# HEAVY METALS AND OTHER POLLUTANTS IN THE ENVIRONMENT

## Biological Aspects

# HEAVY METALS AND OTHER POLLUTANTS IN THE ENVIRONMENT

## Biological Aspects

*Edited by*

**Gennady E. Zaikov, DSc**

**Larissa I. Weisfeld, PhD**

**Eugene M. Lisitsyn, DSc**

**Sarra A. Bekuzarova, DSc**

*Reviewer and Advisory Board Members:*

**Anatoly I. Opalko, PhD, and A. K. Haghi, PhD**

AAP | APPLE ACADEMIC PRESS

Apple Academic Press Inc.
3333 Mistwell Crescent
Oakville, ON L6L 0A2 Canada

Apple Academic Press Inc.
9 Spinnaker Way
Waretown, NJ 08758 USA


Printed in the United States of America on acid-free paper
International Standard Book Number-13: 978-1-77188-437-2 (Hardcover)
International Standard Book Number-13: 978-1-315-36602-9 (CRC Press/Taylor & Francis eBook)



**Library and Archives Canada Cataloguing in Publication**

Heavy metals and other pollutants in the environment : biological aspects / Edited by Gennady E. Zaikov, DSc, Larissa I. Weisfeld, PhD, Eugene M. Lisitsyn, DSc, Sarra A. Bekuzarova, DSc ; reviewer and advisory board members: Anatoly I. Opalko, PhD, and A. K. Haghi, PhD.

Includes bibliographical references and index.
Issued in print and electronic formats.
ISBN 978-1-77188-437-2 (hardcover).--ISBN 978-1-315-36602-9 (PDF)

1. Heavy metals--Environmental aspects. 2. Pollutants--Environmental aspects. I. Zaikov, G. E. (Gennadiĭ Efremovich), 1935-, editor II. Weisfeld, Larissa I., author, editor III. Lisitsyn, Eugene M., author, editor V. Bekuzarova, Sarra A., author, editor

TD196.M4H43 2017 628.5'2 C2017-900591-X C2017-900592-8

**Library of Congress Cataloging-in-Publication Data**

Names: Zaikov, G. E. (Gennadiæi Efremovich), 1935- editor.
Title: Heavy metal pollutants and other pollutants in the environment: biological aspects / editors, Gennady E. Zaikov, DSc. [and three others].
Description: Toronto : Apple Academic Press, 2017. | Includes bibliographical references and index.
Identifiers: LCCN 2017002510 (print) | LCCN 2017004930 (ebook) | ISBN 9781771884372 (hardcover : alk. paper) | ISBN 9781315366029 (ebook)
Subjects: LCSH: Heavy metals--Environmental aspects. | Metals--Environmental aspects. | Pollution--Physiological effect.
Classification: LCC TD427.M44 H42 2017 (print) | LCC TD427.M44 (ebook) | DDC 577.27--dc23
LC record available at https://lccn.loc.gov/2017002510

Apple Academic Press also publishes its books in a variety of electronic formats. Some content that appears in print may not be available in electronic format. For information about Apple Academic Press products, visit our website at **www.appleacademicpress.com** and the CRC Press website at **www.crcpress.com**

# CONTENTS

# LIST OF CONTRIBUTORS

**Rafail A. Afanas'ev**
DSc in Agriculture, Professor, Head of Laboratory, D.N. Pryanishnikov All-Russian Scientific Research Institute of Agrochemistry, 31a, Pryanishnikov St., Moscow, 127550, Russia, Tel.: +79191040585, E-mail: rafail-afanasev@mail.ru

**Zinaida M. Aleschnikova**
DSc in Biology, Head Department, Institute of Microbiology, National Academy of Sciences, 2, Kuprevich St., Minsk, 220125, Belarus, Tel.: +375296385416, E-mail: zaleshhenkov@mail.ru

**Eugene N. Alexandrov**
DSc in Chemistry, Head of Laboratory, Emanuel Institute of Biochemical Physics of Russian Academy of Sciences, 4, Kosygin St., Moscow, 119334, Russia, Tel.: +74950307319, Ee-mail: 28en1937@mail.ru

**Elena A. Alyabysheva**
PhD in Biology, Associated Professor, Mari State University, 1, Lenin Sq., Yoshkar-Ola, 424000, Russia, Tel.: +78362687943, E-mail: e_alab@mail.ru

**Elena Y. Babaeva**
PhD in Biology, Associated Professor, Russian Peoples' Friendship University named after Patrice Lumumba, 6, Miclucho-Maclay St., Moscow, 117198, Russia, Tel.: +79057530081, E-mail: babaevaelena@mail.ru

**Tatyana V. Baranova**
PhD in Biology, Scientific collaborator, Botanical Garden of Voronezh State University, 1, Botanical Garden, Voronezh, 394068, Russia, E-mail: tanyavostric@rambler.ru

**Alan D. Bekmursov**
DSc in Biology, Associated Professor, Department of Environmental Geoscience and Land, K.L. Khetagurov North Ossetian State University, 44-46, Vatutin St., Vladikavkaz, Republic of North Ossetia – Alania, 362025, Russia, Ee-mail: 3210813@mail.ru

**Sarra A. Bekuzarova**
DSc in Agriculture, Honored the Inventor of Russian Federation, Professor, Gorsky State Agrarian University, 37, Kirov St., Vladikavkaz, Republic of North Ossetia – Alania, 362040, Russia, Tel.: +78672362040; North-Caucasian Research Institute of Mountain and Foothill Agriculture, 1, Williams St., Suburban region Mikhailovskoye Vil., Republic of North Ossetia – Alania, 363110, Russia, E-mail: bekos37@mail.ru

**Natalia K. Belisheva**
DSc in Biology, Head of Scientific Department of Medicine-Biological Problems of Human Adaptation in Arctic, Kola Science Centre, Russian Academy of Science, 14A, Fersman St., Apatity, Murmansk region; 184209, Russia, Tel.: +8155579452, +79113039033, E-mail: natalybelisheva@mail.ru

**Oksana I. Bodnar**
PhD in Biology, Associate professor, V.M. Hnatyuk Ternopil National Pedagogical University, 2, M. Krivonos St., Ternopil, 46027, Ukraine, Tel.: +380977821698, E-mail: bodnar_oi@yahoo.com

**Irina I. Chesnokova**
Leading engineer, A.O. Kovalevsky Institute of the Marine Biological Research RAS, 2, Nakhimov Ave., Sevastopol, 299011, Russia, Tel.: +79788271354, E-mail: mirenri@bk.ru

**Tatyana L. Egoshina**
DSc in Biology, Professor, Head of Department of Plant Resources, All-Russian Institute of Game and Fur Farming, 72, Preobrazhenskaya St., Kirov, 610000, Russia; Professor of Cathedra of Ecology and Zoology, Vyatka State Agricultural Academy, 133, Oktyabrsky Ave., Kirov, 610017, Russia, Tel.: +79097166866, E-mail: etl@inbox.ru

**Vasil V. Grubinko**
DSc in Biology, Full professor, Volodymyr Hnatyuk Ternopil National Pedagogical University; Chief of department of General Biology, 2, M. Krivonis St., Ternopil, 46027, Ukraine, Tel.: +380672822463, Ee-mail: v.grubinko@gmail.com

**Vladislav N. Kalaev**
DSc in Biology, Professor, Department of Genetics, Cytology and Bioengineering, Vice Director for Science of B.M. Kozo-Polyansky Botanical Garden; Professor, Botanical Garden of Voronezh State University, 1, Botanical Garden, Voronezh, 394068 Russia, Tel.: +79103450072, E-mail: Dr_Huixs@mail.ru

**Liliya E. Kartyzhova**
PhD in Biology, Leading Scientist, Institute of Microbiology, National Academy of Sciences, 2, Kuprevich St., Minsk, 220141, Belarus, Tel.: +375296385416, E-mail: Liliya_Kartyzhova@mail.ru

**Tatiana Khijniak**
DSc of Biology, Acting Head of Laboratory, Winogradsky Institute of Microbiology, Research Center of Biotechnology of the Russian Academy of Sciences, 33, bld. 2, Leninsky Ave., Moscow 119071, Russia, Tel.: +74991351021, +74991350109, E-mail: mtk.fam@mail.ru

**Katherina V. Kostyuk**
PhD of Biology, Assistant Professors of department of Biology O.O. Bogomolets National Medical University, 13, Taras Shevchenko St., Kyiv, 01601, Ukraine, Tel.: +380930000272, E-mail: kostyuk.katya@gmail.com

**Tatyana B. Kovirshina**
Leading engineer, A.O. Kovalevsky Institute of the Marine Biological Research RAS, 2, Nakhimov Ave., Sevastopol, 299011, Russia, Tel.: +79788307440, E-mail: mtk.fam@mail.ru

**Mikhail V. Kozlov**
PhD of Biology, Researcher, N.M. Emanuel Institute of Biochemical Physics of Russian Academy of Sciences, 4, Kosygin St., Moscow, 119334, Russia, Tel.: +74959397186, E-mail: wer-swamp@yandex.ru,

**Alevtina G. Kudyasheva**
DSc of Biology, Head of Department, Institute of Biology of Komi Scientific Center, Ural Branch of Russian Academy of Sciences, 28, Kommunisticheskaya St., Syktyvkar, 167982, Russia, Tel.: +78212430478, kud@ib.komisc.ru

**Eugene M. Lisitsyn**
DSc in Biology, Assistant professor, Head of Department of Plant Edaphic Resistance, N.V. Rudnitsky Zonal North-East Agricultural Research Institute, 166a Lenin St., Kirov, 610007, Russia; Professor of Department of Ecology and Zoology, Vyatka State Agricultural Academy, 133 Oktyabrsky Av., Kirov, 610017, Russia; +79123649822, E-mail: edaphic@mail.ru

**Angela I. Lutsiv**
PhD, Senior Researcher of the Laboratory of Ecotoxicology, V.M. Hnatyuk Ternopol National Pedagogical University, 2, M. Krivonos St., Ternopil, 46027, Ukraine, Tel.: +380979254150, E-mail: kostyuk.katya@gmail.com

**Genrietta E. Merzlaya**
DSc in Agriculture, Professor, Head of Laboratory, D.N. Pryanishnikov All-Russia Research and Development Institute of Agrochemistry, 31A, Pryanishnikov St., Moscow, 127550, Russia, Tel.: +79623694197, E-mail: lab.organic@mail.ru

**Nataliya N. Roslova**
Senior Engineer, Taras Shevchenko National University of Kyiv, 64/13, Volodymyrska St., Kyiv, 01601, Ukraine, Tel.: +380674012692, E-mail: tereska@bigmir.net

**Irina I. Rudneva**
DSc in Biology, Professor, Leading Scientist, A.O. Kovalevsky Institute of the Marine Biological Research RAS, 2, Nakhimov Ave., Sevastopol, 299011, Russia, Tel.: +79787491704, E-mail: svg-41@mail.ru

**Mariia P. Rudyk**
PhD in Biology, Assistant Lecturer, Taras Shevchenko National University of Kyiv, 64/13, Volodymyrska St., Kyiv, 01601, Ukraine, Tel.: +380504438775, E-mail: rosiente@gmail.com

**Alexey V. Safonov**
PhD in Chemistry, Head of biotechnology and radioecology group, Frumkin Institute of Physical Chemistry Russian Academy of Science, 31, Leninsky Ave., Moscow, 199071, Russia, Tel.: +74953352004, +79169121059, E-mail: alexeysafonof@gmail.com

**Elena V. Sarbayeva**
PhD in Biology, Associated Professor, Mari State University, 1, Lenin Sq., Yoshkar-Ola, 424000, Russia, Tel.: +78362687943, E-mail: sarbaevaev@mail.ru

**Irina A. Shabanova**
PhD in Bioilgy, Associated Professor, Gorsky State Agrarian University, 37, Kirov St., Vladikavkaz, Republic of North Ossetia – Alania, 362040, Russia, Tel.: +79280685702, E-mail: irinashabanova@mail.ru

**Valentine G. Shayda**
Engineer, A.O. Kovalevsky Institute of the Marine Biological Research RAS, 2, Nakhimov Ave., Sevastopol, 299011, Russia, Tel.: +79787491708, E-mail: svg-41@mail.ru

**Viktoriia V. Shepelevych**
PhD in Biology, Researcher, Head of Department, Taras Shevchenko National University of Kyiv, 64/13, Volodymyrska St., Kyiv, 01601, Ukraine, Tel.: +380667817304, E-mail: vshepelevich@ukr.net

**Oksana G. Shevchenko**
PhD of Biology, Leading Scientist, Institute of Biology of Komi Scientific Center, Ural Branch of Russian Academy of Sciences, 28, Kommunisticheskaya St., Syktyvkar, 167982, Russia, Tel.: +78212430478, E-mail: Microtus69@mail.ru

**Lyudmila N. Shikhova**
DSc in Agriculture, Assistant professor, Head of Cathedra of Ecology and Zoology, Vyatka State Agricultural Academy, 133 Oktyabrsky Ave., Kirov, 610017, Russia; +79127213758, E-mail: shikhova-l@mail.ru

**Lyudmila N. Shishkina**
DSc of Chemistry, Professor, Head Department, Emanuel Institute of Biochemical Physics of Russian Academy of Sciences, 4, Kosygin st., Moscow, 119334, Russia, Tel. +74959397186, E-mail: shishkina@sky.chph.ras.ru

**Larysa M. Skivka**
DSc in Medicine, Associated Professor, University Chair, Taras Shevchenko National University of Kyiv, 64/13, Volodymyrska St., Kyiv, 01601, Ukraine, +380445213231, E-mail: realmed@i.com.ua

**Ekaterina N. Skuratovskaya**
PhD in Biology, Researcher, A.O. Kovalevsky Institute of the Marine Biological Research RAS, 2, Nakhimov Ave., Sevastopol, 299011, Russia, Tel.: +79788114046, E-mail: skuratovskaya2007@rambler.ru

**Michail O. Smirnov**
PhD in Biology, Senior Researcher, D.N. Pryanishnikov All-Russian Scientific Research Institute of Agrochemistry, 31A, Pryanishnikov St., Moscow, 127550, Russia, Tel.: +79057966323, E-mail: User53530@yandex.ru

**Anna V. Stanislavchuk**
PhD in Biology, Researcher, Ternopil V. M. Hnatiuk National Pedagogical University, 2, M. Krivonos St., Ternopil, 46027, Ukraine, Tel.: +380673697608, E-mail: anna.stanislavchuk@gmail.com.ua

**Olesya V. Vasilenko**
PhD in Biology, Teacher of higher education institution, Ternopil V. Hnatiuk National Pedagogical University, 2, M. Krivonos St., Ternopil, 46027, Ukraine, Tel.: +380977821698, E-mail: vasylenkoo@mail.ru

**Halina B. Viniarska**
Post-graduate student, Researcher, V. M. Hnatiuk National Pedagogical University, 2, M. Krivonos St., Ternopil, 46027, Ukraine, Tel.: +380977821698, E-mail: viniarska19@gmail.com

**Olga A. Vlasova**
Graduate student, D.N. Pryanishnikov All-Russia Research and Development Institute of Agrochemistry, 31A, Pryanishnikov St., Moscow, 127550, Russia, Tel.: +74990761191, E-mail: cool.vlacova2013@yandex.ru

**Olga L. Voskresenskaya**
DSc in Biology, Professor, Director Institute of Medicine and Science, Mari State University, Russia, 1, Lenin Sq., Yoshkar-Ola, 424000, Russia, Tel.: +78362687946, E-mail: voskres2006@rambler.ru

**Vladimir S. Voskresesenskiy**
PhD, Associated Professor, Mari State University, Russia, 1, Lenina Sq., Yoshkar-Ola, 424000, Russia, Tel.: +78362687943, E-mail: woron-69@rambler.ru

**Larissa I. Weisfeld**
Senior Researcher, Laboratory Gas Analysis and Environmental Toksimetrya, Emanuel Institute of Biochemical Physics of Russian Academy of Sciences, 4, Kosygin St., Moscow, 119334, Russia, Tel.: +79162278685, E-mail: liv11@yandex.ru

**Nadezhda G. Zagorskaya**
Researcher, Institute of Biology of Komi Scientific Center, Ural Branch of Russian Academy of Sciences, 28, Kommunisticheskaya St., Syktyvkar, 167982, Russia, Tel.: +78212430478, E-mail: kud@ib.komisc.ru

**Inga Zinicovscaia**
PhD in Biology, Researcher, Joint Institute for Nuclear Research, 6, Joliot-Curie Str., Dubna, 141980, Russia; The Institute of Chemistry of the Academy of Sciences of Moldova, 3, Academiei Str., Chisinau, 2028, Republic of Moldova, Tel.: +74962163653, E-mail: zinikovskaia@mail.ru

**Olga A. Zubkova**
PhD in Agriculture, Senior researcher, Department of plant edaphic resistance, N.V. Rudnitsky Zonal North-East Agricultural Research Institute, 166a Lenin St., Kirov, 610007, Russia, Tel.: +79226671641, E-mail: edaphic@mail.ru

# LIST OF ABBREVIATIONS

| | |
|---|---|
| AAC | approximately allowable concentrations |
| $AgNO_3$ | silver nitrate |
| Al | aluminum, is a chemical element in the boron group with atomic number 13 |
| $Al^{+++}$ | ion of aluminum |
| ALT | alanin aminotransferase |
| As | arsenic |
| AST | aspartat aminotransferase |
| ATP-ase | adenosine 5'-triphosphatase (ATP hydrolase) |
| B | subsoil horizon: subsurface layer reflecting chemical or physical alteration of parent material |
| Bq/kg | the indicator of the specific activity of natural radionuclides in Becquerel's per 1 kg of dry mass |
| C-mitosis | metaphase |
| c/ha | center/hectare |
| $Ca^{++}$ | ion of calcium |
| CAT | catalase |
| Cd | cadmium is a chemical element with atomic number 48. |
| CDNB | 1-chloro-2,4-dinitrobenzene |
| CINAO | Pryanishnikov All-Russian Scientific Research Institute of Agrochemistry |
| CL | cardiolipin |
| $C_{lab}$ | labile part of total carbon |
| cm | centimeter |
| CO | carbon monoxide |
| Co | cobalt |
| $CO_2$ | carbon dioxide |
| Cr | chromium, chrome |
| Cs | cesium |
| $C_{tot}$ | total organic carbon |
| Cu | copper |
| DAG | diacylglycerol |

| | |
|---|---|
| DNA | deoxyribonucleic acid |
| $E_a$ | activation energy of ion transport |
| EAC | eluvial accumulative coefficient |
| EC | enzyme commission number |
| EDTA | ethylenediaminetetraacetate |
| ETC | electron transport chain |
| Fe | iron |
| g/ha | gram/hectare |
| GPx | glutathione peroxidase |
| GR | glutathione reductase |
| GSH | reduced glutathione |
| GST | glutathione-S-transferase |
| $H^+$ | hydrogen ion |
| HCl | hydrogen chloride |
| Hg | mercury, quicksilver |
| HM | heavy metals |
| IARC | International Agency for Research on Cancer |
| IChL | inducible chemiluminescence |
| K | potassium |
| $K_3Fe(CN)_6$ | potassium ferricyanide, potassium geksatsianoferriat, Gmelin salt, red blood salt |
| KCl | potassium chloride |
| $KH_2PO_4$ | potassium dihydrogen phosphate |
| $K_m$ | Michaelis (Michaelis-Menten) constant |
| LPC | lysoforms of phospholipids |
| LPL | lysophospholipids |
| LPO | lipid peroxidation |
| LSD | Least Significant Difference |
| lx–(lux) | the SI unit of illuminance and luminous emittance, measuring luminous flux per unit area |
| MAC | maximum allowable concentration |
| mg/kg | milligram/kilogram |
| MI | mitotic index |
| mm | millimeter |
| MMCs | melanomacrophage centers |
| Mn | manganese |
| Mo | molybdenum |

| | |
|---|---|
| MPC | maximum permissible concentration |
| N | nitrogen |
| NADPH | nicotinamide adenine dinucleotide phosphate |
| NADPH-GDH | nicotinamide adenine dinucleotide phosphate-glutamate dehydrogenase |
| NaOCl | sodium hypochlorite |
| $NaS_2CN(C_2H_5)_2$ | sodium diethyldithiocarbamate |
| $NaSO_3$ | sodium sulfite |
| NBT | nitroblue tetrasolium |
| NEFA | nonetherified fatty acids |
| $NH_4NO_3$ | ammonium nitrate |
| Ni | nickel |
| Nis | network information services |
| NLMK | Novolipetsk metallurgical complex |
| NPK | complex fertilizer: nitrogen-phosphorus-potassium |
| P | phosphorus |
| PA | phosphatidic acid |
| Pb | lead |
| PC | phosphatidylcholine |
| PE | phosphatidylethanolamine |
| PER | peroxidase |
| $pH_{kcl}$ | the negative logarithm of the hydrogen ion in KCl extract |
| PI | phosphatidylinositol |
| PL | phospholipid and/or phospholipids |
| PM | pathological mitoses |
| PMS | phenazine methosulfate |
| PN | persistent nucleoli |
| PS | phosphatidylserine |
| *r* | coefficient of pair correlation |
| ROS | reactive oxygen species |
| rpm | revolution per minute |
| SanReg | Sanitary Regulations and Norms |
| SChL | spontaneous hemiluminescence |
| SDH | succinate dehydrogenase |
| Se | selenium |
| SM | sphingomyelin |

| | |
|---|---|
| SOD | superoxide dismutase |
| Sr | strontium |
| STRAZ | Software program |
| t/ha | ton/hectare |
| TAG | triacylglycerol |
| TBA | tiobarbituric acid |
| TBA-RS | TBA-reactive substances |
| TCA | trichloroacetic acid |
| U.S. EPA | United States Environmental Protection Agency |
| UFA | unesterified fatty acids |
| $V_{max}$ | maximum speed of penetration |
| Zn | zinc |
| ΣEOPL/ΣPOPL | ratio of the sums of the more easily to the more poorly oxidizable fractions of phospholipids |

# ABOUT THE EDITORS

**Gennady E. Zaikov, DSc**

*Head of the Polymer Division, Emanuel Institute of Biochemical Physics, Russian Academy of Sciences, Moscow, Russia; Professor, Moscow State Academy of Fine Chemical Technology, Russia; Professor, Kazan National Research Technological University, Russia*

Gennady E. Zaikov, DSc, is a Full Professor and Head of the Polymer Division at the Emanuel Institute of Biochemical Physics, Russian Academy of Sciences, Moscow, Russia, and Professor at Moscow State Academy of Fine Chemical Technology, Russia. He is also a Professor at Kazan National Research Technological University, Kazan, Russia. He is also a prolific author, researcher, and lecturer, with over 2700 invited lectures through 2013 (including over 170 in the last three years) at conferences, universities, and research institutes. He has received several awards for his work, including the Russian Federation Scholarship for Outstanding Scientists. He has over 4500 research publications through, including 470 monographic books, 251 volumes, and 14 textbooks. In addition, he is a member of editorial/advisory boards of many books. He has been a member of many professional organizations and on the editorial boards of many international science journals.

**Larissa I. Weisfeld, PhD**

Senior Researcher, N.M. Emanuel Institute of Biochemical Physics, Russian Academy of Sciences, Moscow, Russia

Larissa I. Weisfeld, PhD, is a Senior Researcher at the N.M. Emanuel Institute of Biochemical Physics, Russian Academy of Sciences in Moscow, Russia, and a member of the N.I. Vavilov Society of Geneticists and Breeders. She is the author of about 300 publications in scientific journals and conference proceedings, as well as the coauthor of a work on four new cultivars of winter wheat. Her research interests concern the basic problems of chemical mutagenesis, cytogenetics, and the other ecological problems. She has worked as a scientific editor at the publishing house Nauka (Moscow) and of the journals *Genetics* and *Ontogenesis*. She was an author and coeditor of the books *Ecological Consequences of Increasing Crop Productivity: Plant Breeding and Biotic Diversity*; and *Biological Systems, Biodiversity, and Stability of Plant Communities*; and *Temperate Crop Science and Breeding: Ecological and Genetic Studies.*

**Eugene M. Lisitsyn, DSc**
*Assistant Professor, N.V. Rudnitsky North-East Agricultural Research Institute, Russian Academy of Sciences; Professor of Ecology and Zoology, Vyatka State Agricultural Academy, Kirov, Russia*

Eugene M. Lisitsyn, DSc in Biology, is an Assistant Professor at the N.V. Rudnitsky NorthEast Agricultural Research Institute, Russian Academy of Sciences in Kirov, Russia, and a Professor of ecology and zoology at Vyatka State Agricultural Academy, also in Kirov, Russia. He is a member of the N. I. Vavilov Society of Geneticists and Breeders. Dr. Lisitsyn is the author of about 230 publications in scientific journals and conference proceedings, as well as the co-breeder of oats variety Krechet. His research interests concern the basic problems of plant adaptation to environmental stressors such as soil acidity, aluminum and heavy metals, and the other ecological problems. He has published chapters in several books in Apple Academic Press, including *Barley: Production, Cultivation and Uses*; *Biological Systems, Biodiversity and Stability of Plant Communities*; *Chemical and Structure Modification of Polymers*; *Materials Chemistry: A Multidisciplinary Approach to Innovative Methods*; *Temperate Crop Science and Breeding: Ecological and Genetic Studies*, among others.

**Sarra A. Bekuzarova, DSc**

*Professor and Head, Laboratory of Plant Breeding, Republic of North Ossetia-Alania; Professor, Gorsky State University of Agriculture, Professor at North-Ossetia State University, Vladikavkaz, Republic of North Ossetia-Alania, Russia*

Sarra A. Bekuzarova, DSc in Agriculture, is a Deserved Inventor of Russia and has been awarded the Medal of Popova. She is a Professor and Head of the Laboratory of Plant Breeding and Feed Crops of long-term seed grower of fodder crops in North Caucasus Institute of Mountain and Foothill Agriculture of the Republic of North Ossetia – Alania. She is also a Professor at the Gorsky State University of Agriculture, Vladikavkaz, Republic of North Ossetia Alania, Russia, as well as Professor at L.N. Kosta Khetagurov North Ossetia State University, Vladikavkaz, Republic of North Ossetia Alania, Russia. She is a prolific author, researcher, and lecturer. She is corresponding member of the Russian Academy of Natural Sciences, and a member of the International Academy of Authors of the Scientific Discoveries and Inventions, the International Academy of Sciences of Ecology; Safety of Man and Nature; the All Russian Academy of Nontraditional and Rare Plants; and the International Academy of Agrarian Education. Dr. Bekuzarova is also a member of the editorial boards of several scientific journals. She was an author and a coeditor of several books, including *Ecological Consequences of Increasing Crop Productivity: Plant Breading and Biotic Diversity*; *Biological Systems, Biodiversity and Stability of Plant Communities*; and *Temperate Crop Science and Breeding: Ecological and Genetic Studies*.

# PREFACE

The commonly known triad of global civilization development joins global historical processes into three consequent stages: *traditional civilization* with traditional culture of unconscious use of the complex of natural resources; *industrial civilization* with conscious but merciless exploitation of atmospheric, water, soil, mineral resources, as well as of flora and fauna; and *information civilization of the "third wave,"* when the exploitation of natural resources, almost as merciless as before, is accompanied by the meaningful analysis of possible risky consequences of anthropogenic activity and by the resistance of civil society.

The stage of traditional civilization is rooted in the periods of Paleolithic and Mesolithic ages (early and middle stone ages) with their low population density, nomadic and/or semi-nomadic life style, absence of mechanisms or any other tools of radical environment destruction, when the transition from barbarity to civilization indeed happened. Localities, disturbed by the relatively limited anthropogenic load, naturally restored themselves during few decades, sometimes during two or three centuries. Further developing of new territories by humans and their more effective use are known in the history under the name of "Neolithic revolution," when plant cultivation started. Anthropogenic load on the environment increased with permanently accelerating rate, but it did not reach globally alarming proportions.

The stage of industrial civilization, which started since the end of XVIII century, promised to the humankind the overcoming of the dependence on the forces of the nature and of the permanent threat of famine. During further centuries the massive introduction of science in industrial production and agriculture took place, which led to the use of steam engine; creation of a series of engines (water and steam turbines, combustion engine); development of railways and sea shipping; invention of radio, telegraph, telephone; creation of automobile and airplane; development of electric energy. The technical progress evoked new branches of production, namely chemical, electrical industry, etc. Computing and industrial automation, creation of artificial materials, technology of use of atom properties originated from this period. Consequently destructive loads on the environment increased, and by the end of XXth century they began to acquire globally-threatening character. The understanding of risks of technogenic activity of humans by the public in developed countries radically differs from the aspirations of the population of developing countries, as well as from USSR and other states with totalitarian regimes, with their chronic deficit of food. The civil society of most of European countries considers the environmental pollution by industrial, in particular radioactive, waste, as well as the contravention of the status of nature reserves, disappearance of any biological species, from ethic point of view, mainly negatively, and it effectively oppose the government. However the majority of the population of developing countries approves the deforestation aimed to free the territory for "intensive" agriculture or for industrial or housing construction, because they relate it with their eternal aspirations to well-being.

The stage of information civilization of so-called "third wave" started in seventies of the last century, when technologically advanced sectors of the world community started the transition to the post-industrial economy. The economic activity in the agrarian sector is now, as well as always before, related with the production of sufficient amount of food. However, now the limited accessibility of natural resources, in particular of agricultural land, becomes the limiting factor. In the industrial economy the economical activity is mainly aimed to produce goods. However, there the capital remains the main limiting factor, both for investments in the production itself, as well as for ensuring the purchasing power of the population. As for information economy, economical activity there consists mainly in the objective analysis of situation, forecasting of development trends, preparing of reliable information and its application aimed to increase effectiveness of all other production forms and thus create more material economical activity. The inexhaustibility of information resources creates unique possibility to apply information for the service of the whole world. There are following limiting factors there: availability of knowledge in the world information space, professional level of information consumers and the position of state, on which the possibilities of information exchange, as well the effectiveness of the use of obtained information depend.

With all the variety of views on the historical development of humankind, the considerable contribution of Soviet scientists in the world science is recognized by all researchers independently on their political position. The question, which scientific achievements took place “due to” and which ones “in spite of” the totalitarian regime of this time, requires separate analysis.

Without going into details of commonly known achievements of the Soviet science, let’s notice that under conditions of totalitarian regime and strictly controlled mass media ecological consequences of ambitious plans of “nature transformation” under loud populist slogan «all for a person, all in the name of a person!» did not become a topic of basic research. Moreover, even ecological disasters, which happened, remained unknown to the population of the former USSR, as well as, in many cases, for the world public. The world knew about them only in the cases of pollutants emission out of the territory of the communist state, as, e.g., under the accident at Chernobyl nuclear power plant. As a result of the Chernobyl disaster, considerable pollution affected not only the Gomel and Mogilev regions of Belarus, Kiev and Zhitomir regions of Ukraine, Briansk region of Russia, but also a series of European countries. Samizdat (as underground publication of the literature forbidden in USSR was called) informed also about many pre-Chernobyl accidents with emissions of high amounts of radioactive substances, including accidents at Leningrad NPP (1975), Beloyarsk (1978, Sverdlovsk region), Armenian (1982, Medzamor), Zaporozhe (1984, Energodar town, Ukraine) and other nuclear power plants. Progressive Soviet scientists and writers warned about the ecological impact of flooding of enormous territories due to the construction of hydro plants, of the pollution of agricultural lands by untreated wastes of giant industrial factories, and of the contamination of water resources by sewages of these factories including unique freshwater Russian Lake of Baikal. They informed about the danger of the transformation of the Aral Sea, which was sometime the fourth lake in the world by area, into a salted desert because of the use of the Amudaria and Syrdaria rivers, which disembogue into it, for the irrigation of cotton and rice fields in the arid lands of Kazakhstan and Uzbekistan, as well as about many other consequences of ignoring ecological safety issues in USSR. However, in those times the ignoring of ecological inspections requirements was always justified by "public interest" of accomplishing plans at any cost.

After all, an opportunity appeared to demand the owner of a private company provide the ecological safety without infringement of the public interest, to force him to invest in the construction treatment plants of in order to protect atmospheric, water, soil resources, as well as flora and fauna. Indeed, these resources condition the well-being of mankind and finally the survival of the *Homo sapiens* species itself.

The ecological expertise and the forecasting of industrial activities impacts on the environment are performed for the projects related with the creation of new enterprises, as well as with the reconstruction of existing ones. However, the owners of ecologically dangerous enterprises, the same as in former times directors of Soviet plants and factories, continue to ignore any ecological measures, which require investments. The real brake of environment protecting measures, which was masked in the Soviet period by the "public interest", now developed even more under the patronage of corrupted officials from one side and because of enormous possibilities for the owner from other side. The last one can "buy" the approval of an official by spending only a small part of excess profit that he gets. Attempts of blocking of environmentally hazardous enterprises, attempted by social activists, are often stopped by law enforcement authorities and even—often are punished by corrupted courts.

In connection with the above, theoretical and applied studies performed by Russian specialists and their colleagues from some other post-Soviet countries concerning the questions of occurrence, distribution, ecological risk assessment and behavior of heavy metals and other pollutants in the system soil-plant-human, as well as results of the environment monitoring, represent considerable interest not only to scientists in biology, ecology and agriculture, but also for businessmen searching for objects for investment in the industrial or agrarian sectors of economy in the post-Soviet space.

—***Anatoly I. Opalko***
***Larissa I. Weisfeld***

# INTRODUCTION

The proposed book *Heavy Metal Pollutants and Other Pollutants in the Environment: Biological Aspects* is dedicated to the questions of occurrence and behavior of heavy metals in the complex system soil–plant–animal–human.

"Heavy metals" is the most widely recognized term for the large group of chemical elements with an atomic density above 6 $g/cm^3$. They are often called "trace elements" because their concentrations in the parent rocks of the earth's crust are less than 100 mg/kg. All heavy metals are toxic for living organisms when present in excess, but some of them are essential for the normal growth in small concentrations. So sometimes they are referred as "potentially toxic elements." Heavy metals could be important constituents of living organisms and many nonliving substances in the environment.

Heavy metals represent one of the most widespread type of pollutants. The problems connected with their negative influence on living systems and, in particular, on the person, have arisen with the beginning of technical revolution though till 1950–1960 their scales were not realized by the public. For example, in the USA the Federal law on a state policy in the field of environment has been accepted in 1970. With acceptance of similar laws an ecological situation has changed sharply for the better; reduction of impurity of biosphere by heavy metals has begun on a global scale. Reduction of aerial emissions from stationary sources and means of transportation has occurred by especially high rates.

However, in many countries of the world there were the places strongly polluted as a result of the previous human activity. So, the large amounts of metal pollutants has got to soils of the Western Europe from second half of XIX century; for example, in France by 1996 some hundreds places have been revealed strongly polluted by heavy metals. Such soils, for many years incorporated considerable amounts of pollutants, demand studying and land improvement. It concerns both Russia and other countries of the former Soviet Union.

Modern and most active pollution of soils goes by a hydrogenous way, especially in agriculture sphere. For example, if by means of the filters established at the enterprises, it is possible to detain up to 95% of aerial pollutants than sewage are cleared much worse—three quarters of them is not cleared completely. Thus impure sewage is widely used for irrigation in the droughty countries, one of which is China.

One more serious source of heavy metals is sediment of sewage of city water drains. Sediments of sewage are considered perspective for fertilizer of agricultural and wood soils as sources of organic substance, phosphorus and nitrogen.

It is important to realize that the soil is both a sink of metal and also its natural source. Studies of heavy metals content within whole ecosystems have indicated that many areas contain anomalously high concentrations of these elements. The factors controlling total content of heavy metals and their bioavailability in soils are of great importance for estimation of its damage level both for human toxicology and agricultural productivity. By acting as a sink for metal the soil own to properties of its some compounds (like organic matter) also functions as a filter protecting the vegetation from inputs of potentially harmful metals.

It is important to emphasis that in many part of the world agricultural production is limited by deficiencies of such "essential" heavy metals as Zn, Cu, Mn, and Co. Iodine, boron and iron are also important deficit micronutrients but are not defined as heavy metals. Their absence limits growth of microorganisms, plants and animals. Human health has been seriously affected by deficiencies of these heavy metal micronutrients. So, it is desirable to measure and control concentrations of these elements in the environment.

This book, *Heavy Metal Pollutants and Other Pollutants in the Environment: Biological Aspects*, covers the important results in research of heavy metals in soils, microorganisms, plants, and foodstuff.

Special efforts were made to invite experts to contribute chapters in their own areas of expertise. Since the discussed area is very broad, no one can claim to be an expert in all discussed questions; collective contributions in solving of this problem are better than a single author's presentation for a book.

In *Part 1 (Heavy Metals in Soils: The Factors Influencing Their Accumulations in Plants)* the data showing distribution of heavy metals in soils and parent rocks of natural and farmlands is cited, and also the factors influencing their bioavailability, such as the content of organic matter, soil acidity, entering of mineral fertilizers and sewage are considered. Researchers from the North-East Agricultural Research Institute (Kirov, Russia) have shown that in the north of the European Russia background levels of the heavy metals content do not exceed level of maximum allowable concentration, but are above level world clark. This data testifies to the raised regional geochemical background of these elements. Statistically significant seasonal variability of the content of mobile forms of these elements in 1.5–10 times is shown by the same authors. The system of organic matter of arable soil is unstable and is more dynamical, than in soils of forest biocenoses. The minimal values are characteristic to beginning of a growing season, and maximum for its second half.

Authors from Mari State University (Yoshkar-Ola, Russia) present the data about pollution by heavy metals of soils of the urbanized territories, its influence on assimilation apparatus of wood plants and specify in the important role of root systems as barriers on uptake of soil pollutants by aboveground parts of plants. In many countries of the world considerable attention have received heavy metals in sewage sludge and agricultural fertilizers both mineral and organic.

Scientists from the All-Russia Institute of Agrochemistry (Moscow, Russia) on the basis of the long-term researches recommend to observe regulations of sewage application under agricultural crops including doses and periodicity of entering into soil in order to avoid contamination of environment and agricultural production with heavy metals containing in many kinds of sediments.

The cleaning of soil from oil contamination is nowadays a global problem. The application of chemical substances for the neutralization of hydrocarbons brought more harm than good. Hydrocarbons disturb gas exchange, air temperature, kill living organisms, disturb ecology not only in the soil, but also in surrounding atmosphere. The article wrote by scientists from Gorsky State Agrarian University (Vladikavkaz, Republic of North Ossetia – Alania) and N.M. Emanuel Institute of Biochemical Chemistry (Moscow, Russia) presents an overview of the current state of pollution and attempts of soil restoration, avoidance of chemicals application, and soil environment enrichment by biological methods using plants, biological preparations, and mineral fossils. The concrete experiment of living systems application is presented.

Influence of soil heavy metals on growth and development of wild-growing plants is shown in *Part 2 (Impact of Heavy Metals on Vegetation)*. Researchers of Gorsky State Agrarian University and the North Ossetia State University (Vladikavkaz) for definition of toxicity of soils carried out an estimation of level of their pollution by means of the plants-indicators possessing ability to absorb heavy metals. They show intensity of accumulation of heavy metals by different species and varieties of plants during growing season. Scientists have found out that level of accumulation of heavy metals in plants of legume grasses is specific not only to species, but also varieties. Depending on degree of accumulation of lead, cadmium, copper, zinc, and nickel in different phases of development of the plants, the conclusion is made about toxicity of soils and actions are planned for its decrease.

On the basis of long-term research researchers of the North-East Agricultural Research Institute and All-Russian Institute of Game and Fur Farming (Kirov, Russia) have noted a considerable divergence between standard and observed confinedness of species of wild-growing plants to soils with various acidity levels and the content of heavy metals. Authors explain this fact by a poor studying of species biology on all extent of their areal. Their data is also according to conclusions of authors from Vladikaukaz about essential distinction in specific accumulation of toxicants, adding them with indications on threshold levels of the content of various metals in soil. At the same time grain crops (oats and barley) feel toxic effect of heavy metals only at their concentration considerably exceeding level of maximum allowable concentration.

Authors from the Botanical garden of the Voronezh State University (Voronezh, Russia) have presented results of cytologic studying of wood plants from the areas pure and polluted by heavy metals. Results of cytological study show genome instability of *Betula pendula* and *Rhododendron ledebourii* seed progeny, collected in the ecologically polluted areas. Cytogenetic responses of indigenous and introduced species are similar in nature, which is shown as the increase of mitotic index due to the increase of the proportion of cells at prophase, the presence of cells with mitotic pathologies and similar disturbances of mitosis.

In Kola Scientific Centre of the Russian Academy of Sciences (Apatity, Russia) the comparative characteristic of effects of interaction of some heavy metals (Pb, As, Hg, Cd, Ni, Cr, and V) with cellular structures is reported. Ions of heavy metals may induce some mutations in living organisms. It is noticed that the more widely is the spectrum of the infringements caused by action of metal at cellular level, the more plural character has pathology on organism level, that is the original projection of microlevel on macrolevel and on the contrary takes place. Therefore width of a spectrum of infringements at one level of the organization makes it possible to judge infringements at other level of the organization. These phenomena support toxic and carcinogenic activity of metals, which could take the role of initiators, as well as promoters of carcinogenesis.

Especially interesting are investigations on influence of various heavy metals on soil microorganism activity. *Part 3 (Effects of Different Pollutants on Algae, Fungi, and Soil Microorganisms)* of the offered book is devoted disclosing of these questions. Researchers from Ternopol National Pedagogical University (Ternopol, Ukraine) and National Medical University (Kiev, Ukraine) have shown that accumulation of ions of $Mn^{2+}$, $Zn^{2+}$, $Cu^{2+}$ and $Pb^{2+}$ by the cells of *Chlorella vulgaris* Beij is fluctuating and is determined by the concentration of ions in the environment and duration of its action on the cells. There are four stages: the stage of protective self-isolation of cells as a result of the primary stress response; the stage of the active accumulation as a result of decrease in resistance and destruction of outer membrane; the stage of inhibition of the accumulation as a result of formation of the secondary concentric membrane; the stage of uncontrolled accumulation as a result of destruction of the secondary concentric membrane. The process of a concentric double membrane system formation is universal and does not depend on the nature of the toxicant (biogenic $Mn^{2+}$, $Zn^{2+}$ and $Cu^{2+}$, toxic $Pb^{2+}$).

In work with algae *Chlorella* spent by authors from Ternopol National Pedagogical University (Ternopol, Ukraine) it is shown that increased content of selenium and zinc in *Chlorella* cells by adding them as selenite 10.0 mg/dm$^3$ and zinc ions 5.0 mg/dm$^3$ within 7 days modified cell metabolism through activation of photosynthetic systems, adaptive adjustment of energy metabolism, and antioxidant protection that increased physiological and biochemical status of cells with simultaneous accumulation of selenium and zinc in macromolecules, mostly in lipids.

Scientists from Institute of Microbiology, National Academy of Sciences (Minsk, Belarus) present results of researches on the selection of *Rhizobium galegae* strains resistant against disinfectants, herbicides, and oil pollution. Their mutant forms keeping symbiotic properties at joint application with disinfectants, herbicides and providing effective symbiosis with a host plant under extreme conditions are selected.

Biosorption of hexavalent heavy metals (chromium and uranium) by using such types of microorganisms as bacteria, algae, and fungi is reviewed in study of Moldavian and Russian microbiologists (Institute of Chemistry, Chisinau, Republic of Moldova, Institute of Physical Chemistry and Electrochemistry, Moscow, Russia and Winogradsky Institute of Microbiology, Moscow, Russia). Authors show that microorganisms are an efficient and potential class of bio-sorbents for the removal of hexavalent chromium and uranium from industrial wastewater. The microorganisms contain a variety of functional groups responsible for metal adsorption, which allows producing cheap and effective bio-sorbents for large-scale application.

At last in *Part 4 (Risks of Environment Pollution by Heavy Metals for Animals and Possibility of Foodstuff Protection From Pollutants)* there are two research papers deal with effect of heavy metals on higher organisms – fish and animals. In mutual research of scientists from Komi Scientific Centre (Syktyvkar, Russia) and N.M. Emanuel Institute of Biochemical Chemistry (Moscow, Russia) comparative analysis of the early and long-term biological consequences under the low-intensity $\gamma$-radiation action at low dose and combined action of $\gamma$-radiation and lead nitrate at different concentrations was performed by using of the morpho-physiological parameters and the separate indices of the lipid peroxidation in the functionally distinct tissues of mice. They found that the direction and scale of changes of studied parameters in studied tissue or organ differ significantly. Such changes have complex nonlinear character, depending on dose and time duration after the radiation action.

In second article reactions of sea fishes' liver to environment pollution was studied by biologists from A.O. Kovalevsky Institute of Marine

Biological Research (Sevastopol, Russia) and Institute of Biology of Taras Shevchenko National University (Kyiv, Ukraine). They found out that influence of anthropogenic pollution of environment on fishes is shown in development of histopathological deviations and oxidizing stress. These damages promote strengthening of activity of immune and anti-oxidation protective systems. The same reactions underlie organism adaptations to unfavorable living conditions and can serve as informative indicators for working out of monitoring programs.

The authors hope that this book, *Heavy Metal Pollutants and Other Pollutants in the Environment: Biological Aspects*, is to be used as a reference book for the environmental professional. Professors, students, and researchers in environmental, chemical, and public health and science will find valuable educational materials here.

**Eugene M. Lisitsyn, DSc,**
*Assistant Professor at the N.V. Rudnitsky North-East Agricultural Research Institute, Russian Academy of Sciences (Kirov, Russia), as well as Professor of Cathedra of Ecology and Zoology at Vyatka State Agricultural Academy (Kirov, Russia)*

# PART I

# HEAVY METALS IN SOILS: THE FACTORS INFLUENCING THEIR ACCUMULATIONS IN PLANTS

## CHAPTER 1

# HEAVY METALS IN SOILS AND PARENT ROCKS OF NATURAL LANDS IN NORTH-EAST OF EUROPEAN RUSSIA

LYUDMILA N. SHIKHOVA

*Vyatka State Agricultural Academy, 133 Oktyabrsky Prospect, Kirov, 610017, Russia, E-mail: shikhova-l@mail.ru*

## CONTENTS



### ABSTRACT

The data on the content and profile distribution of total and mobile forms of heavy metals (HM) (Mn, Fe, Cu, and Ni) in the basic types of soils and parent rocks of the Kirov region is presented in the article. It is established that background levels of content of the elements do not exceed their critical volume; however as a whole they are some above than in soils of the central and western regions of Russia. Profile distribution of total forms of HM conforms to well-known laws. Distribution of mobile forms of HM has no accurate regularity.

## 1.1 INTRODUCTION

Heavy metals – widespread components of emissions of many enterprises of different industries and transport concern to the most dangerous pollutants of atmospheric air, soil, waters, vegetation, and further – animals and human. The Kirov region is included into first ten subjects of the Russian Federation on a ratio of agricultural areas with excess of maximum allowable concentration on heavy metals [1]. Existing researches on problem of HM in soils of the Kirov region do not give complete representation about the contents and features of distribution of some chemical elements. Frequently they are received by different methods. Therefore, researches on the given theme are extremely relevant.

The important problem at realization of monitoring of soils pollution by heavy metals is definition of criterion of comparison of pollution degree. As a rule, soil pollution by that or other element is judged by comparing its content to value of maximum allowable concentration (MAC). However, it is not always justified. Geochemical and geomorphological features of soils of exact regions lead to situation when background concentration of some elements in soils and plants of these regions exceed MAC values [2]. Besides, at studying of their pollution by heavy metals there is a question on values of the MAC itself. In references and statutory acts rather essential differences in values of MAC for many elements take place [3, 4]. Biogeochemical features of territories impose sometimes restrictions on use of this indicator, and existing norms of MAC consider not all the factors influencing behavior of HM.

Determination of MAC for mobile forms of elements is even more difficult. Comparing of obtained data on them is very difficult because of use of different methods of extraction and estimation of mobile fractions of HM by different researchers. But the content of mobile fractions of

HM can give the first information on the beginning of negative processes in soil. We agree with opinion of some researchers that in this situation regional background concentration of HM, especially of their mobile forms, can represent itself as criterion of comparison [5–7].

Therefore, a main aim of the given work was definition of levels of background content of HM in soils and parent rocks of the Kirov region.

## 1.2 MATERIALS AND METHODOLOGY

The Kirov region is in a northeast part of the European Russia and borders on PredUral. The basic part of its territory belongs to South taiga subband of a taiga zone. The most northern part concerns a subband of middle taiga, and a southern part – to a zone of mixed coniferous-broad-leafed forest.

The most widespread soils in region – podzolic, sod-podzolic, gray wood and alluvial soddy soils, basically of heavy granulometric composition [8] were investigated. Soil samples were selected in background territories. The territories removed from the nearest large settlements and asphalted roads not less than on 3–5 km were considered as background. The great part of samples is selected from top organogenic horizons. For the characteristic of the profile content and distribution of heavy metals soil samples were selected from soil profiles by genetic horizons to 70–150 cm depth. Estimation of soil features was conducted with widespread methodic: exchangeable acidity – after Sokolov, $pH_{KCl}$ – potentiometrically, content of organic carbon – after Tjurin [9].

Description of soil horizons see Appendices in article Lyudmila N. Shikhova, Olga A. Zubkova, and Eugene M. Lisitsyn “Dynamics of Organic Matter Content in Sod-Podzolic Soils Differ in Degree of Cultivation” in this book.

Definition of total content of HM in soils is spent in small number of samples. Mobile fractions of HM defined in acetic-ammonia buffer solution, pH 4.8. Content of HM was estimated after corresponding processing by a method of atomic-absorption spectrophotometry [10, 11].

Statistical data processing is spent on personal computer with use of software package “STATGRAPHICS *Plus* for Windows 5.1.”

## 1.3 RESULTS AND DISCUSSION

### *1.3.1 LEAD (PB)*

Lead content in lithosphere is about 16 mg/kg of soil. Lead is concentrated in acid magmatic rocks and clay deposits where its concentration fluctuates within 10–40 mg/kg [12].

In blanket loams from different areas of the Kirov region the amount of total Pb varies from 5.0 to 43.0 mg/kg with average level equal to 25.9 mg/kg (Table 1.1). Eluvium-deluvium of Perm deposits has the least content of total lead (in Kotelnitch and Orlov districts of Kirov region – 7.8 and 9.0 mg/kg, accordingly) as well as light granulometric mother rocks. Mobile forms of Pb are predominant in loamy soils (0.45–3.16 mg/kg) then in sandy soils (0.45–2.22 mg/kg). As a whole, levels of content of total and mobile Pb in basic mother rocks of the region can be characterized as moderate lying within the limits of concentration characteristic for sedimentary rocks.

The lead content in soils is inherited from mother rocks. Concentration of Pb in top horizons of different types of soil varied within 3–189 mg/kg.

**TABLE 1.1** Content of Some Heavy Metals in Basic Mother Rocks of Kirov Region (mg/kg)

| | Pb | | Cd | | Zn | | Cr | |
|---|---|---|---|---|---|---|---|---|
| **Soils** | **Total** | **Mobile** | **Total** | **Mobile** | **Total** | **Mobile** | **Total** | **Mobile** |
| Blanket loams | 25.9 | 1.49 | 0.92 | 0.12 | 55.0 | 1.70 | 140.6 | 1.59 |
| Eluvia-deluvium of Perm deposits | 8.4 | 7.1 | — | 2.4 | — | 1.0 | — | 1.8 |
| Eluvia-deluvium of Cretaceous deposits | — | 0.84 | — | 0.15 | — | 3.50 | — | 0.95 |
| Fluvio-glacial and old-eluvium deposits | 2.66 | 1.04 | — | 0.11 | 15.7 | 2.64 | 6.7 | 0.86 |
| Modern eluvia deposits | — | 1.15 | — | 0.20 | — | 3.13 | — | 0.78 |

Note: "-" total content of element is not determined

Average values on types of soil make 32 mg/kg. On a global scale average concentration of Pb in topsoil is estimated in 25 mg/kg [12]. In soils of Russian Plain the background content of an element varied from 2.6 to 43 mg/kg [13]. Content of total lead in investigated soils of Kirov region is close to that values (Table 1.2).

Arable horizons of loamy agricultural soil content about 26–30 mg/kg of lead. In literature [14] there are some lower values: regional natural background of Pb content was 12.0 mg/kg and in arable soils – 14 mg/kg. This clash of opinions is explained with the fact that we paid much attention to soils of heavy granulometric content. In top part of soil profile there is biogenic accumulation of Pb. Sometimes in eluvial horizons content of the element is higher. Such a character of profile distribution of Pb is noted by other investigators [15–20].

Mobile forms of Pb consists only little part of total content of an element. The greater part of lead complexes has low or very low solubility. But in acid soils essential part of Pb (II) up to 10–70% exists in exchangeable state basically as organic-mineral complexes [21].

On data of [22] acetic-ammonia solution with pH 4.8 extracts only little amounts of Pb from background sod-podzolic soils – about 5–7% of total Pb. In our investigations ratio of mobile Pb in total lead content varied in high degree. In top humгы-accumulated part of soil profile it is the greatest (4–23%) as well as in lowest horizons adjoining to mother rocks (12–22%).

In horizons of forest ground litter on podzolic and sod-podzolic soils of Kirov region content of mobile Pb is about 1 mg/kg (Table 1.3).

Ground litters of loamy soils content significantly more lead than ground litters of sandy-loam soils. There is not any regularity in distribution of lead on profile of forest soils because in acid and strong acid media Pb migrates actively within soil profile and could be transferred beyond its bounds as chelate composition.

Maximum of Pb content within soil profile is not always noted in organogenic horizons. Eluvial accumulative coefficient (EAC) in ground litter horizon varies from 0.2 till 13.4. Loamy forest soils content significantly more mobile forms of lead than sandy-loam soils.

Content of mobile forms of lead in humus horizons of arable and meadow sod-podzolic soils are not differ significantly and consists on

**TABLE 1.2** Content of Heavy Metals in Different Horizons of Sod-Podzolic and Loamy Podzolic Soils in Kirov Region (mg/kg of Soil)

| Soil horizon | Pb | | Cd | | Zn | | Cr | |
|---|---|---|---|---|---|---|---|---|
| | Total | Mobile | Total | Mobile | Total | Mobile | Total | Mobile |
| Arable soil | | | | | | | | |
| $A_p$ | 26–37 | 1.28–2.67 | 0.88–1.11 | 0.04–0.13 | 58.6–105.3 | 2.0–5.0 | 120–161 | 3.1–10.5 |
| $A_{he}$ | 17–37 | 1.25–3.33 | 0.78–1.05 | 0.10–0.21 | 48.0–58.4 | 2.0–2.2 | 115–128 | 1.9–4.2 |
| AB | 15–39 | 0.40–2.69 | 0.79–1.07 | 0.04–0.16 | 52.5–74.0 | 0.6–2.0 | 112–130 | 1.8–6.6 |
| B | 26–32 | 0.40–3.20 | 0.74–0.99 | 0.01–0.11 | 59.0–66.1 | 0.6–2.8 | 133–156 | 0.9–10.1 |
| BC | 23–30 | 1.11–3.11 | 0.93–1.20 | 0.01–0.13 | 37.1–83.5 | 0.2–2.8 | 134–172 | 0.9–5.8 |
| C | 20–38 | 1.00–4.25 | 0.76–1.19 | 0.08–0.16 | 58.5–79.1 | 0.4–1.0 | 143–172 | 0.8–5.0 |
| Meadow | | | | | | | | |
| $O_e$ | — | 2.15 | — | 0.11 | — | 4.21 | — | 3.0 |
| $A_h$ | 20 | 2.00 | 0.70 | 0.29 | 58.6 | 2.05 | 110 | 2.0 |
| $A_{he}$ | 16 | 1.44 | 0.86 | 0.10 | 51.6 | 2.05 | 149 | 1.1 |
| AB | 29 | 1.68 | 0.86 | 0.05 | 48.5 | 2.05 | 131 | 1.1 |
| B | 15 | 3.26 | 0.77 | 0.02 | 55.3 | — | 109 | 1.0 |
| Forest | | | | | | | | |
| $O_i$ | 24–37 | 2.42–6.75 | 0.81–1.03 | 0.13.0.23 | 54.9–78.2 | 2.0–3.1 | 127–171 | 1.7–8.3 |
| $A_{he}$ | 22–37 | 2.42–3.20 | 0.83–1.03 | 0.05–0.19 | 43.9–78.2 | 0.4–2.8 | 128–171 | 1.7–8.2 |
| AB | 20–27 | 1.42–2.62 | 1.02–1.08 | 0.01–0.04 | 35.7–62.6 | 0.6–2.8 | 125–151 | 1.7–4.8 |
| B | 20–33 | 1.42–2.62 | 0.72–1.08 | 0.05–0.08 | 45.5–46.4 | 2.0–3.4 | 72–168 | 1.7–4.8 |
| BC | 20–35 | 2.41–3.20 | 0.66–1.02 | 0.07–0.12 | 39.2–47.4 | 2.0–3.1 | 71–168 | 1.7–4.0 |

Note: "-" total content of element is not determined.

**TABLE 1.3** Content of Mobile Forms of Heavy Metals (mg/kg) in Organogenic Soil Horizons in Kirov Region of Russia

| Types of soils | | Pb | Cd | Zn | Cr |
|---|---|---|---|---|---|
| Ground litters of forest podzolic and sod-podzolic soils | Sandy-loam soils | 0.87±0.07 | 0.15±0.02 | 3.53±0.20 | 1.26±0.10 |
| | Loamy soils | 1.22±0.16 | 0.09±0.01 | 4.78±0.37 | 1.57±0.29 |
| Sod-podzolic soils | Arable soils | 2.04±0.16 | 0.11±0.02 | 3.36±0.24 | 3.12±0.49 |
| | Meadows, fallows | 1.67±0.34 | 0.14±0.02 | 3.79±0.45 | 1.42±0.21 |
| Gray forest soils | | 1,56±0,18 | 0.10±0.01 | 2.82±0.25 | 2.69±0.53 |
| Eluvia-sod soils | Sod horizons | 1.52±0.23 | 0.15±0.03 | 4.50±0.41 | 0.95±0.21 |
| | Humus horizons | 0.96±0.18 | 0.18±0.05 | 3.87±0.30 | 1.49±0.25 |

average 1.95±0.14 mg/kg (see Table 1.3). Profile distribution of an element in arable and meadow soils has weakly expressed biogenic-accumulating character. In some cases top part of soil profiles is depleted in mobile lead in compare with lower horizons. In mother rock increasing of content of mobile forms of Pb is expressed more clearly. Minimum content of Pb is noted in eluvium horizons.

Content of Pb and regularity of its profile distribution in gray forest soils is the same as in sod-podzolic soils. But humus horizons of gray forest soils content significantly less mobile Pb – 1.56±0.18 mg/kg (see Table 1.3).

Eluvia-sod soils had less mobile Pb than sod-podzolic and gray forest soils – 1.16±0.15 mg/kg on average. Its maximum content is characteristic for top part of soil profiles. In lower sandy deposits content of mobile Pb is minimum. In such soils EAC for lead is higher than 1 as a rule.

### *1.3.2 CADMIUM (CD)*

Average concentration of cadmium in lithosphere varies about 0.10 mg/kg; in sedimentary rocks – about 0.30 mg/kg [12, 23].

Content of total cadmium in mother rocks of Kirov region is some higher than world clark of the element and it's known concentrations in sedimentary rocks (see Table 1.1). In investigated samples of loamy rocks variability of a concentration of the element is 0.71–1.09 mg/kg.

Content of mobile Cd in mother rocks is little and loamy rocks are differing insignificantly from sandy rocks. The highest concentrations of mobile Cd are noted in alluvia sediments.

Content of Cd in soil are determined by soil type in great degree. More often limits of variability of total Cd content in un-pollinated soils are about 0.01–0.70 mg/kg; in podzolic and sod-podzolic soils – 0.70–2.31 mg/kg; in gray forest soils – 0.65 mg/kg [15, 24]. Concentration of the metal in humus horizon of sod-podzolic loamy soils of European Russia is near 0.14 mg/kg [23].

Content of total cadmium in soils of investigated territory of European North-East is near its content in soils of neighbor regions but some higher than in Russia as a whole (see Table 1.2). In different horizons of sod-podzolic and podzolic loamy soils cadmium content varies from 0.66 till 1.11 mg/kg. Distribution of total Cd within soil profile has biogenic-accumulating character as a rule [18–20]. However, biogenic accumulation of total Cd in investigated soil profiles is noted not always or weakly expressed; some authors explain that fact by significantly migration of Cd downstream along profile in soils of humid landscapes [12]. Content of total Cd as well as its mobile forms is lowering with depth [5, 25]. So it is eluvial-illuvial re-distribution of total Cd in soil profiles.

It is known that high concentration of organic matter of humate composition, silty fractions, hydroxides of iron and aluminum, neutral or alkaline pH of medium promote fixing and accumulating of Cd in soil in slow-moving state. Perhaps complexes of secondary minerals with organic matter and hydroxides of Fe and Al play important role in Cd accumulation [26]. Cd has highest mobility and availability to plants in acid soils with low pH level, low base exchange capacity, fulvatic type of humus, and flushing water regime [18, 27]. In acid soils Cd forms soluble organic-mineral complexes with organic matter and so is highly mobile [12, 28].

In different horizons of soils of Kirov region content of mobile cadmium varies from 0.01 till 0.30 mg/kg. The highest one is in accumulating-eluvial part of soil profile and in mother rock. The lowest content was in illuvial horizons.

As a rule ratio of mobile Cd in total Cd content is not more than 10% [19, 22]. In our case it varies in wide limits. Its maximum is characteristic for upper part of profile (from 10 up to 40%) that is linked with existence of organic matter of fulvatic type. In illuvial part of profile ratio of mobile cadmium decreased to 1–11%, and increased again near mother rock.

Content of mobile Cd in organogenic horizons of investigated podzolic and sod-podzolic soils is very little (see Table 1.3). Forest ground litters of sandy soils content significantly higher amounts of mobile Cd than loamy soils. Heightened content of the element is noted in lowest accumulating-eluvial and eluvial horizons. As a rule the highest amounts of EAC in upper horizons are characteristic for loamy soils. In illuvial horizons increasing in content of mobile fractions of Cd is not noted.

Sod-podzolic soils of the region have lowest content of mobile cadmium (see Table 1.3). Its content in humus horizons of meadow and fallows soils is some higher than in arable soils. Similar lower content of mobile Cd in plow horizons of sod-podzolic soils (0.09–0.28 mg/kg) was noted earlier by other researchers [29, 30]. The lowest content of mobile Cd had plow horizons of soils in eastern and north-eastern districts of the region (Afanasevo, Zuevka, and Falenki districts) – 0.05–0.07 mg/kg that is explained by heavy granulometric soil composition and higher content of sesquialteral oxides in mother rocks that decrease Cd mobility.

Maximum concentration of mobile Cd in arable soils is in eluvial horizons (EAC is about 4–6 units), and illuvial part of soil profile has minimum content of the element almost always. In meadow soils humus horizons have maximum mobile Cd, but illuvial horizons – minimum one.

In humus horizons of gray forest soils Cd content does not differ from sod-podzolic soils (see Table 1.3). Profile distribution of the element and EAC values are similar to that in sod-podzolic soils too.

The highest content of mobile Cd is characteristic to sod and humus horizons of floodplain soils (see Table 1.3). In middle part of its profiles content of the element is decreased but increased again near mother rock – alluvial deposits. In rare cases it is noted increasing of mobile Cd content in illuvial horizons.

### 1.3.3 ZINC (ZN)

Average clark of Zn in lithosphere is about 50 mg/kg. According to our data content of total Zn in loamy mother rocks of Kirov region varies from 43 to 68 mg/kg (see Table 1.1). These levels are close to data [29] for sedimentary rocks. The highest content of total zinc among the investigated rocks is characteristic for surface loamy soils of Murashi, Afanasevo, and Falenki districts of Kirov region (59.2; 56.9; and 67.9 mg/kg, respectively).

High content of total Zn is not mean high content of mobile forms of the element. Correlation coefficient between content of total and mobile forms of the element in loamy mother rocks is insignificant ($r = 0.16$). These rocks have only little amount of mobile Zn.

Water-glacial and ancient-alluvial sediments have the lowest content of total Zn [31]. However, they have some time more mobile forms of the element than loamy sediments. Perhaps this fact is linked with minimum content of silty fractions and of organic matter, which are basic absorbent of zinc [12]. Considerable amount of mobile zinc is noted in modern alluvial deposits.

Arable soils of Russia content near 34.4 mg/kg of total Zn as a whole. In central regions of Non-Chernozem Zone of Russia content of mobile Zn reaches 1.6 mg/kg [32].

Total zinc content in sod-podzolic soils of Kirov region may be characterized as heightened (see Table 1.2). Its content within investigated soil profiles is not less than 40 mg/kg according to soil horizon. Zn is sharply expressed element of soil biogenesis. Its biogenic accumulation in upper parts of soil profile is very characteristic as well as its depletion in podzolized horizons. Second significant maximum of its content is noted in illuvial parts of profiles.

Mobile zinc consists 0.24–9.46% of total Zn content in investigated sod-podzolic soils of Kirov region (see Table 1.2). Ratio of mobile Zn may increases with soil depth as well as decreases. There is not any regularity in change of ratio of mobile zinc within soil profile.

Analysis of content of mobile Zn within some types and sub-types of soil has shown that degree of leaching or podzolizing of soil has significant effect on mobility of the element. Amount of mobile zinc in sod-podzolic and gray forest soils is increased naturally in accordance with increasing of their degree of podzolizing; in sod-carbonate soils – with increasing of their leaching degree [31]. There are some evidences about increasing of Zn mobility in over-moistened soils [33]. Mineral colloids and organic matter especially complex compounds play important role in migration or fixing of zinc in soils.

Zinc is biophyle element; its maximum content is in upper soil horizons. However, content of mobile zinc in organogenic horizons of most spread soils of Kirov region is rather low (see Table 1.3). Forest ground litters have only 3.93±0.19 mg/kg of mobile zinc; ground litters of loamy forest soils have significantly more Zn than sandy soils that is linked with originally high content of Zn in their mother rocks. Distribution of mobile zinc within profiles of forest soils has accumulating character as a rule. Zn has highest EAC values among all investigated elements but seasonal dynamics of Zn content may disturbs this regularity. Eluvial horizons of podzolic soils are depleted with Zn.

Organogenic horizons of arable and meadow soils have significantly more mobile zinc than forest ground litters (see Table 1.3). Soil of eastern, north-eastern, and southern districts of Kirov region have insignificantly more mobile zinc. There is biogenic accumulation of the element in upper horizons that is testified with high values of EAC. Eluvial-illuvial re-distribution is noted not always because of significant seasonal dynamics and periodical anthropogenic influence on arable soils. Total regularity of profile distribution of zinc in meadows and arable soils is the same as in forest soils.

Content of mobile Zn in humus horizons of gray forest soils is lower than in sod-podzolic soils. Profile distribution of the element has also biogenic-accumulating character; depletion of podzolic horizons with zinc is not noted. EAC values in humus horizons of gray forest soils is higher than 1.

Floodplain soils have high content of mobile zinc not only in organogenic horizons but in whole soil profile (see Table 1.3). But its distribution within profile has not any regularity. EAC values in humus horizons vary within wide limits.

Basic soils of Kirov region have low Zn content especially of mobile compounds as a whole.

### 1.3.4 CHROMIUM (CR)

Clarke of chromium in Earth crust is about 83–100 mg/kg. Soils have near 20 mg/kg of the element in average. Chromium content in mother rocks of investigated region of Russia lays within limits characteristic for sedimentary rocks (see Table 1.1). Content of total form of the element in loamy deposits varies from 102 to 166 mg/kg. Water-glacial deposits (in Orichi district of Kirov region) contain 6.7 mg/kg Cr. Highest amount of total chromium is noted for surface loamy soils in Afanasevo, Falenki, and Zuevka districts of Kirov region. This fact may be connected with their participation in forming of Perm rocks as well as glacial deposits brought from Ural Mountains. Glacial deposits formed at the expense of mother rocks of different regions keep geochemical specificity of original territory after transition into Kirov region. Material brought into Russian Plain from Ural Mountains has some higher concentration of Cr in compare with soils brought from Baltic Shield [34].

There is a little mobile form of Cr in loamy soils: its amount varies from 0.56 to 4.8 mg/kg, i.e., about 0.5–3 5 of total Cr content. Light soil-forming rocks contain significantly lower amount of mobile Cr than soils of heavy granulometric composition. There is not significant correlation between content of total and mobile forms of the element. Heightened amount of mobile chromium is estimated in mother rocks of Afanasevo and Malmyzh districts that may be explained by the same reasons as for total Cr. Concentration of mobile Cr in alluvial deposits of floodplain soils does not differ from other types of soil.

Most part of chromium is in form $Cr^{3+}$; its compounds in soils are considered highly stable. But form $Cr^{6+}$ is highly unstable and mobilized easy in acid as well as alkali soil. Behavior of Cr is determined by soil pH and redox potential. It is known that concentration of Cr in soil is depended from its content in mother rocks [12]. However, variability of concentration of the element in arable and eluvial horizons is governed with soil-forming processes [35]. Cr content in soils is depended first of all with granulometric composition, mineralogical composition, and degree of humification. It increases at heightened of content of humus and silty fraction. Cr concentration has positive correlation with silt content [34, 36].

Biogenic accumulation with organic matter is characteristic for Cr distribution in humus horizons. Second maximum of total Cr content is noted in illuvial horizons [25]. Podzolic process leads to depletion of upper horizons with the element.

Data on the chromium content in soils of Russia is rather diverse. Soils of an arable land of the Russian Federation as a whole contain less than 50 mg/kg of total trivalent Cr.

The content of mobile chromium in soils is insignificant. For example, in sod-podzolic light-clay arable soils of the Ramensky agrochemical experimental station (Moscow Region) contains 0.41–0.85 mg/kg Cr (1M HCl-solution). At weighting of granulometric structure the amount of mobile Cr increases to 0.36–4.8 mg/kg [37]. Sod-podzolic gley and gleyey soils of the Smolensk-Moscow Height contain nearby 40–50 mg/kg of total chromium and 0.4–2.7% of the mobile Cr. The element content varies considerably on years [38].

In sod-podzolic soils of the Kirov region the amount of total chromium fluctuates within 71–171 mg/kg in various soil horizons. Biogenic accumulation of an element in organogenic horizons and eluvial-illuvial differentiation of its content in a profile is accurately shown. The presented means of concentration of total chromium are close to the data resulted on an ecological-geochemical card of soils of the Kirov region (1996). Proceeding from it, the natural regional background makes 84 mg/kg of soil; in arable horizons of agricultural soils there is about 111 mg/kg Cr; in technogenic landscapes – near 95 mg/kg.

Under our data mobile chromium makes 0.5–7.0% of total for different horizons. The amount of mobile Cr is maximum in the top part of a profile and is minimum in eluvial part. However, accurate laws in its distribution on profiles are not observed. Raised concentrations of Cr are often marked in other horizons (see Table 1.2). The greatest share of mobile chromium from total is also observed in the top part of a profile.

Organogenic horizons of soils of region are characterized by the low content of mobile forms of an element (see Table 1.3). In forest ground litter contains nearby 1.36±0.11 mg/kg Cr on the average. Ground litters of forest loamy soils contain more mobile chromium than ground litters of sandy soils though it is doubtful.

In profiles of heavy soils the element content is considerably above than in easy soils. Both for granulometrically light and for heavy soils maximal contents are marked in different horizons. But more often the highest concentration are characteristic for humus-accumulative and illuvial parts of profiles. It confirms that Cr accumulates basically in silty fraction and in complexes with organic matter. EAC in the top horizons of forest soils seldom falls below 1 but fluctuates within 0.1–6.2 units.

Sod-podzolic arable soils contain in arable horizon significantly more mobile chromium than forest ground litters (see Table 1.3). Element distribution in soil profiles of arable soils is more uniform frequently without clearly expressed maximum in illuvial horizons. In arable horizons the maximum content of mobile chromium is marked. EAC is within limits 0.1–5.7. The same features can be noted in sod-podzolic soils under meadow vegetation.

Grey forest soils on the content of mobile chromium in organogenic horizons do not differ from sod-podzolic one (see Table 1.3). The maximum content of an element in a profile is on humus horizons. EAC is practically always more than 1 and fluctuates from 0.8 to 4.9. Illuvial accumulation is absent practically.

Floodplain soils contain a minimum amount of mobile chromium in humus horizons among the studied soils. EAC in humus horizons fluctuates from 0.2 to 4.5. Illuvial chromium accumulation is most characteristic for floodplain soils.

### *1.3.5 MANGANESE (MN)*

Manganese is one of the most widespread microelement. Average clark of total Mn in lithosphere makes about 850–900 mg/kg [12].

Among loamy mother rocks of the Kirov region the least amount of total Mn contains in blanked loams of the western and northwest districts of the region (Podosinovets, Murashi, Darovskoy districts) (Table 1.4). Acid aluminosilicate moraine of Scandinavia and Kola Peninsula rather poor with microelements took considerable participation in formation of these rocks [39].

**TABLE 1.4** Content of Some Heavy Metals (mg/kg) in Basic Mother Rocks of Kirov Region

| Soils | Mn | | Fe | | Cu | | Ni | |
|---|---|---|---|---|---|---|---|---|
| | Total | Mobile | Total* | Mobile | Total | Mobile | Total | Mobile |
| Blanket loams | 818.3 | 39.8 | 6.55 | 160.4 | 55.0 | 1.70 | 140.6 | 1.59 |
| Eluvia-deluvium of Perm deposits | 656 | 70.2 | — | 194.2 | 25.4 | 4.25 | 64.7 | 5.3 |
| Eluvia-deluvium of Cretaceous deposits | — | 26.4 | — | 240.9 | — | 3.10 | — | 0.18 |
| Fluvio-glacial and old-eluvium deposits | — | 30.0 | — | 220.9 | 5.2 | 1.70 | 22.8 | 0.56 |
| Modern eluvia deposits | — | 57.9 | — | 304.6 | — | 1.95 | — | 0.49 |

Note: * – for total Fe means are in % not in mg/kg; "-" total content of element is not determined.

In blanked loams of east, northeast and central districts of the region (Afanasevo, Uni, Zuevka districts) content of Mn is higher – about 860–890 mg/kg of total and 50–60 mg/kg of mobile Mn. Acid magmatic rocks of Northern Ural Mountains, New Land Isla and the Timansky range containing higher amounts of total manganese have served as an initial material for formation of glacial deposits in these parts of the region. Besides, Paleozoic bed-rocks rather rich in microelements and leaving close to a day surface near Vyatsky Uvaly and the Verkhnekamsky Height took part in formation of mother rocks of PredUralje [34, 40]. Raised content of Mn is noted in Nizhny Novgorod region, in Tatarstan, in Chuvashiya, and Udmurtiya in rocks relative on genesis [31].

In light rocks there are significantly less amounts of mobile Mn. The high average content of mobile manganese is characteristic for alluvial sediments. Soils substantially inherit level of the total content of an element in mother rocks. Fluctuations in Mn content in soils can reach from several tens to 2000 and more mg/kg [12]. Maximum allowable concentrations for total manganese make 1000–1500 mg/kg, for mobile – 60–80 mg/kg [41].

The greatest means of content of total manganese are characteristic for acid soils. Researchers mark exclusively high biogenic accumulation of an element in eluvial landscapes [19, 31, 42].

The maximum levels of total Mn content in soils are also characteristic for east and northeast districts of the region. For mobile fractions of Mn such law is not observed (Table 1.5). Mobile Mn makes very small part from the total. Its share as a rule is above in the top and bottom parts of a profile.

For all studied soils biogenic-accumulative accumulation of an element and eluvial-illuvial differentiation of a profile in Mn content is characteristic. Apparently, character of this redistribution depends on degree of soil podzolization. More accurate laws of profile distribution of an element are shown for total Mn. For mobile forms they are rather conditional: maximal contents are observed in eluvial horizons as well. Possibly, at change of oxidizing conditions to reducing one in podzolic horizon the element can segregate in numerous small complexes which give high level of content at the analysis [40].

As a whole, average values of the content of mobile manganese in organogenic horizons of the studied soils can be characterized as raised

**TABLE 1.5** Content of Heavy Metals (mg/kg of soil) in Different Horizons of Sod-Podzolic and Loamy Podzolic Soils in Kirov Region

| Soil horizon | Mn | | Fe* | | Cu | | Ni | |
|---|---|---|---|---|---|---|---|---|
| | Total | Mobile | Total (%) | Mobile | Total | Mobile | Total | Mobile |
| *Arable soil* | | | | | | | | |
| $A_p$ | 1137 | 28.8 | 3.33 | 185.0 | 60.0 | 2.79 | 52.3 | 0.22 |
| $A_{he}$ | 1015 | 15.1 | 4.80 | 177.5 | 41.9 | 2.78 | 64.0 | 0.65 |
| AB | 756.3 | 18.3 | 5.00 | 170.9 | 33.1 | 2.43 | 67.7 | 0.92 |
| B | 883.3 | 17.5 | 5.66 | 167.0 | 51.9 | 3.40 | 73.7 | 1.01 |
| BC | 762.3 | 22.4 | 5.79 | 169.6 | 41.8 | 3.37 | 77.0 | 0.90 |
| C | 897.7 | 54.5 | 6.16 | 166.9 | 44.8 | 2.89 | 77.3 | 1.29 |
| *Meadow* | | | | | | | | |
| $O_e$ | — | 17.9 | — | 280.2 | — | 1.43 | — | 1.16 |
| $A_h$ | 804 | 17.8 | 4.58 | 280.1 | 38.9 | 1.58 | 63 | 0.72 |
| $A_{he}$ | 442 | 17.9 | 3.79 | 282.0 | 39.8 | 1.60 | 49 | 0.51 |
| AB | 518 | 26.8 | 4.43 | 270.0 | 34.5 | 1.59 | 51 | 0.70 |
| B | 796 | 27.0 | 4.36 | 201.5 | 31.5 | 1.61 | 63 | 0.72 |
| *Forest* | | | | | | | | |
| $O_i$ | — | 54.5 | — | 131.1 | — | 2.68 | — | 0.66 |
| $A_{he}$ | 824 | 26.7 | 3.85 | 148.9 | 30.7 | 2.88 | 41.7 | 0.65 |
| AB | 672 | 20.6 | 4.93 | 149.7 | 25.5 | 2.39 | 48.0 | 0.69 |
| B | 772 | 27.7 | 5.33 | 145.3 | 32.6 | 2.34 | 84.7 | 1.09 |
| BC | 660 | 39.2 | 5.57 | 126.9 | 29.1 | 3.06 | 64.0 | 0.95 |

Note: * – for total Fe means are in % not in mg/kg; "-" total content of element is not determined

(Table 1.6). The maximum values of the content are received for forest ground litters. And ground litters of light soils accumulate significantly more mobile Mn than heavy soils. Organogenic horizons of floodplain soils differ with the high content of Mn too. The least content of mobile manganese is characteristic for arable sod-podzolic soils. In meadow and fallow sod-podzolic soils significantly more Mn collects in humus horizons than on an arable land. Grey forest soils are intermediate. The correlation analysis has shown low frequently not significant coefficients of correlation between the content of mobile forms of manganese and degree of soil acidity as well as with content of organic carbon ($r$ = 0.3–0.4).

**TABLE 1.6** Content of Mobile Forms of Heavy Metals (mg/kg) in Organogenic Soil Horizons in Kirov Region of Russia

| Types of soil | | Mn | Fe | Cu | Ni |
|---|---|---|---|---|---|
| Ground litters of forest podzolic and sod-podzolic soils | Sandy-loam soils | 74.80±5.90 | 200.4±10.0 | 1.63±0.13 | 0.59±0.03 |
| | Loamy soils | 58.15±5.80 | 174.5±12.6 | 1.72±0.21 | 0.67±0.07 |
| Sod-podzolic soils | Arable soils | 38.32±3.55 | 180.9±11.0 | 2.42±0.19 | 0.41±0.07 |
| | Meadows, fallows | 66.51±10.25 | 174.0±22.5 | 2.15±0.32 | 0.68±0.24 |
| Gray forest soils | | 56.09±4.04 | 277.8±21.8 | 2.17±0.23 | 0.26±0.03 |
| Eluvia-sod soils | Sod horizons | 86.77±11.55 | 224.3±49.2 | 2.05±0.35 | 0.40±0.07 |
| | Humus horizons | 48.39±5.66 | 235.8±24.8 | 2.18±0.30 | 0.65±0.14 |

### *1.3.6 IRON (FE)*

Iron takes the second place among metals on the content in earth crust after aluminum. Its clark is made by 4.65% [12]. The iron content as heavy metal in mother rocks is great up to the whole percent. According to Ref. [43] content of $Fe_2O_3$ in moraine loams and sandy loams from different districts of the Kirov region fluctuates within 3.82–6.77% (2.61–4.63 for Fe); in fluvio-glacial sand and sandy loams – 1.16–3.96% (0.79–2.71 for Fe); in blanked carbonate-free loams and clays of different districts of the region – 3.13–9.87% (2.14–6.75 for Fe); in eluvia of the Perm bedrocks depending on granulometric structure – 1.74–9.68% (1.19–6.62 for Fe). Similar values of total Fe content for similar rocks in some districts of the region are resulted in works of other authors [29, 44]. These values are some higher than resulted in Ref. [24] for soils and mother rocks as a whole. High values of the total content of iron are received in our researches (see Table 1.4). The highest values – 7.21 and 7.14% Fe are noted in rocks of east districts (Afanasevo and Zuevka). In northwest Murashi district blanked loams contain less iron (4.52%).

Thus, all researchers in the given region mark the greatest content of Fe in eluvia and eluvia-deluvium of Perm carbonate rocks or in sediments in which formation the given rocks took part. The Perm rocks are characterized by heavy granulometric structure and the considerable content of non-silicate forms of iron. Total content of $Fe_2O_3$ makes up to 12%, in silty fraction – up to 14% [43].

The content of mobile iron in mother rocks has no accurate geographical laws. Significantly higher amount of an element is noted only in blanked loess loams in zone of mixed coniferous-broad leaved forests in a southern part of the region. As well as for manganese, significantly higher values of the content of mobile Fe are characteristic for fluvio-glacial and old-eluvium deposits. But the highest values are noted in modern alluvial sediments.

The iron content in soils is substantially inherited from mother rocks; and its distribution on a soil profile and within one horizon reflects a direction and features of soil-forming process. Eluvial-illuvial and accumulating-eluvial types of distribution are presented in podzolic and sod-podzolic soils. For iron it is characteristic formation of congestions or new formations. As a part of these new formations there is a considerable part of other heavy metals [24].

The behavior of iron in soils in high degree depends on an oxidation-reduction mode of soil. Behavior of oxidic and protoxidic iron essentially differ depending on soil pH. Complexes of trivalent iron ($Fe^{3+}$) are inactive. In reduction conditions oxidic iron passes in protoxidic with formation of soluble complexes. $Fe^{2+}$ is most movably at pH < 8–8.5 [24]. Real solubility of $Fe^{3+}$ in a soil solution is considerably higher than theoretical that is caused by accompanying reactions. Iron forms easily complexes with many organic compounds. Stability of these complexes is various essentially depend on pH value. Sharp increase of the content of mobile forms of iron sometimes toxic for plants causes development of anaerobic conditions. Acid organic matter raises mobility of iron too [12, 24].

The content of total iron in the studied soils essentially does not differ from other regions of the European part of Russia (see Table 1.5). The minimum amount of total Fe in the top part of a profile and increase of its content is characteristic at approach to mother rocks. The eluvial-accumulative coefficient (EAC), as a rule, is less than 1 in the top part of a profile especially in eluvia horizons and increases in illuvial horizons.

The opposite tendency is traced in the content of mobile iron; its amount is higher in the top horizons. The share of mobile iron from the total is insignificant: no more than 1% in the top horizons and decreases at approach to mother rocks.

Organogenic horizons of soils of the region contain significant amounts of mobile iron (see Table 1.6). The lowest values of the content of an element are noted in forest ground litters and humus horizons of sod-podzolic soils.

In soil profiles maxima and minima of the content of mobile iron are observed in different horizons. EAC fluctuates in considerable limits from 0.3 up to 3.3. In loamy soils dynamics of the profile content is smoother than in the sandy soils. EAC in the majority of horizons is close to 1. Its insignificant increase is marked in eluvial parts of profiles.

Smoother distribution of the content of mobile iron on depth of a profile in comparison with meadow and fallow soils is characteristic for arable sod-podzolic soils.

Humus horizons of gray forest soils differ from organogenic horizons of other soils with the greatest content of mobile Fe (see Table 1.6). Profile distribution of an element is more uniform. Rather weak accumulation of mobile iron is marked in humus and illuvial horizons.

Element distribution in profiles of floodplain soils is non-uniform (see Table 1.6). In a great number of cases maxima are marked in the top parts of profiles. EAC fluctuates in different horizons from 0.2 up to 5.3.

### *1.3.7 COPPER (CU)*

Average clark of copper in lithosphere makes 100 mg/kg. For the copper content in rocks the same laws are characteristic as for manganese. In mother rocks it depends basically on quantity of silty fraction [20, 45]. Eluvia of Perm both Triassic clay and loams differ with the high content of copper (about 50 mg/kg). The content of mobile forms of an element in these rocks on the contrary is the least. The amount of mobile forms of copper fluctuates from 10 to 33%. Its content is highest in loess loams; much less – in eluvia of Perm clays and loams. In process of movement on the east of the European Russia the total content of copper increases; the amount of mobile copper increases also up to 5–7 mg/kg [40].

According to our data, mother rocks of east, northeast and southern districts of the region contain more total copper than western and northwest districts. There is not any accurate law in the content of mobile copper. It is possible to note only raised content of Cu in blanked loams of Afanasevo district. As a whole total amount of copper in the studied loamy rocks are close to the element content in similar rocks of surrounding regions; but in sand and sandy loams it is slightly less.

Clark of copper in soils is about 20 mg/kg [40]. The average background content of copper in soils fluctuates within 6–100 mg/kg. High content of clay fraction and organic matter concerns to the factors increasing the content of copper in soil [12].

Mobility of copper in soil as well as its availability to plants depends on many reasons. Water-soluble salts of copper mainly with low-molecular organic matters as well as absorbed by colloids' surface copper are mobile [12, 46]. Mobility of copper especially in the form of complex with organic acids increases in anaerobic media [33, 45].

On an ecological-geochemical map of soils of the Kirov region [14] background values of the content of total forms of copper in soils are estimated within limits of 50 mg/kg; on the average about 28 mg/kg and for arable horizons of farmlands – 37 mg/kg. It is higher than for the western and central regions of Russia. The data of our researches is close to these values also (see Table 1.5). They are higher than world clark of a metal. The high content of total Cu exceeding MAC (66 mg/kg) in some organogenic and illuvial horizons and rocks is noted in Afanasevo and Zuevka districts. In the studied soils the eluvial-illuvial differentiation of profiles on the content of total copper and dependence on biogenic factors is well traced. Usually the copper content increases considerably in the top horizons of soils following the content of humus [5]. However, in soils with a washing mode relation between content of humus and Cu within a profile is not always found out [20].

The content of mobile copper does not exceed 4–5 mg/kg with fluctuations from trace amount up to 5 mg/kg. Mobile copper makes a small part from the total. The greatest share of mobile copper from the total content, as a rule, is dated to eluvia parts of profile – from 4 to 20% in different soils (see Table 1.5). In mother rocks there is near 1.5–7%; in organogenic horizons – 3–4%. The share of mobile copper in gleyey horizons is raised.

In organogenic horizons of soils of the Kirov region the average content of mobile forms of copper is low (see Table 1.6). In forest ground litter it makes about 1.5 mg/kg. Arable sod-podzolic soils differ slightly from forest and meadow soils on content of total copper (see Table 1.6). However, there are significantly more mobile fractions of an element in organogenic horizons of arable soils (see Table 1.5).

As a whole, distribution of total copper on profiles has biogenic-accumulative character. Carrying out of copper from podzolic horizons and accumulation in illuvial is marked [47]. However, accumulation of its mobile forms in forest ground litters can be absent. The eluvial-accumulative coefficient is often less than 1. There is no any law in distribution of an element on profiles. Copper content is much higher in samples of soils from east and northeast districts (Falenki, Zuevka, and Afanasevo) both in heavy and in easy on granulometric structure.

Grey forest soils do not differ significantly from sod-podzolic soils on amount of mobile forms of an element (see Table 1.6). Element distribution in profiles has more expressed biogenic-accumulative character. In all studied profiles EAC for humus horizons is more than 1.

The majority of authors mark the high content of microelements, especially their mobile fraction, in floodplain soils [19, 40]. According to our data, floodplain soils are not distinguished significantly with higher content of mobile fractions of copper in organogenic horizons (see Table 1.6). Copper distribution in a profile has poorly expressed biogenic-accumulative character. Eluvial-illuvial redistribution of copper is marked seldom.

### *1.3.8 NICKEL (NI)*

The average content of nickel in lithosphere is about 80 mg/kg, in soils – 40 mg/kg. Loamy mother rocks of the Kirov region contains nearby 50–80 mg/kg total Ni (see Table 1.4) that is close to the element content in loamy rocks of other regions of the European Russia. In granulometrically light deposits the nickel content is essentially lower (10–45 mg/kg). The highest content of total nickel is marked in blanked loams of the northeast of the Kirov region (Afanasevo district) – 81.3 mg/kg. The content of mobile nickel is in limits 0.29–1.10 mg/kg for loamy rocks, and 0.28–1.11 mg/kg for easy rocks that is does not depend on granulometric structure unlike the total content. Modern alluvial sediments do not differ practically from other rocks by the content of mobile forms of nickel.

The nickel content in soils depends in many respects on its content in mother rocks. However, level of its concentration in the top layer of soils depends on soil-forming processes and technogenic pollution. In the top horizons of soils nickel is basically in connection with organic matter including as a part of easily soluble chelates [45, 48]. However, Ni connected with oxides of iron and manganese also makes a considerable part of the total content of an element [49].

Element distribution on a soil profile depends on the content of organic matter, amorphous oxides, and clay fraction [12, 45].

The content of total Ni in the studied soils of the region is in limits of 30–106 mg/kg for different soil horizons (see Table 1.5). Close values are resulted on an ecological-geochemical map of the Kirov region: a natural regional background is 33 mg/kg; in arable lands – 47 mg/kg; in technogenic landscapes – 42 mg/kg. Much more total Ni contains in soils of Zuevka and Afanasevo districts. Practically in all studied profiles element accumulation in illuvial horizons and depletion of podzolic parts with it is traced accurately. Arable horizons often contain an element in smaller amounts than eluvial horizons.

The content of mobile Ni fractions in the studied soils is insignificant. The eluvial-illuvial differentiation of a profile on the content of mobile nickel is marked. The share of mobile nickel makes 0.3–4.29% of the total content and, as a rule, increases with depth. Increase of a share of mobile nickel of total is sometimes marked in humus horizons.

As a whole, the content of mobile nickel in organogenic horizons of soils of region is low (see Table 1.6). The greatest content of an element is noted in forest ground litter. And its content in ground litter of loamy and sandy soils does not differ significantly. Sod-podzolic arable and meadow soils of the region contain less mobile nickel in organogenic horizons.

Studying of distribution of mobile Ni on a profile of sod-podzolic soils testifies to its weak enough accumulation in humus and arable horizons. EAC fluctuates from 0.1 to 2.7. However, sometimes the coefficient rises up to 20 that occur, obviously, as a result of anthropogenous pollution. More essential accumulation of an element is marked in illuvial horizons of the majority of soils.

Grey forest soils contain significantly less mobile nickel in humus horizons than sod-podzolic soils. Increase in the content of an element in illuvial parts is also characteristic for profile distribution at weak enough accumulation in humus horizons. EAC in humus horizons is in limits 0.2–1.8 without exceeding 1.0 more often.

Floodplain soils of the region contain significantly more mobile Ni in organogenic horizons in comparison with sod-podzolic and gray forest soils. EAC in humus horizons varies from 0.5 to 2.8. It is not established any laws in profile distribution.

## 1.4 CONCLUSIONS

The content of total and mobile forms of heavy metals in soils and mother rocks of the Kirov region does not exceed critical sizes; but as a whole it is above than values of world clarks both the content in soils and rocks of the western and central regions of Russia.

Content of heavy metals increases at direction from the West by the East as increase of participation of Paleozoic deposit rocks in process of soil formation. In certain cases in background territories in east part of the region high concentration of total and mobile Cu, Mn, Ni are revealed exceeding MAC that testifies to the raised regional geochemical background of these elements.

Distribution of total forms of elements in soil profiles submits to the laws known for similar soils of other regions. For mobile forms of heavy metals owing to their mobility and seasonal dynamics biogenic accumulation and eluvial-illuvial differentiation in profiles are expressed poorly.

Correlation dependences between content of heavy metals and soil properties (acidity, content of organic carbon) are mostly insignificant and doubtful probably as a result of a high spatial and seasonal variation of the content of elements.

For similar regions with lithologic variability of mother rocks more detailed working out is necessary of MAC levels depending on their geochemical features.

## KEYWORDS


- **cadmium**
- **chromium**
- **copper**
- **gray-forest soils**
- **iron**
- **lead**
- **manganese**
- **nickel**
- **podzolic soils**
- **zinc**


## REFERENCES


1. State (national) Report on Condition and Use of Land in Russian Federation for 1995, Moscow: Russlit Publ., 1996, 120 p. (in Russian).
2. Chernykh, N. A., Milashchenko, N. Z., Ladonin, V. F. Ecological-and-toxicological aspects of soil pollution with heavy metals. Moscow: Agroconsalt, 1999, 176 p. (in Russian).
3. Ilin, V. B. An estimation of existing ecological specifications of content of heavy metals in soil. Agrochemistry. 2000, №9, 74–79 (in Russian).
4. Stepanova, M. D. Approaches to estimation of soil and plant pollution with heavy metals. Chemical elements in system "soil-plant." Novosibirsk: Nauka ("Science" in English), 1981, 92–105 (in Russian).
5. Ilin, V. B. Heavy metals in system "soil-plant." Novosibirsk: Nauka ("Science" in English), 1991, 150 p. (in Russian).
6. Matveev, Yu. M., Sheptukhov, V. N., Reshetina, T. V. State of soil cover of Russia. Protection of Natural Environment. Soils. Moscow: Publishing Office of Research Institute of Nature Protection, 2001, 161–165 (in Russian).
7. Nagabedan, I. A., Minkina, T. M., Nazarenko, O. G. Certification of soils of ground areas. Agrokhimichesky Vestnik ("Agrochemical Herald" in English). 2003, 2, 25–26 (in Russian).
8. Tjulin, V. V., Gushchina, A. M. Features of Soils of Kirov Region and Their Use at Intense Agriculture. Kirov: Publishing House Vyatka, 1991, 94 p. (in Russian).
9. Arinushkina, E. V. Handbook on Chemical Analysis of Soils. Moscow: Publishing House of Moscow University, 1970, 488 p. (in Russian).

10. Methodical indications for estimation of microelements in soils, fodders, and plants by atomic-absorption spectroscopy. Moscow: CINAO (Central Institute of Agrochemical Operation of agriculture) Publ., 1995, 95 p. (in Russian).
11. Methodical instructions by definition of heavy metals in soils of agricultural lands and plant growing production. Moscow: Ministry of Agriculture of Russian Federation. 1992, 61 p. (in Russian).
12. Kabata-Pendias, A. Trace Elements in Soils and Plants. Fourth Edition. Boca Raton, Florida. CRC Press. 2010, 548 p.
13. Zyrin, N. G., Chebotaryova, N. A. About forms of copper, zinc, and lead complexes in soils and their availability to plants. Content and Form of Microelements Complexes in Soils. Moscow: Publishing House of Moscow University, 1979, 350–386 (in Russian).
14. Ecological-geochemical map of soils of the Kirov region. Saint-Petersburg: All-Russian Geological Research Institute, 1996 (in Russian).
15. Zolotareva, B. N., Scripnichenko, I. I., Geletjuk, N. I., Sigaeva, E. V., Piunova, V. V. Content and distribution of heavy metals (lead, cadmium, and mercury) in soils of European USSR. Genesis, fertility and amelioration of soils. Pushchino: Scientific Center of Academy of Sciences, 1980, 77–90 (in Russian).
16. Foner, H. A. Anthropogenic and natural lead in soils in Israel. Israel Journal of Earth Sciences. 1993, 42(1), 29–36.
17. Krosshavn, M., Steinnes, E., Varskog, P. Binding of Cd, Cu, Pb and Zn in soil organic matter with different vegetational background. Water, Air, and Soil Pollution. 1993, 71, 185–193.
18. Titova, V. I. Optimization of plant nutrition and ecological-agrochemical estimation of fertilizers use on soils with high contents of exchangeable phosphorus. DSc Thesis. Saint-Petersburg, 1998, 340 p. (in Russian).
19. Izerskaya, L. A., Vorob'eva, T. S. Heavy metal compounds in alluvial soils of the Middle Ob valley. Eurasian Soil Science. 2000, 33 (1), 49–55.
20. Gorjunova, T. A. Heavy metals (Cd, Pb, Cu, Zn) in soils and plants of south-west part of Altai Territory. Siberian Ecological Journal. 2001, 2, 181–190 (in Russian).
21. Ponizovsky, A. A., Mironenko, E. V. Mechanisms of lead (II) absorption by soils. Pochvovedenie ("Soil Science" in English). 2001, 4, 418–429 (in Russian).
22. Obukhov, A. I., Lurie, E. M. Regularities of heavy metal distribution in soils of sod-podzolic sub-zone. Geochemistry of Heavy Metals in Natural and Technogenic Landscapes. Moscow: Moscow State University Publ., 1983, 55–62 (in Russian).
23. Zink and Cadmium in the Environment. Ed. V. V. Dobrovolsky. Moscow: Nauka ("Science" in English), 1992, 200 p. (in Russian).
24. Orlov, D. S. Soil Chemistry. Moscow: Moscow State University Publ., 1985, 376 p. (in Russian).
25. Tsiganjuk, S. I. Influence of long-term application of phosphorus and lime fertilizers on accumulation of heavy metals in soil and plant products. PhD Thesis. Moscow: AllRussian Institute of Fertilizers and Agrochemistry, 1994, 126 p. (in Russian).
26. Levy, R., Francis, C. W. Adsorption of Cadmium by Synthetic and Natural Organo-Clay Complexes. Geoderma. 1976, 15(5), 361–370.

27. Gavrilova, I. P., Bogdanova, M. V., Samonova, O. A. Experience of Square Estimation of Degree of Russian Soil Pollution by Heavy Metals. Herald of Moscow University. Soil Science. 1995, 1, 48–53 (in Russian).
28. Dovbysh, S. A. Forms of Heavy Metals in Natural and Technogenic Polluted Chernozem Soils of Altai Priobe. Abstract of PhD Thesis. Barnaul: Altai State Agrarian University, 2000, 19 p. (in Russian).
29. Prokashev, A. M. Soils with Complex Organic Profiles of South of Kirov Region. Kirov: Publishing House Vyatka, 1999, 174 p. (in Russian).
30. Yulushev, I. G. System of Fertilizers Use in Crop Rotation. Kirov: Publishing House Vyatka, 1999, 154 p. (in Russian).
31. Kuznetsov, N. K. Microelements in Soils of Udmurtiya. Izhevsk: Publishing House of Udmurt University, 1994, 285 p. (in Russian).
32. Ermolaev, S. A., Sychev, V. G., Plyushchikov, V. G. Agrochemical and Agro-Ecological State of Russian Soils. Plodorodie ("Fertility" in English). 2001, 1, 4–9 (in Russian).
33. Berdyaeva, E. V. Influence of perennial use of sewage and lime on fractional content of copper and zinc in sod-podzolic sandy-loam soil. Herald of Moscow University. 2001, 2, 24–29 (in Russian).
34. Dobrovolsky, V. V. Geography of microelements. Global distribution. Moscow: Mysl' ("Thought" in Russian), 1983, 272 p. (in Russian).
35. Sinkevich, E. I. Content of chromium in mother rocks and soils of eastern Fennoscandy. Ecological Functions of soiLs in Eastern Fennoscandy. Petrozavodsk: Publishing House of Karelian Scientific Center of Russian Academy of Sciences. 2000, 65–75 (in Russian).
36. Andersson, A. J. The distribution of heavy metals in soil material as influenced by the ionic radius. Swedish Journal of Agricultural Researches. 1977, 7, 79–83.
37. Karpova, E. A., Potatueva, Yu. A. Influence of perennial use of liquid complex and solid composite fertilizers on heavy metal content in sod-podzolic soil and plants of oats and vetch. Pochvovedenie ("Soil science" in English). 2003, 2, 45–49 (in Russian).
38. Samonova, O. A., Kasimov, N. S., Kosheleva, N. E. Space-Temporal Variability of Heavy Metal Content in Sod-Podzolic Soils of South Taiga. Herald of Moscow University. Soil Science. 2000, 2, 20–26 (in Russian).
39. Yakushevskaya, I. V., Kovda, V. A. Geochemistry of microelements in soils and crust of weathering. Basics of Doctrine of Soils. Moscow: Nauka ("Science" in English), 1973, 199–231 (in Russian).
40. Microelements in soils of USSR. Editors, V. A. Kovda, N. G. Zyrin. Moscow: Publishing House of Moscow University, 1981, 252 p. (in Russian).
41. List of Maximum Allowable Concentration (MAC) and Approximately Allowable Concentrations (AAC) of Chemical Matters in Soil. Hygienic Norm 6229–91. Moscow: Publishing office of Goscomsanepidnadzor of Russia, 1993, 6 p. (in Russian).
42. Lipkina, G. S. The content of mobile Mn, Zn, Fe in sod-podzolic loamy soils and their influence on a crop in the conditions of intensive fertilizing. Bulletin of All-Russian Institute of Fertilizers and Agrochemistry. 1991, 108, 8–13 (in Russian).
43. Tjulin, V. V. Soils of Kirov Region. Kirov: Publishing House Vyatka, 1976, 288 p. (in Russian).

44. Shikhova, L. N. Influence of Drying Amelioration on Features of Sod Gley Soils on Eluvia-Deluvium of Perm Carbonate Rocks. PhD Thesis. Moscow: Moscow State University, 1995, 160 p. (in Russian).
45. Kovda, V. A. Bio-Geochemistry of Soil Cover. Moscow: Nauka ("Science" in English), 1985, 262 p. (in Russian).
46. Dobrovolsky, V. V., Alishchukin, L. V., Filatova, E. V., Chupahina, R. P. Migration forms of heavy metals of soil as the formation factor of metals mass-flows. Heavy metals in environment: Materials of International Symposium. Pushchino, 15–18 October, 1996, Pushchino, 1997, 5–14 (in Russian).
47. Grishina, A. V., Barinov, V. N., Ivanov, V. F. The ecologic-agrochemical characteristic of heavy metals in agriculture. Bulletin of All-Russian Institute of Fertilizers and Agrochemistry. 2001, 115, 125–126 (in Russian).
48. Bloomfield, C., Greenland, D. J., Hayes, M. H. B. The Chemistry of Soil Processes. New York: John Wiley and Sons, 1981, 463 p.

# CHAPTER 2

# SEASONAL DYNAMICS IN CONTENT OF SOME HEAVY METALS AND MICROELEMENTS IN ARABLE SOILS OF TAIGA ZONE OF EUROPEAN RUSSIA

LYUDMILA N. SHIKHOVA[1] and EUGENE M. LISITSYN[1, 2]

[1]*Vyatka State Agricultural Academy, 133 Oktyabrsky Prospect, Kirov, 610017, Russia*

[2]*North-East Agricultural Research Institute, 166-a Lenin St., Kirov, 610007, Russia, E-mail: shikhova-l@mail.ru*

## CONTENTS



## ABSTRACT

Research of seasonal variability of the content of mobile forms of heavy metals (Cd, Pb, and Cr) and microelements (Cu, Zn, and Mo) in arable soils of a taiga zone of the Kirov region of Russia is conducted. For research podzolic and sod-podzolic soils on various rocks have been chosen: on eluvia-deluvium of Perm clays; on blanket carbonate-free loams; on blanket loams of middle taiga subband. In each geographical place two variants of use of soil in agriculture were investigated – without entering of ameliorants and fertilizers and at long-term entering of mineral fertilizers. Results of research have shown that studied arable horizons of sod-podzolic loamy soils of the Kirov region contain 2.04±0.16 mg/kg of mobile lead at fluctuations from 0.75 to 5.8 mg/kg that does not exceed maximum allowable concentration (MAC) for this element. Levels of fluctuation of the content of lead in soils during a growth season is a lot of below MAC and do not differ essentially depending on studied geographical locations. Arable horizons of sod-podzolic loamy soils contain 0.11±0.02 mg/kg of mobile Cd at fluctuations from 0.02 to 0.81 mg/kg. During the separate periods of a growth season its content can exceed MAC. In variants with application of fertilizers level of the content of an element is higher that can be connected with acidifying action of high doses of mineral fertilizers. The average content of mobile chrome in arable horizons is 3.12±0.49 mg/kg at fluctuations from 0.18 to 11.10 mg/kg; in some arable soils of the region excess of MAC is observed. Dynamics of the content of mobile chrome in arable soils of the region has similar character not only on different locations but also in different soil horizons. The maximum content of chrome is marked in the beginning of a season (May–June). The minimum is at the end of plant growth. Distinctions in soils on degree of cultivation do not influence significantly character of its dynamics. The content of mobile copper makes 2.42±0.19 mg/kg at scope of a variation of 0.70–4.80 mg/kg. A variation of the content during a growth season is from 0.5 to 5.0 mg/kg. Arable horizons of loamy sod-podsolic soils contain 3.36±0.24 mg/kg of mobile Zn at fluctuations from 0.85 to 8.11 mg/kg. Seasonal fluctuations of the content of zinc are essential; at entering of high doses of fertilizers the content of mobile zinc in a profile decreases. During a season supplying of soil with zinc available to plants can vary from low to high level. The average content of mobile Mo in arable horizons makes 0.28±0.02 mg/kg at fluctuations from 0.11 to 0.56 mg/kg that testifies to average and high degree of supply of soils with the microelement. Scope of fluctuations Scope of fluctuations of the contents of one or another element during a season reaches 1.5–2 times. The minimum of its content in arable horizons is observed in summer months at relative maxima in the spring and in the fall.

## 2.1 INTRODUCTION

Microelements play the important role in obtaining of a high and qualitative yield of agricultural crops. Being in plants in trace amounts microelements carry out the important functions as a part of enzymes, hormones, and vitamins. For example, it is known that Cu is essential element yet it may be toxic to both humans and animals when its concentration exceeds the safe limits and its concentration in some human tissues such as thyroid can be change dependent on the tissue state including cancerous or noncancerous [1, 2].

The content of microelements in soil is studied for a long time. However, mechanisms of their behavior are studied insufficiently. In the majority of researches total content of microelements are considered though only their mobile forms can give authentic representation about amount of microelements accessible to plants. Works devoted to studying of change of the content of microelements in soils during a growth season are not enough. However, such researches can give more exact picture of security of soils by microelements in concrete phases of development of plants and also supervise their ecological safety.

As well as the majority of microelements heavy metals (HM) presented in plants in small amounts can also carry out the important functions in organisms, being a part of biologically active substances. However, excess of these concentrations specific to each chemical element and/or a plant species causes negative consequences for live organisms. Pollution of agricultural soils in Europe by heavy metals is highly important ecological problem. Heavy metals are one of the most prevalent agents causing public health problems, as crops grown on heavy metal contaminated soils are used for human diet. Moreover some soil organisms, vitally important for soil health and fertility, are sensitive to heavy metal stress and biological diversity of soil is reduced by heavy metal contamination [3, 4]. On the other hand some ubiquitous heavy metals, especially Pb, have limited availability for plant uptake due to complexation with solid soil fractions [5].

For example, in every contaminated soil examined by Leštan and Grčman [6], no or a very small fraction of total soil Pb was present in a form directly available to plants. Therefore, labile (exchange) forms of HM represent the greatest danger as are characterized by high biochemical activity. By their toxicity action such metals as lead, cadmium, and chromium are belong to first (Pb, Cd) and second (Cr) class of danger according to All-Union State Standard [7].

The numerous facts of environmental contamination by heavy metals as a result of dispersion of industrial emissions through atmosphere or in the form of a waste (slags, sludge) and the polluted process waters are described in the literature [8–12]. The Kirov region is included into first ten subjects of the Russian Federation on a quota of farmland with excess of MAC (maximum allowable concentration) on heavy metals [13]. The great number of the large industrial enterprises of metallurgical, chemical and building branches operates now and operated in the recent past in territory of the Kirov region. HM make a considerable share of emissions of these enterprises [14]. For example, annual receipt of lead and cadmium into environment, according to Ref. [15] makes 451.64 and 4.05 tons, accordingly.

Mostly, chemical pollutions are found in the form of combinations and mixtures of some contaminants in the environment and in between soil pollution is a multielements problem in many areas, which are caused by heavy metals [16]. Environmental contamination with HM has as a rule not continuous, but local character, concentrating basically rounds big cities, settlements, railway stations, and highways [17]. It is especially dangerous that the great bulk of the population lives here and the considerable part of agricultural production are made here too. Arable soils are subject to high risk of pollution with HM. Long-term application of mineral fertilizers leads to accumulation of metals in soil [18–21]. Cadmium input is 2 times more than its carrying out by cultural plants. It leads to excess of MAC of metals in soils and plants. In production of plant industry of some regions concentration of cadmium, nickel, lead and chrome is already above their maximum concentration limit [22]. Use of technics for soil processing can also promote enrichment of arable soils with HM. Reduction of volumes of liming has led to acidity increase on the considerable areas of arable soils and, as consequence, to increase in HM mobility [23].

Many researchers studying toxicity of heavy metals for agricultural crops list them in a following order on toxicity: Cd>Zn>Cu>Pb or Cd>Ni>Cu>Zn>Cr>Pb [24, 25]. Soil pollution with HM cannot be shown long time owing to high buffer ability of soils [26]. The content of mobile fraction of HM in this case can give the first information on the beginning of negative processes [17].

At studying of degree of pollution of object it is accepted to compare concentration of pollutants in it to values of MAC. However, values of MAC are rather conditional. Biogeochemical features of territories impose sometimes restrictions on use of this indicator; and existing norms of MAC consider not all the factors influencing behavior of HM. Therefore, at studying of TM content use of indicators of their background content would be more correct. Besides, it will allow to note the initial stages of negative processes of HM accumulation in soils.

HM content as well as other properties of soil essentially varies during a growth season. Therefore, complex studying of the content mobile fraction of HM in soils as well as seasonal dynamics of their content will allow to supervise reliably an ecological condition of soil during the various periods of a growth season.

## 2.2 MATERIALS AND METHODOLOGY

Studying of seasonal dynamics of chemical elements (Pb, Cd, Cr, Cu, Zn, and Mo) was spent in 2010–2014 in territory of the Kirov region of the Russian Federation. The Kirov region locates in a northeast part of the European Russia. The basic part of territory of the region enters into a South taiga subband. The most northern part concerns a subband of a middle taiga, and a southern part – to a zone of mixed coniferous-broad-leafed forest. Objects of research: (i) Sod-podzolic loamy soil on eluvia-deluvium of Perm sediments (an experiment field of North-East Agricultural Research Institute (NEARI), 10 km from Kirov city). There are two variants of soil treatment – without application of fertilizers and ameliorants (control) and at long entering of $N_{100}P_{100}K_{100}$ (test). (ii) sod-podzolic loamy soil on blanket carbonate-free loams (Falenki district of the Kirov region, 130 km on the east from Kirov). There are two variants of soil – poor-cultivated strongly acid soil and well-cultivated soil under grain-grass crop rotation. (iii) Podzolic loamy soil on blanket loams of middle taiga subbands (Podosinovets district of the Kirov region, 180 km on the northwest from Kirov). Soil samples from three top horizons ($A_p$, $A_{he}$, AB) were selected 5 times per growth season. The soils were place in the labeled plastic bags. They were air dried, mechanically ground and sieved through 2 mm sieve. Estimation of mobile forms of studied chemical elements is spent with 1.0 M acetate-ammonia extract with pH 4.8 by method of atom-absorption spectrophotometry [27, 28].

## 2.3 RESULTS AND DISCUSSION

### *2.3.1 LEAD (PB)*

Recently lead involves a great attention as one of the main components of chemical pollution of environment and as an element toxic for plants and human. Though there is no data that lead is vital for plants there are indications on stimulating action on the last of some salts of lead at low concentration.

The most part of compounds of lead differs with low or very low solubility. In sod-podzolic soils mobile lead makes no more than 5–7% of total [29]. In acid and strongly acid media lead actively migrates within a soil profile in connection with organic matter.

In arable horizons of sod-podzolic soils of the Kirov region the share of mobile Pb in the total makes 4–10% [30]. As whole, arable horizons of sod-podzolic loamy soils of the region contains 2.04±0.16 mg/kg of mobile lead at fluctuations from 0.75 to 5.80 mg/kg. Thus, in agricultural soils of natural territories level of the content of an element does not exceed MAC for this element – 6 mg/kg [31].

Levels of fluctuation of lead content in soils during a growth season is also a lot of below MAC and do not differ essentially depending on locations (Figure 2.1).

The lowest concentrations of Pb are characteristic for all studied variants at the end of a growth season. Maximum content as a rule is more strongly pronounced in the beginning of a season. Most plausible reason of spring maxima of the content of an element is occurrence of time

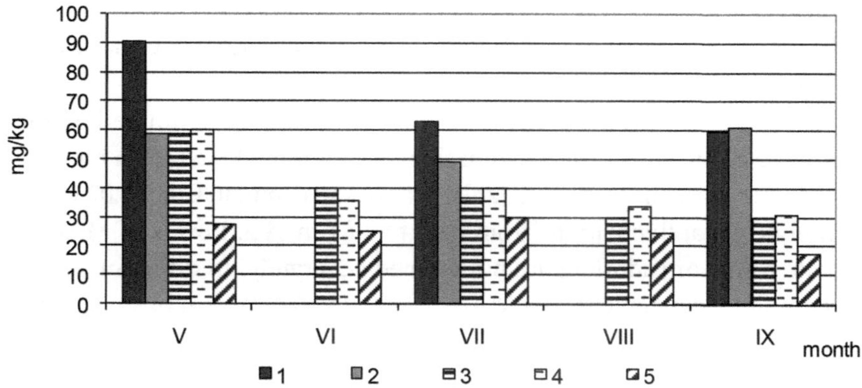


**FIGURE 2.1** Seasonal dynamics of mobile Pb in arable horizons of sod-podzolic soils in Kirov region of Russian Federation. 1 – Control soil, Kirov; 2 – test soil, Kirov, N100P100K100; 3 – poor-cultivated soil, Falenki district; 4 – well-cultivated soil, Falenki district; 5 – podzolic loamy soil, Podosinovets district.

reducing conditions which promote increase in mobility of iron, manganese, and, hence, lead as its considerable part in soil is connected with these elements. Increase of Pb mobility can be caused also by strengthening of a mineralization of organic matter in the beginning of a season. The most part of this element is also connected with organic matter in soil. For example, high negative coefficients of correlation between the content of mobile Pb and the content of total organic carbon in soils of Podosinovets and Falenki districts testify to increase in Pb mobility at reduction of amounts of organic carbon.

Thus, the obtained data testifies that in arable soils there are significant seasonal fluctuations of Pb content, which do not exceed critical level and do not represent danger from the ecological point of view.

### *2.3.2 CADMIUM (CD)*

Cadmium is not a necessary element for a plant metabolism. For human and animals cadmium represents cumulative poison. Researchers notice that Cd is 2–20 times more toxic than other metals.

Cadmium has the greatest mobility and availability to plants in acid soils with low pH level, low capacity of absorption, fulvatic structure of humus, and washing water mode [32, 33]. Thus, in podzolic and sod-podzolic soils the share of mobile cadmium in total, as a rule, does not exceed 10% [29, 34].

In our researches the content of mobile Cd in total in arable horizons of sod-podzolic soils of the region fluctuates within 4–15%. On the average, arable horizons of sod-podzolic loamy soils contain 0.11±0.02 mg/kg of mobile Cd at fluctuations from 0.02 to 0.81 mg/kg. Levels of MAC for mobile forms of cadmium are not defined accurately. In different sources it is possible to meet levels of MAC from 0.2 to 0.6 mg/kg. In some arable soils the content of mobile Cd in arable horizons exceeds all known specifications [30].

High mobility of an element in soils of podzolic type causes by absence of accurate laws of an element behavior in seasonal dynamics (Figure 2.2).

Dynamics of the content of mobile Cd is various in soils from different districts of Kirov region. The least amplitudes of fluctuation are characteristic for soils on eluvia-deluvium of the Perm sediments near Kirov city. Thus in a variant with entering of high doses of fertilizers level of the content of an element is higher than in the control that is probably connected with acidifying action of high doses of mineral fertilizers. The significantly high positive coefficient of correlation between value of exchange acidity and Cd content ($r = 0.81$) testifies to it. Character of dynamics in

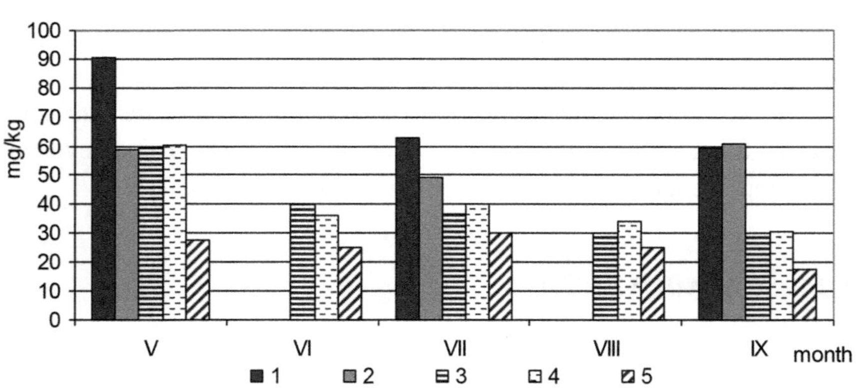


**FIGURE 2.2** Seasonal dynamics of mobile Cd in arable horizons of sod-podzolic soils in Kirov region of Russian Federation. 1 – Control soil, Kirov; 2 – test soil, Kirov, N100P100K100; 3 – poor-cultivated soil, Falenki district; 4 – well-cultivated soil, Falenki district; 5 – podzolic loamy soil, Podosinovets district

soils on blanket carbonate-free loams of Falenki district differs for poor-cultivated and well-cultivated soils. At relative increase in the content in arable horizons of both soils at the beginning of a season and a minimum in the middle of a season, on poor-cultivated field the sharp increase in the content is observed in the fall; on the well-cultivated field it is not observed. In podzolic soil on blanket loams under clover crops (Podosinovets district) the cadmium content is minimum in the beginning and the middle of a season and sharply rises in the fall to the values exceeding all known specifications of MAC. Probably, it is connected with plowing of clover after-grass in the end of summer and increase in the content of labile humus. In the bottom horizons of the studied soils character of dynamics and levels of fluctuation of the content of an element are almost identical to that in arable soils.

Thus, the content of mobile cadmium in arable podzolic and sod-podzolic soils fluctuates considerably during a growth season that is caused by presence in them of the whole complex of the factors promoting increase of Cd mobility. The constant control of its content in soil is necessary as during the separate periods of a growth season the content of this HM can exceed MAC values.

### *2.3.3 CHROME (CR)*

Chrome is considered a toxic element for plants and animals. However, there is some data about positive influence of its small concentration on growth of some plants, about its participation in a glucose and cholesterol metabolism.

The behavior of chrome in soils depends on its content in mother rocks, content of humus and silt, pH, oxidation-reduction potential. The mobile form of Cr makes under the different data from 0.5 to 4% of total amount. Its content depends on a hydrothermal mode and varies essentially on years [35].

In arable sod-podzolic loamy soils of the Kirov region the share of mobile Cr makes 1.3–7.0% of the total content. The average content of mobile chrome in arable horizons is about 3.12±0.49 mg/kg at fluctuations from 0.18 to 11.10 mg/kg. Level of MAC for mobile Cr makes 6.0 mg/kg

[31]. That is, in some arable soils especially from east and northeast districts of the region MAC excess is observed. In our opinion it is connected with participation in formation of soils of sedimentary mother rocks of Paleozoic which in these districts leave close to a surface [30]. In the bulk arable soils the content of mobile Cr is insignificant.

Dynamics of the content of mobile chrome in sod-podzolic loamy arable soils of the region has similar character not only on different locations but also in different horizons of the same soil profile. The maximum content of chrome is marked in the beginning of a season (May-June). The minimum is noted at the end of growth season. Distinctions in soils on degree of cultivation do not influence essentially character of dynamics (Figure 2.3).

Exception is the variant of podzolic soil of the most northern district – Podosinovets. Here gradual increase of the content of an element from the beginning by end of the growth season was marked. Obviously, such behavior of an element is connected with concrete soil conditions. For example, for the cultivated soils of Podosinovets and Falenki districts significant negative correlations with pH values (–0.89–(–0.50)) are received. That is mobility of an element increases at soil acidification. Reduction of

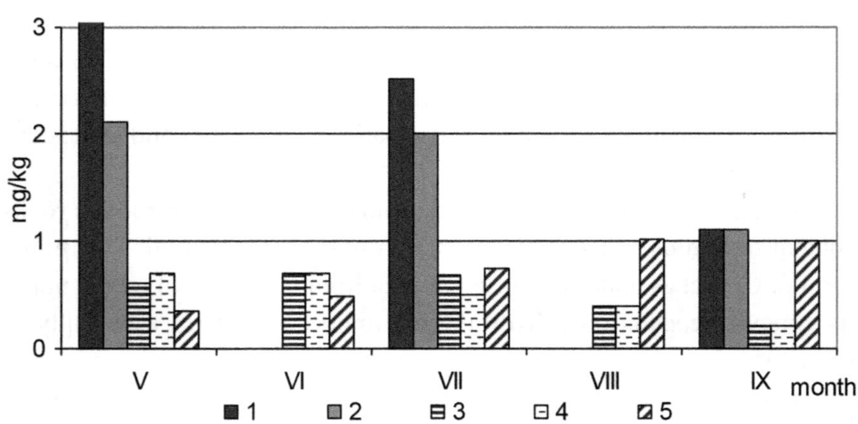


**FIGURE 2.3** Seasonal dynamics of mobile Cr in arable horizons of sod-podzolic soils in Kirov region of Russian Federation. 1 – Control soil, Kirov; 2 – test soil, Kirov, N100P100K100; 3 – poor-cultivated soil, Falenki district; 4 – well-cultivated soil, Falenki district; 5 – podzolic loamy soil, Podosinovets district.

the content of humus carbon promotes clearing of chrome from organo-mineral complexes and to increase in its mobility to what the high negative coefficient of correlation (–0.92) between these parameters in podzolic soil testifies (Podosinovets district).

The total level of the content of an element is higher in the soil formed on eluvia-deluvium of the Perm mother rocks (fields near Kirov city) that is caused by higher content of chrome in these rocks.

Thus, arable soils of the Kirov region as a whole are characterized by the low content of mobile chrome. The raised content of an element is characteristic for the soils generated on the Perm rocks or with their participation. For seasonal dynamics of an element presence of a spring maximum is characteristic. The increase in acidity and reduction of the content of organic matter promote strengthening of mobility of chrome.

### 2.3.4 COPPER (CU)

Copper compounds play the important physiological role participating in photosynthesis, breath, a metabolism of carbohydrates, nitrogen, and proteins. The content of mobile copper in soil fluctuates largely from 2 to 33% of the total content. The share of mobile copper in eluvial horizons is raised. In arable horizons it is less and makes 3–15% [30].

In arable horizons of sod-podzolic soils in the northeast of the European Russia the amount of mobile copper consist about 2.42±0.19 mg/kg in average at scope of a variation of 0.70–4.80 mg/kg. The variation of the content during a growth season in investigated soils makes from 0.5 to 5.0 mg/kg.

The raised content of mobile copper is characteristic for sod-podzolic soils on eluvia-deluvium of the Perm sediments (Figure 2.4) – about 3–5 mg/kg. The raised content of not only total, but also mobile copper in the soils generated on rocks of the Perm age or at their participation was marked by researchers in other regions of east part of the European Russia also [36].

Character of dynamics of the content of mobile copper differs in different types of arable soils. Sharp reduction of Cu content during a season is observed in soil of a control variant on eluvia-deluvium of Perm rocks. A

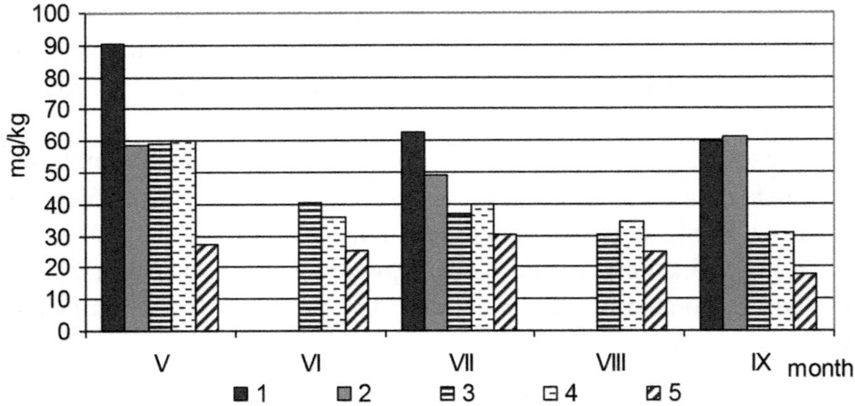


**FIGURE 2.4** Seasonal dynamics of mobile Cu in arable horizons of sod-podzolic soils in Kirov region of Russian Federation. 1 – Control soil, Kirov; 2 – test soil, Kirov, N100P100K100; 3 – poor-cultivated soil, Falenki district; 4 – well-cultivated soil, Falenki district; 5 – podzolic loamy soil, Podosinovets district.

spring maximum of the content is above CAM level for mobile copper – 3 mg/kg [31]. In loamy soils it is possible to explain such fact by increase in mobility of an element in anaerobic conditions created by spring overwetting of an arable layer. The increase in mobility of copper in anaerobic conditions is noted by other researchers also [37]. In a variant with input of high doses of fertilizers the content of copper during a season is raised and is rather stable.

Arable soils on blanket carbonate-free loams of Falenki district are characterized by the low content of mobile copper – less than 2 mg/kg. Depending on level of cultivation character of dynamics of Cu content is distinguished. In arable horizon of poor-cultivated soils rather uniform low amounts of copper is observed in first half of growth season; but is sharp (almost in 2 times) raises in the fall. On the well-cultivated field the maximum amount of an element is marked in the beginning of a season with sharp reduction in the middle and in the end. Such dynamics is characteristic for all horizons of both fields. Spring top dressing of a winter rye by ammoniac saltpeter and gradual dissolution of fertilizer leads to increase of the content of copper in the end of a season. It is known that entering of ammoniac and nitrate fertilizers promotes increase of Cu mobility [38, 39]. However, acid conditions brake processes of nitrification in the beginning of season on poor-cultivated field and they develop enough only by the end of season. Besides, in the end of a season consumption of an element by plants is sharply reduced after harvesting. On the well-cultivated field at light acidic conditions entering of ammoniac saltpeter in the spring makes active these processes at once as leads to increase of mobility of copper in the beginning of a season. The spring maximum on the well-cultivated field can be caused by anaerobic conditions at over-wetting of an arable layer by spring precipitations and thawed snow. Entering of lime and phosphoric fertilizers at cultivating of sod-podzolic soil as well as essential carrying out of copper with a crop reduce amount of an element on the well-cultivated field during the summer-fall period [40].

Sod-podzolic arable soil on blanket loams from Podosinovets district contains also the insignificant amount of mobile fractions of copper. Relative maxima are noted in the spring and in the fall. During the summer period perhaps as a result of intensive consumption by a clover the element content is minimum (about 0.5 mg/kg) that can testify to its lack.

Thus, the content of copper accessible to plants in arable soils essentially varies during a growth season. Character of dynamics depends from both the concrete soil-climatic conditions and amount of inputted fertilizers and from an element consumption level in different crops.

### *2.3.5 ZINC (ZN)*

Zinc compounds in certain concentration are necessary for processes of a metabolism of carbohydrates and proteins in plants. They are a part of some many enzymes.

The content of mobile zinc in the region soils is insignificant and makes 1.2–9.7% of the total content. Arable horizons of sod-podzolic loamy soils contain 3.36±0.24 mg/kg of mobile Zn at fluctuations from 0.85 to 8.11 mg/kg.

The most important factors governing solubility of zinc in soil solution are the content of clay minerals, hydroxides of Fe and Al, pH value, degree of alkalinity and podzolizity [32].

Seasonal fluctuations of the content of zinc are essential (Figure 2.5). In arable horizons of soils of a South taiga subband (Kirov, Falenki district) the element content is maximum in spring and the beginning of summer and decreases by the end of a season; especially sharp it is in soil on eluvia-deluvium of Perm rocks (fields near Kirov city). On this site it is accurately visible that at entering of the high doses of fertilizers in comparison with the control the content of mobile zinc in a profile decreases obviously as a result of carrying out with the high crop.

At the same time arable horizons of poor-cultivated and well-cultivated fields on blanket sediments (Falenki district) do not differ significantly on amount of mobile zinc and character of its dynamics. Dynamics of the content of an element in arable horizon of the most northern variant of arable soils on blanket loams (Podosinovets district) differs sharply. The maximum of the content of mobile Zn forms is observed in the beginning of summer. After summer recession the amount of zinc increases again in the fall. Absolute values of the content of an element is higher than in other studied soils and according to the specifications accepted in the Russian Federation given soil can be characterized [41] as highly supplied with the given microelement. The reasons of the high content of zinc are possibly connected with climatic features of year of research and with physiological features of grown crop – a clover. It is known that a clover absorbing

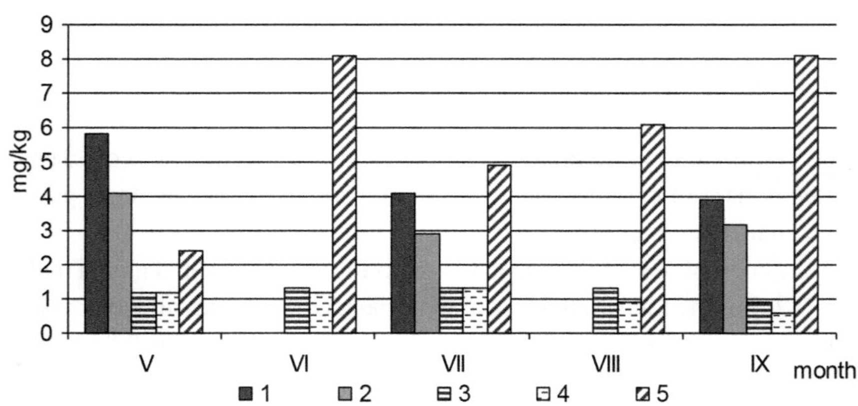


**FIGURE 2.5** Seasonal dynamics of mobile Zn in arable horizons of sod-podzolic soils in Kirov region of Russian Federation. 1 – Control soil, Kirov; 2 – test soil, Kirov, N100P100K100; 3 – poor-cultivated soil, Falenki district; 4 – well-cultivated soil, Falenki district; 5 – podzolic loamy soil, Podosinovets district.

considerable amount of the bases during its growth promotes acidification of soils. It can lead to increase of mobility of many elements including zinc. In this soil significant correlation between the content of mobile zinc and the content of labile organic carbon ($r = 0.82$) is noted.

Thus, the content and dynamics of mobile zinc during a growth season is influenced by primary riches an element of rocks on which the soil is formed as well as climatic features of a season and physiological features of grown agricultural crop. During a season supply of soil by zinc accessible to plants can vary from low to the high level.

### 2.3.6 MOLYBDENUM (MO)

Molybdenum is a vital microelement for plants. It is a part of enzyme systems of nitrogenase and nitrate-reductase, oxidases and Mo-ferrodoxin, which accelerate various reactions untied among them.

On the average soils of the European Russia contain about 0.16 mg/kg of mobile Mo. In northeast and east regions of the European Russia values of content fluctuate largely but, as a rule, do not exceed 0.25 mg/kg (Figure 2.6). The average content of mobile Mo in arable horizons of

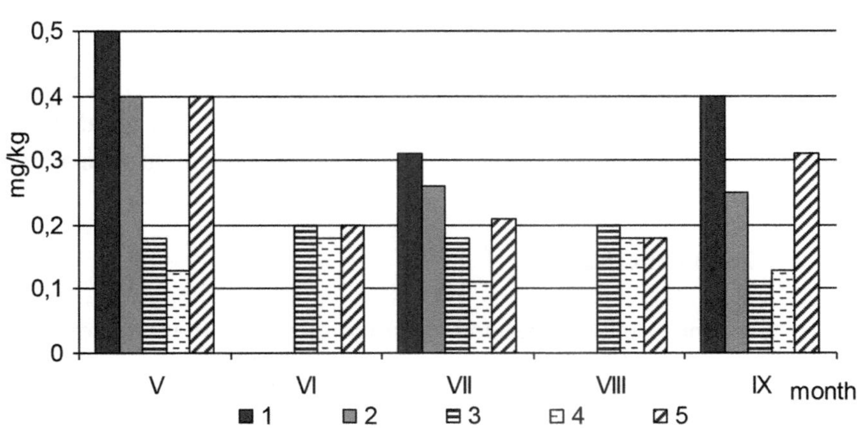


**FIGURE 2.6** Seasonal dynamics of mobile Mo in arable horizons of sod-podzolic soils in Kirov region of Russian Federation. 1 – Control soil, Kirov; 2 – test soil, Kirov, N100P100K100; 3 – poor-cultivated soil, Falenki district; 4 – well-cultivated soil, Falenki district; 5 – podzolic loamy soil, Podosinovets district.

sod-podzolic soils of the Kirov region makes 0.28±0.02 mg/kg at fluctuations from 0.11 to 0.56 mg/kg that testifies to average and high degree of supplying of soils with a microelement. Mobile forms of an element make from 4 to 33% of the total. Their share is highest in the top horizons and in mother rocks [30].

The factors increasing mobility of molybdenum in soil are: increase in concentration of anions OH–, $PO_4^{3-}$, and COO– in a soil solution, entering into soil of calcium ions, a mineralization of organic matter. The factors reducing mobility of molybdenum in soil and its availability to plants are: increase in concentration of cation $H^+$, enrichment of soil with $Al(OH)_3$ and $Fe(OH)_3$, presence of exchange $Al^{3+}$, surplus of cations $Fe^{3+}$, $Mn^{2+}$, entering of physiologically acid mineral fertilizers without liming, molybdenum fixing by acid humus forms containing sesquioxides [38].

Scope of fluctuations in the content of an element during a growth season reaches 1.5–2 times. The minimum content in arable horizons is observed in summer months at relative maxima in the spring and in the fall (see Figure 2.6). Such fluctuations in the content are caused by intensive absorption of Mo by plants in active growth period and by soil processes. High level of the content of an element during all season is marked in sod-podzolic soil on eluvia-deluvium of the Perm rocks. In a variant with entering of high doses of fertilizers level of the content of Mo is lower than in the control variant that is caused as entering of physiologically acid fertilizers reducing availability of Mo and with higher carrying out of an element with a crop. Increase in degree of cultivation of soils on blanket carbonate-free loams initially poor in the given element leads to decrease in level of its content as a result of carrying out of an element with a crop.

## 2.4 CONCLUSIONS

1. Content of mobile fractions of microelements in arable soils of a taiga zone varies significantly (by several times) during a growth season. The minimum content of mobile microelements is characteristic in most cases for the middle of a growth season – the period of their intensive consumption by plants.
2. Raised contents of mobile Cu, Zn, and Mo despite fluctuations during a season are characteristic for the soils generated on eluvia-deluvium of the Perm sediments.

3. For the soils generated on blanket carbonate-free loams lower degree of supply by microelements is characteristic; their additional entering especially during the periods of intensive growth and formation of a crop of agricultural crops is necessary.
4. Content and character of seasonal dynamics of microelements in arable soils is influenced by soil-climatic conditions a concrete growth season and by physiological conditions of grown crop.
5. Non-polluted loamy arable soils of the Kirov region contain on the average insignificant amounts of mobile Pb, Cd, and Cr. In some cases in arable horizons of soils essential excess of MAC on these metals connected with geochemical features of territories (Cr) and, possibly, with technogenic influence (Pb, Cd) is revealed.
6. Content of mobile HM varies significantly during a growth season by 1.5–10 times.
7. Maximum content of all studied elements is characteristic with rare exception to beginning of a growth season. The minimum is on the middle or the end of growth.
8. During the separate periods of a growth season (spring, fall) the content of mobile HM can increase sharply coming nearer to levels of MAC and above that can represent danger to agricultural plants.
9. The control over level of the content of mobile fractions of HM is necessary in agricultural soils especially during the spring period.

## KEYWORDS

- **cadmium**
- **chrome**
- **copper**
- **lead**
- **mobile form of element**
- **molybdenum**
- **podzolic soil**
- **seasonal dynamics**
- **zinc**

## REFERENCES


1. Yang, X. E., Long, X. X., Ni, W. Z., Ye, Z. Q., He, Z. L., Stoffella, P. J. Assessing copper thresholds for phytotoxicity and potential dietary toxicity in selected vegetable crops. Journal of Environmental Science and Health. Part, B. 2002, 37, 625–635.
2. Yaman, M., Akdeniz, I. Sensitivity enhancement in flame atomic absorption spectrometry for determination of copper in human thyroid tissues. Analytical Sciences. 2004, 20, 1363–1366.
3. Dahlin, S., Witter, E., Martensson, A., Turner, A., Baath, A. Where's the limit? Changes in the microbiological properties of agricultural soils at low levels of metal contamination. Soil Biology and Biochemistry. 1997, 29, 1405–1415.
4. Giller, K. E., Witter, E., McGrath, S. P. Toxicity of heavy metals to microorganisms and microbial processes in agricultural soils: a review. Soil Biology and Biochemistry. 1998, 30, 1389–1414.
5. Rieuwerts, J. S., Thornton, I., Farago, M. E., Ashmore, M. R. Factors influencing metal bioavailability in soils: preliminary investigations for the development of a critical loads approach for metals. Chemical Speciation and Bioavailability. 1998, 10, 61–75.
6. Leštan, D., Grčman, H. Speciation of lead, zinc and cadmium in contaminated soils from Mežica valley. Zbornik Biotehniške fakultete Univerze v Ljubljani. Kmetijstvo. Agricultural issue. 2001, 77, 205–214.
7. Nature protection. Soils. Classification of Chemicals for Pollution Control. GOST 17.4.1.02–83. Moscow. Standartinform. 2008. 9 p.
8. Nasrabadi, T., Nabi Bidhendi, G. R., Karbassi, A. R., Hoveidi, H., Nasrabadi, I., Pezeshk, H., Rashidinejad, F. Influence of Sungun copper mine on groundwater quality, NW Iran. Environmental Geology. 2009, 58, 693–700.
9. Diviš, P., Machat, J., Szkandera, R., Dočekalova, H. In situ measurement of bioavailable metal concentrations at the downstream on the Morava River using transplanted aquatic mosses and DGT technique. International Journal of Environmental Research and Public Health. 2012, 6 (1), 87–94.
10. Okuku, E. O., Peter, H. K. Choose of Heavy Metals Pollution Biomonitors: A Critic of the Method that uses Sediments total Metals Concentration as the Benchmark. International Journal of Environmental Research and Public Health. 2012, 6(1), 313–322.
11. Machado, S., Rabelo, T. S., Portella, R. B., Carvalho, M. F., Magna, G. A. M. A study of the routes of contamination by lead and cadmium in Santo Amaro, Brazil. Environmental Technology. 2013, 34, 559–571.
12. Egoshina, T. L., Shikhova, L. N., Lisitsyn, E. M., Zhiryakov, A. S. Accumulation of heavy metals in aquatic systems having different degree of contamination. Problems of Regional Ecology. 2007, 2, 17–23. (In Russian).
13. State (national) Report on Condition and Use of Land in Russian Federation for 1995. Moscow. Russlit Publ., 1996. 120 p. (in Russian).
14. Saet Yu.E., Raevich, B. A., Yanin, E. P. Geochemistry of Environment. Moscow. Nedra Publishing, 1990, 335 p.

15. Burkov, N. A. Applied Ecology. Kirov. Publishing House Vyatka, 2005, 271 p. (in Russian).
16. An, Y. J., Kim, Y. M., Kwon, T. I., Jeong, S. W. Combined effect of copper, cadmium, and lead upon Cucumis sativus growth and bioaccumulation. Science of the Total Environment. 2004, 326, 85–93.
17. Ilin, V. B. Estimation of mass-flow of heavy metals in system soil-agricultural crop. Agrochemistry. 2006, 3, 52–59 (in Russian).
18. Oyedele, D. J., Asonugho, C., Awotoye, O. O. Heavy metals in soil and accumulation by edible vegetables after phosphate fertilizer application. Electronic Journal of Environmental Agricultural Food Chemistry 2006, 5(4), 1446–1453.
19. Guzman, E. T. R., Regil, E. O., Gutierrez, L. R. R., Albericl, M. V. E., Hernandez, A. R., Regil, E. D. Contamination of corn growing areas due to intense fertilization in the high plane of Mexico. Water, Air and Soil Pollution. 2006, 175, 77–98.
20. Mendes, A. M. S., Duda, G. P., Nascimento, G. W. A., Silva, M. O. Bioavailability of Cadmium and Lead in a Soil Amended With Phosphorus Fertilizers. Scientia Agricola. 2006, 63, 328–332.
21. Dissanayake, C. B., Chandrajith, R. Phosphate Mineral Fertilizers, Trace Metals and Human Health. Journal of the National Science Foundation of Sri Lanka. 2009, 37(3), 153–165.
22. Dobrovolsky, V. V. Zink and cadmium in the environment. Moscow. Nauka Publ., 1992, 200 p. (in Russian).
23. Efimov, V. N., Ivanov, A. I. Hidden Degradation of Well-Cultivated Sod-Podzolic Soils of Russia. Agrochemistry. 2001, 6, 5–10.
24. Yagodin, B. A., Govorina, V. V., Vinogradova, S. B., Zamarajev, A. G., Chapovskaya, G. V. Accumulation of Cd and Pb by some agricultural crops on sod-podzolic soils of different degree of cultivation. Reports of Timiryazev Agricultural Academy. 1995, 2, 85–100 (in Russian).
25. Zyrin, N. G., Sadovnikova, L. K. (Ed.) Chemistry of heavy metals, arsenic, and molybdenum in soils. Moscow. Publishing House of Moscow University, 1985, 208 p. (in Russian).
26. Chernykh, N. A., Milashchenko, N. Z., Ladonin, V. F. Ecological-and-Toxicological Aspects of Soil Contamination with Heavy Metals. Moscow. Agroconsalt Publ., 1999, 148 p. (in Russian).
27. Arinushkina, E. V. Handbook on Chemical Analysis of Soils. Moscow. Publishing House of Moscow University, 1970, 488 p. (in Russian)
28. Methodical Indications for Estimation of Microelements in Soils, Fodders, and Plants by Atomic-Absorption Spectroscopy. Moscow. CINAO Publ., 1995, 95 p. (in Russian)
29. Obukhov, A. I., Lurie, E. M. Regularities of heavy metal distribution in soils of sod-podzolic sub-zone. Geochemistry of Heavy Metals in Natural and Technogenic Landscapes. Moscow. Moscow State University Publ., 1983, 55–62. (in Russian).
30. Shikhova, L. N., Egoshina, T. L. Heavy Metals in Soils and Plants of North-East of European Part of Russia. Kirov. North-East Agricultural Research Institute, 2004, 263 p. (in Russian).

31. List of maximum allowable concentration (MAC) and approximately allowable concentrations (AAC) of chemical matters in soil. Hygienic Norm 6229–91. Moscow. Publishing Office of Goscomsanepidnadzor of Russia, 1993, 6 p. (in Russian).
32. Kabata-Pendias, A. Trace Elements in Soils and Plants. Fourth Edition. Boca Raton Florida, CRC Press. 2010, 548 p.
33. Titova, V. I. Optimization of plant nutrition and ecological-agrochemical estimation of fertilizers use on soils with high contents of exchangeable phosphorus. DSc Thesis. Saint-Petersburg. 1998, 340 p. (in Russian).
34. Izerskaya, L. A., Vorobeva, T. S. Heavy Metal Compounds in Alluvial Soils of the Middle Ob Valley. Eurasian Soil Science. 2000, 33(1), 49–55.
35. Voloshin, E. I. Monitoring of Cr in soils of Middle Siberia. Agrochemical Herald. 2001, 2, 29–31 (in Russian).
36. Microelements in Soils of USSR. Kovda, V. A., Zyrin, N. G. (Eds.) Moscow. Moscow State University Publ., 1981, 242 p. (in Russian).
37. Berdyaeva, E. V. Influence of Long-Term Application of Sludge and Lime on Fraction Content of Copper and Zinc in Sod-Podzolic Loamy Soil. Herald of Moscow State University. 2001, 2, 24–29 (in Russian).
38. Peive, Ya. V. Soil Biochemistry. Moscow. Selkhozgiz, 1961. 421 p. (in Russian).
39. Potatueva, Yu. A., Khlystovsky, A. V., Yanchuk, I. A., Kornienko, E. F., Optimakh, V. P., Ryabova, A. N. Microelements in plants and soils at systematic use of mineral fertilizers, manure, and lime. Agrochemistry. 1984, 6, 83–92 (in Russian).
40. Tkachenko, V. M., Vdovenko, O. P. Increasing of efficiency of microelements in complex of phosphorus microfertilizers. Increasing of efficiency of fertilizers using and soil fertility in Ukrainian USSR. Kharkov Agrarian Institute Publ., 1985, 87 (in Russian).
41. Kuznetsov, N. K. Microelements in Soils of Udmurtiya. Izhevsk. Udmurt University Publ., 1994. 287 p. (in Russian).

CHAPTER 3

# DYNAMICS OF ORGANIC MATTER CONTENT IN SOD-PODZOLIC SOILS DIFFER IN DEGREE OF CULTIVATION

LYUDMILA N. SHIKHOVA,[1] OLGA A. ZUBKOVA,[2] and EUGENE M. LISITSYN[1,2]

[1]*Vyatka State Agricultural Academy, 133 Oktyabrsky Prospect, Kirov, 610017, Russia*

[2]*North-East Agricultural Research Institute, 166-a Lenin St., Kirov, 610007, Russia, E-mail: shikhova-l@mail.ru*

## CONTENTS



## ABSTRACT

Data of two-year observation over seasonal dynamics of the content and reserves of the total and labile carbon of humus substances of an arable variant of podzolic soils of a southern taiga are presented. It is revealed that arable soils contain insignificant amount of humus carbon; its content and reserves essentially vary not only on depth of a soil profile but also during a growth season. The content and reserves of humus organic carbon essentially vary in a profile of arable soils during all term of observation within 0.58–1.11% (of soil weight) and 31.49–45.58 t/ha accordingly. In transitive eluvial horizon AB the content and reserves of the total carbon is much less (from 0.17 to 0.66%). The horizon B contains insignificant amount of carbon – from 0.15 to 0.30%. The maximum content of labile carbon is characteristic for arable horizon (from 0.010 to 0.222%). In horizon AB the content fluctuates from 0.016 to 0.160%. Character of dynamics of labile carbon in horizon AB is close to that in arable horizon. The content maximum is marked in the beginning of a season, a minimum – in the end. In horizon B it is noted sharp fluctuations of labile carbon content. As a whole the carbon content in this horizon is minimal (from 0.030 to 0.094%) and varies rather smoothly during a season. The top layers of soil are characterized by rather constant reserves of carbon with small amplitudes of fluctuation during a season. The thicker is considered layer, the higher amplitude of fluctuation of the content of carbon during a season. The general tendency of change of reserves of labile compounds of carbon in both years of research is decrease in reserves from the beginning of the growth period up to the end with different in amplitude fluctuations of the content. In horizon of a forest ground litter (0–5 cm) of forest biogeocenosis reserves stocks of organic matter have made 42.94–50.25 t/ha on the average for a season. During the various periods of a growth season significant wavy fluctuations of carbon reserves are observed. In the beginning of a growth season, carbon reserves of a ground litter made 38.81 t/ha. To end of a season reserves have increased by 20.1% and have made 45.76–48.68 t/ha. The maximum value of reserves of organic matter in a ground litter is noted in the second 10-day period of June (55.64 t/ha). The minimum has been noted in the end of May and has made 26.47 t/ha. The system of organic matter in a forest ground litter is stable and possesses dynamic balance. In the beginning of the growth period the minimum reserves of carbon in horizon $O_i$ of forest soil – 6.52 t/ha are noted. The maximum value of reserves of organic carbon in the given horizon (21.86–25.15 t/ha) has been noted in the second 10-day period of June. In the end of the period of observation reserves of organic carbon have made 12.57–15.35 t/ha that significantly exceeds values in the season beginning. Horizon $A_{he}$ differs by the greatest amplitude of fluctuation of carbon reserves. In the beginning of a season it was equal to 13.15–23.03 t/ha, in the end – 18.24–29.41 t/ha. The maximum value of reserves is noted in the middle of September, it has made 52.21 t/ha whereas the minimum reserve is noted in the third 10-day period of May and has made 16.64 t/ha. In the heaviest horizon B the greatest reserves of organic carbon of humus substances are contained. In the beginning of a season of researches they made 36.27–43.71 t/ha, in the end – 31.62–33.48 t/ha. Reserves of the total carbon in this horizon in the end of a season were significantly reduced by 23.4%. The minimum values of reserves of organic carbon of humus are characteristic as a rule to beginning of a growth season, maximum – for second half of growth season. Despite considerable fluctuations of carbon reserves during the growth periods, in the beginning and the end of seasons of research its values did not differ statistically that can be a sign of stability of all system of soil organic substance. Stability of system remains in annual dynamics too.

## 3.1 INTRODUCTION

Threat of global ecological crisis dictates necessity of studying of laws of chemical elements circulations in natural and anthropogenic disturbed ecosystems. The climate change problem is connected with damage of a global carbon cycle. For last century there was a strengthened increasing of a greenhouse effect in atmosphere.

The natural non-anthropogenic-disturbed ecosystems are characterized by the balanced, almost closed circulation of chemical elements. Anthropogenous transformation of natural ecosystems leads to breakdown of biogeochemical cycles of elements. A carbon cycle is one of the major for Earth's biosphere. Its breakdown as a result of human activity will lead to global environmental problems, climate change, and degradation or reorganization of whole biosphere.

Soil organic carbon is one of key links of a global carbon cycle [1]. Organic carbon can define as property of soils, modes, content of compounds of nutritive elements available for plants, and as a whole stability of whole ecosystem. Organic carbon of a soil component is the enormous geochemical accumulator, the store and the keeper of a solar energy and information on a terrestrial surface.

Availability of heavy metal for plants is determined not only by heavy metal contents in soil but also by soil pH value, organic matter and clay contents, and use of different fertilizers [2, 3]. These soil parameters can significantly change mobile fraction of heavy metals at the same level of their total amount [4–6]. Besides soil pH, which is a key parameter, the content and mainly the quality of soil organic matter can influence the availability of heavy metals in soil. Some researchers showed that amendment of contaminated soils with organic matter reduced bioavailability of heavy metals [7]. High organic matter content was reported to decrease concentrations of Cd and Ni in soil solution [8]. Soil organic matter has been of interest in studies of heavy metal sorption by soils because organic ligands are known to form stable complexes with heavy metals [9, 10]. The content of organic matter affects speciation of heavy metals in soil [11]. Dissolved organic-matter molecules contain functional groups (e.g., COOH and phenol-OH) capable of complexion metals [12, 13]. The metal–organic matter interaction has various and complex consequences both on the solubility, mobility and bioavailability of metals [14] and on soil organic matter turnover [15]. Humic substances represent a significant proportion of total organic carbon. The predominant fraction of humus substances is humic acids that are very active in interacting with inorganic contaminants [16]. Excellent sorption properties of humic acids that depend on their chemical structure were reported by many authors [17, 18]. Thus high content of organic matter in soil can serves as the ways to exclude heavy metals from the trophic chain.

It is considered that the major suppliers of carbon in atmosphere are natural forest ecosystems, in particular boreal, and agriculture. Forest ecosystems in the course of photosynthesis absorb carbon dioxide of atmosphere, which is one of the basic greenhouse gases. Boreal forests of Eurasia play the important role in accumulation of atmospheric carbon. It is known that boreal forests owing to climatic features of territories of their growth are capable to deposit carbon not only in a live and dead biomass, but also in wood and vegetative detritus, humus, and soil.

Hence, the estimation of reserves of carbon in forest ecosystems causes a particular interest and demands careful studying. During our long-term researches individual questions of dynamics of the content and reserves of carbon in South taiga forest and arable podzolic soils have been studied [19–21]. Logic continuation of these works is studying of dynamics of reserves of various forms of carbon in soils of a sub-band of a southern taiga, especially in a forest ground litter, which is the basic source of organic carbon in soil of forest biogeocenoses. In the scientific literature there is not enough data on the similar problems. Meanwhile, in biological circulation of carbon, nitrogen and ash elements, and also in soil formation processes in forest ecosystems the forest ground litter has important value. It is known that entry of the vegetative rests on a soil surface in forest ecosystems occurs non-uniformly and their specific accessory varies strongly on months. Amount of inputting tree waste, its structure and intensity of decomposition substantially define character of formation of a ground litter, morphology and properties of soil.

At organization and development of farming agriculture and cattle breeding large areas of natural ecosystems are collapsed. The artificial agricultural systems created on their place are characterized by incompleteness of biogeochemical cycles, notable losses of elements, including carbon. The soil is the major and integral element of any land ecosystem; serves as a habitat of soil flora and fauna; is enriched by nutritious elements for plants; and, as a whole, provides stability of ecosystems. Ploughing up of soils leads to strengthening of a mineralization of soil organic substance; and alienation of vegetative mass with a crop interferes with entry in soil and humification of the fresh vegetative rests [22]. Soils lose carbon that, in turn, conducts to strengthening of greenhouse effect.

Despite presence of the data about carbon reserves in soils of various ecosystems there is not full picture of the process [23]. Data on carbon reserves in soil and their dynamics during a season are fragmentary.

The work purposes are: (i) the estimation of the content and reserves of organic carbon in arable soil, as well as in horizon of ground litter and in soil component of a forest ecosystem of a sub-band of a southern taiga of the Kirov region; and (ii) studying of their dynamics during a growth season.

## 3.2 MATERIALS AND METHODOLOGY

### *3.2.1 DYNAMICS OF THE CONTENT AND RESERVES OF HUMUS CARBON IN ARABLE PODZOLIC SOILS*

Object of research is the arable variant of podzolic soil. The object is in a South taiga sub-band, in territory of the Kirov region in 15 km to the south of a Kirov city. The key site where soil samples were collected is located in an average part of a slope to a brook (1–2°) of a southeast exposition. Researches were spent during seasons 2009 and 2010 from April till October inclusive. Soil type is arable sandy podzolic, on fluvio-glacial sandy sediments spread by loamy sediments.

Soil samples were collected from three top horizons of $A_p$ (0–20 cm), AB (20–40 cm), B (40–80 cm) with a soil drill in five repeats. For studying of dynamics of carbon compounds soil samples were collected in the beginning of season – once a week, further – once in 2–4 weeks, depending on weather conditions. Soil samples were prepared for the analysis by standard methods.

Next parameters of soil were defined: humidity (a thermostat-weight method); volume density and content of the total carbon of humus – by Tyurin's method stated in [24]; content of labile organic carbon – according to [25].

### *3.2.2 DYNAMICS OF RESERVES OF ORGANIC MATTER IN A FOREST GROUND LITTER AND SOIL LAYER OF A FOREST TAIGA ECOSYSTEM OF SOUTH TAIGA BIOGEOCENOSIS*

Object of research is the soil component of forest biogeocenosis presented by a 60–80 year-old wood sorrel spruce forest which is in a South taiga sub-band in territory of the Kirov region. Dominating species is European spruce (*Picea abies* L.). Soil is sandy podzolic on fluvio-glacial sandy sediments spread by loamy sediments. A water supply is mixed; a mode of humidifying – washing.

Three registration sites 10×10 meters each have been put in 2009–2012 for studying of dynamics of carbon reserves. Samples of a forest ground litter were collected in five repeats in the beginning of season – once a week, further – once in 2–4 weeks depending on weather conditions. Next parameters were defined in samples of ground litter prepared by the standard methods: field humidity and losses at burning [24].

In a soil layer from three top horizons $O_i$ (0–5 cm), $A_{he}$ (5–20 cm), and B (20–80 cm) soil samples were collected in five repeats by means of a soil drill. Soil samples were prepared for the analysis by standard methods. Next parameters were defined: humidity (a thermostat-weight method), volume density, content of total organic carbon on Tyurin's method [24]; content of labile organic carbon [25].

The obtained primary data was processed by means of methods of dispersive and correlation analyzes with application of a package of standard programs Excel and Statistica 10. On the basis of received data reserves of the total and labile carbon were counted up on horizons and layers of a soil profile.

## 3.3 RESULTS AND DISCUSSION

### *3.3.1 DYNAMICS OF CONTENT AND RESERVES OF HUMUS CARBON IN ARABLE PODZOLIC SOILS*

The soil organic matter plays the important role not only in fertility of arable soils, but also in circulation of carbon and other chemical elements in agro-biocenoses. It is rather difficult to estimate the content and reserves of organic carbon in soils as they vary essentially during a growth season and depend not only on soil factors, but also on climatic features of year, features of cultivated crops, agricultural technicians, etc.

As our researches have shown the content and, accordingly, reserves of humus organic carbon ($C_{tot}$) vary essentially in a profile of arable soils during all term of observation.

In a profile of arable soil the greatest content and reserves of $C_{tot}$ are characteristic for arable horizon.

During a season of 2009 the content and reserves of humus carbon varied in limits 0.76±0.15 – 1.10±0.07% (from soil weight) which is correspond to 31.49–45.58 t/ha.

The minimum content is noticed in spring that is connected with absence of coming of fresh organic matter for humification, and weak humification of available organic remains because of low soil temperature (Figure 3.1). The carbon content rises in summer, especially in the beginning, and in an early autumn when optimum hydrothermal conditions for humification are created.

In transitive eluvial horizon AB the content and reserves of $C_{tot}$ is much less (from 0.17±0.02 to 0.66±0.03%). Dynamics of the content of carbon during a growth season is significant; character of dynamics is complex (see Figure 3.1). But, as the basic source of organic carbon in this horizon is its coming from overlying arable horizon during the separate periods of a growth season then negative correlation (minus 0.77 at P = 0.95) is marked between $C_{tot}$ contents in horizons $A_p$ and AB. In other words, after snow thawing $C_{tot}$ coming in horizon AB from arable horizon with snowmelt runoff and atmospheric precipitations.

The horizon B contains insignificant amount of carbon – from 0.15±0.01 to 0.30±0.06%. Change of its content during all season as a whole is insignificant statistically; amplitudes of fluctuations are insignificant too.

It is possible to consider 2009 as an average on weather parameters. 2010 was characterized by abnormal weather conditions during the summer period, high air temperatures and absence of precipitations.

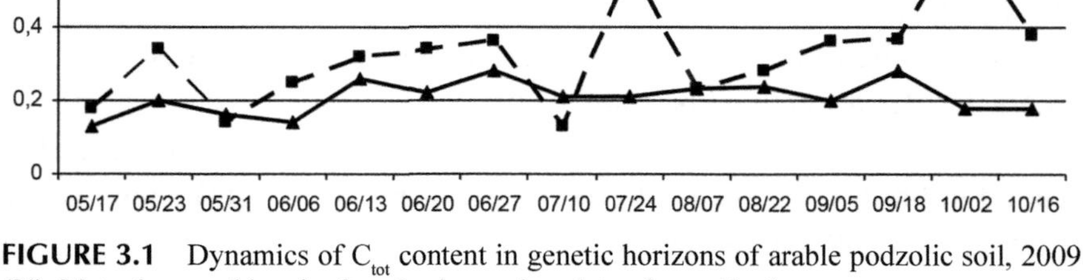


**FIGURE 3.1** Dynamics of $C_{tot}$ content in genetic horizons of arable podzolic soil, 2009 (%) (Note: here and hereinafter abscissa axis – data of sampling).

The humus carbon content and its reserves in arable horizon of soil in 2010 is less, than in previous – from 0.58±0.04 to 1.11±0.06%. Changes of the content during a growth season are significant. Character of dynamics differs from previous year. The maximum content and its greatest fluctuations were in spring period (Figure 3.2). In a period of summer drought fluctuation of humus $C_{tot}$ content are not significant, and the content per se is slightly also. The minimum of the content of humus carbon is noted on an early autumn.

Such character of dynamics is caused by features of transformation of organic matter under extreme weather conditions. At the minimum soil humidity and high air temperature there is a braking of humification processes and the content of organic carbon varies slightly. Obviously, in such conditions organic carbon is presented by compounds strongly connected with a mineral part of soil or poorly decayed organic rests [26].

Underlying horizons of a soil profile contain insignificant amounts of $C_{tot}$ and its fluctuation are absent practically (are not statistically significant) during all 2010 growth season except for the spring period.

The system of soil organic matter is very complex. There are compounds of different structure, complexity and functional activity. It is conditionally accepted to divide soils organic matter into stable and labile parts [27, 28]. The border between them is rather conditional and depends

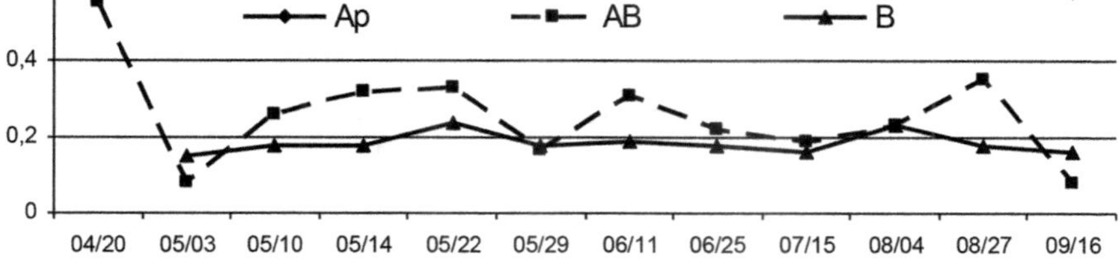


**FIGURE 3.2** Dynamics of content of $C_{tot}$ in genetic horizons of arable podzolic soil, 2010 (%).

on methods of extraction and studying of these parts. If the stable "conservative" component of organic matter is rather well studied, the labile part is studied poorly and the uniform standardized techniques of its extraction and studying do not exist.

Labile components play the important role in soil processes; they define and correct many major soil modes – oxidation-reduction, acid-alkaline and others. Labile components are the most accessible source of nutritive elements for microorganisms and plants. Besides, labile compounds participate in creation and maintenance of balance of system of specific humic matters of soil.

In parallel with definition of the content and reserves of the humus total carbon the content and reserves of labile components of soil organic matter ($C_{lab}$) have been defined.

During two growth seasons significant fluctuations of the content and reserves of labile carbon in all three-soil horizons are observed.

In 2009 the maximum content of $C_{lab}$ was characteristic for arable horizon (from 0.010±0.004 to 0.189±0.021%). During a season statistically significant fluctuations of the content and reserves of labile carbon are noted (Figure 3.3). The maximum values of the content are observed in the beginning of season (in spring and in the beginning of summer) that is caused by increase of microbiological activity at occurrence of favorable levels of temperature and soil humidity and active processing of organic matter.

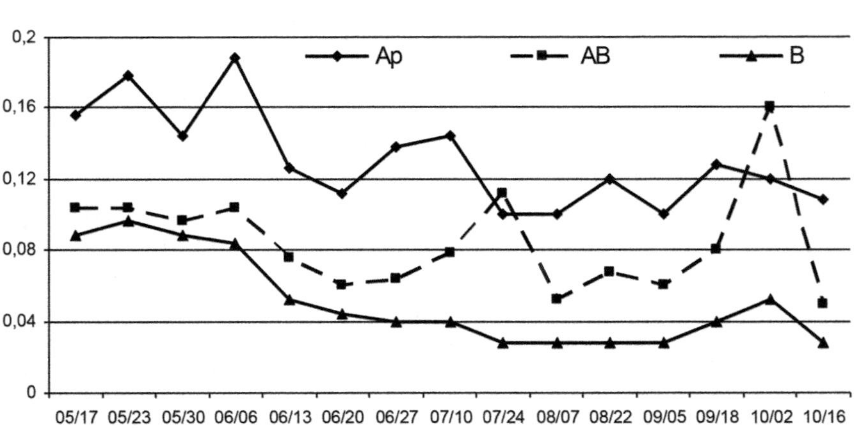


**FIGURE 3.3** Dynamics of $C_{lab}$ content in genetic horizons of arable podzolic soil, 2009 (%).

In the presence of considerable fluctuations during a growth season the tendency of reduction of the $C_{lab}$ content is observed to the end of a season of observing. It, obviously, is connected also with active microbiological destruction of labile fractions of humus and with active transformation of labile compounds in more conservative, poorly mobile compounds.

In underlying horizons AB and B the content of labile compounds of carbon is lower. In horizon AB the content fluctuates from 0.050±0.009 to 0.160±0.007%. Character of dynamics of $C_{lab}$ in horizon AB is close to that in arable horizon. The content maximum is marked in the beginning of a season, a minimum – in the end. In horizon B it is noted sharp fluctuations of the content of labile carbon. As a whole the carbon content in this horizon is minimal (from 0.030±0.004 to 0.094±0.007%) and varies during a season rather smoothly.

During a season of 2010, despite climatic anomalies, a maximum of $C_{lab}$ content is marked in arable horizon also (Figure 3.4). In this horizon the maintenance of $C_{lab}$ fluctuates from 0.059±0.003 to 0.222±0.025%. Thus amplitudes of fluctuation on sampling terms are higher than in 2009. The maximum of the content and the maximum amplitudes of fluctuations are observed in the beginning of a growth season.

At relative decreasing of the content of humus total carbon in comparison with previous year, the increase in $C_{lab}$ content is probably explained

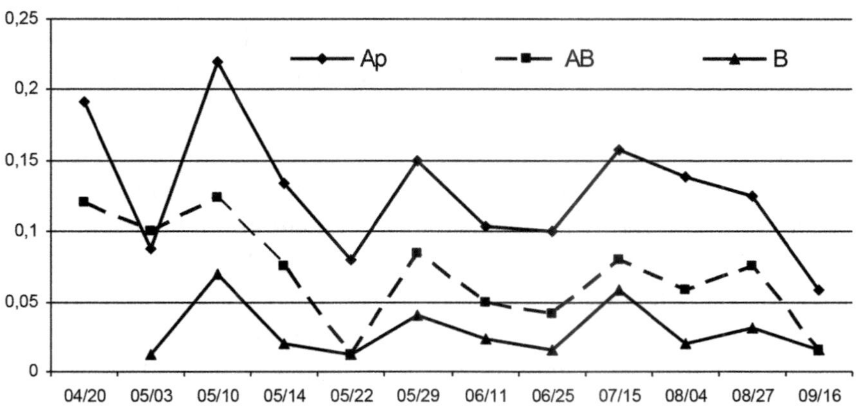


**FIGURE 3.4** Dynamics of $C_{lab}$ content in genetic horizons of arable podzolic soil, 2010 (%).

by chemical destruction of more composite conservative part of humus under extreme weather conditions.

The content of $C_{lab}$ in horizons AB and B is much lower (from 0.016±0.005 to 0.126±0.018% and from 0.015±0.003 to 0.073±0.022%, accordingly). Dynamics of its content during a growth season is expressed distinctly and close to character of its dynamics in arable horizon.

For all surveyed soil horizons the general trend of reduction of $C_{lab}$ content from the beginning of a season by the end is swept up.

Unfortunately, for two years of researches it was not possible to reveal accurate significant dependences between the content of different forms of humus carbon during a season and some soil parameters. Such soil parameters as a mode of humidity, acidities, a temperature mode per se are dynamical in time too and also depend from weather and biocenotic factors. In a real environment joint action of many factors as a result does not give an accurate picture of their influence on system of soil organic matters.

On the basis of studying of the content of carbon in arable soil, its reserves have been counted up on horizons and on profile layers (Tables 3.1 and 3.2).

According to the content, the greatest reserves of humus carbon are intrinsic in arable horizon. Change of reserves during a season repeats dynamics of the carbon content.

If to consider dynamics of reserves on soil layers than the top layers are characterized by rather constant reserves of carbon with small amplitudes of fluctuation during a season (Figures 3.5 and 3.6). The thicker is a considered layer the higher amplitude of fluctuation of the carbon content during a season.

The general tendency of change of labile carbon compounds reserves in both years of research – decrease in reserves from the beginning of growth period up to the end with fluctuations of content different in amplitude. Abnormal on weather conditions 2010 year differs by smaller reserves of labile carbon of humus.

Dynamics of reserves of the total humus carbon on soil layers differs in 2009 and 2010. If in 2009 from the beginning by the end of a growth season, despite fluctuations, the carbon content grows gradually, in 2010 the content falls. It is especially appreciable at higher capacity of a layer.

**TABLE 3.1** Reserves of Humus Total Carbon ($C_{tot}$) and Carbon of Labile Organic Matters ($C_{lab}$) in Some Horizons of Arable Podzolic Soil, 2009 (t/ha)

| Date | 05/17 | 05/23 | 05/30 | 06/06 | 06/13 | 06/20 | 06/27 | 07/10 | 07/24 | 08/07 | 08/20 | 08/05 | 09/18 | 10/02 | 10/16 |
|---|---|---|---|---|---|---|---|---|---|---|---|---|---|---|---|
| Reserves of $C_{tot}$ | | | | | | | | | | | | | | | |
| $A_p$ | 31.91 | 32.74 | 34.81 | 39.78 | 43.10 | 36.05 | 40.61 | 39.37 | 36.47 | 40.61 | 41.03 | 45.58 | 41.44 | 31.49 | 44.76 |
| AB | 4.90 | 9.50 | 4.03 | 7.20 | 8.93 | 9.22 | 9.50 | 3.74 | 16.42 | 6.34 | 8.06 | 10.08 | 10.37 | 19.01 | 10.66 |
| B | 7.61 | 10.66 | 8.12 | 7.61 | 13.20 | 10.66 | 14.72 | 10.66 | 10.66 | 11.67 | 12.69 | 10.15 | 15.23 | 9.14 | 9.14 |
| Reserves of $C_{lab}$ | | | | | | | | | | | | | | | |
| $A_p$ | 6.26 | 7.38 | 6.01 | 7.83 | 5.47 | 4.64 | 5.55 | 5.93 | 4.14 | 4.14 | 4,97 | 4.14 | 5.39 | 4.97 | 4.56 |
| AB | 3.02 | 3.00 | 2.74 | 2.93 | 2.19 | 1.70 | 1.90 | 2.16 | 3.17 | 1.44 | 2,02 | 1.73 | 2.30 | 4.61 | 1.44 |
| B | 4.2 | 4.77 | 4.52 | 4.36 | 2.74 | 2.39 | 2.03 | 1.98 | 1.52 | 1.52 | 1.52 | 1.52 | 2.03 | 2.54 | 1.52 |

**TABLE 3.2** Reserves of Humus Total Carbon ($C_{tot}$) and Carbon of Labile Organic Matters ($C_{lab}$) in Some Horizons of Arable Podzolic Soil, 2010 (t/ha)

| Date | 04/20 | 05/03 | 05/10 | 05/14 | 05/22 | 05/29 | 06/11 | 06/25 | 07/15 | 08/04 | 08/27 | 09/16 |
|---|---|---|---|---|---|---|---|---|---|---|---|---|
| Reserves of $C_{tot}$ | | | | | | | | | | | | |
| $A_p$ | 48.07 | 35.22 | 34.40 | 46.00 | 37.30 | 28.18 | 33.15 | 34.40 | 31.49 | 35.64 | 33.15 | 24.04 |
| AB | 26.78 | 2.59 | 7.49 | 8.93 | 8.93 | 4.61 | 8.35 | 6.34 | 5.18 | 6.34 | 9.50 | 2.30 |
| B | | 7.26 | 9.14 | 9.14 | 11.67 | 8.63 | 9.64 | 9.14 | 7.82 | 11.17 | 9.14 | 7.61 |
| Reserves of $C_{lab}$ | | | | | | | | | | | | |
| $A_p$ | 8.04 | 3.73 | 9.20 | 5.59 | 3.19 | 6.26 | 4.27 | 4.10 | 6.51 | 5.80 | 5.10 | 2.44 |
| AB | 3.51 | 2.85 | 3.63 | 2.19 | 0.81 | 2.45 | 1.44 | 1.21 | 2.25 | 1.64 | 2.19 | 0.46 |
| B | | 1.42 | 3.70 | 1.37 | 0.76 | 2.28 | 1.32 | 1.02 | 3.15 | 1.32 | 1.73 | 0.81 |

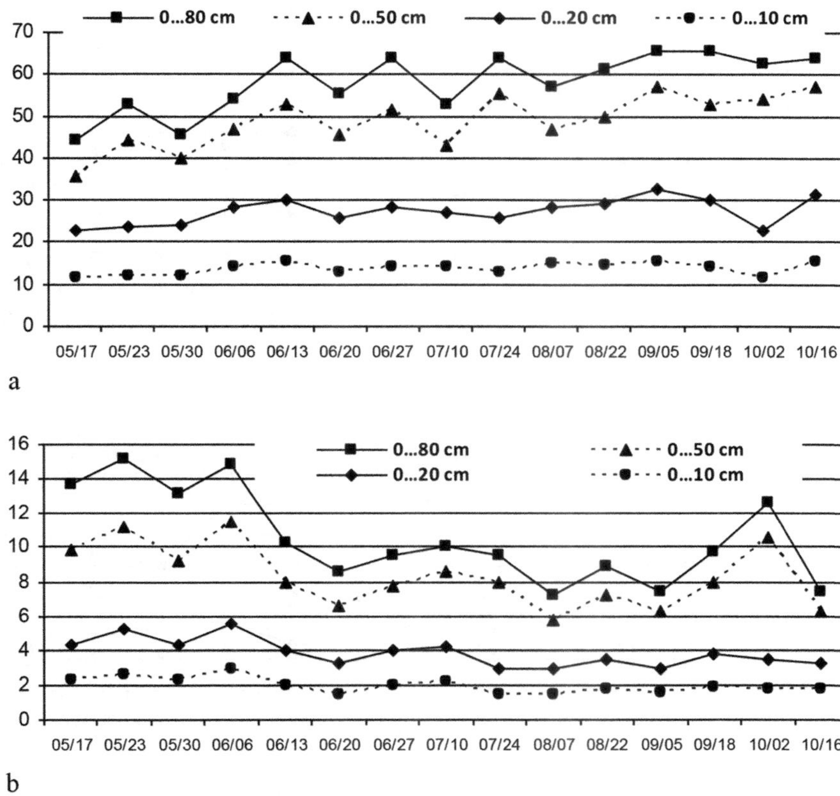


**FIGURE 3.5** Dynamics of reserves of total (a) and labile (b) humus carbon on soil layers, 2009 (t/ha).

### *3.3.2 DYNAMICS OF RESERVES OF ORGANIC MATTER IN A FOREST GROUND LITTER OF SOUTH TAIGA BIOGEOCENOSIS*

The general scheme of formation of an organic part of aboveground cover inseparably linked with biological circulation of matters and soil formation for which the most essential unit is a decomposition of dead organic rests accompanied by simultaneously going processes: accumulation of a primary organic material, decomposition, microbial synthesis, humification, and mineralization [29]. Such unit in South taiga biocenosis is forest

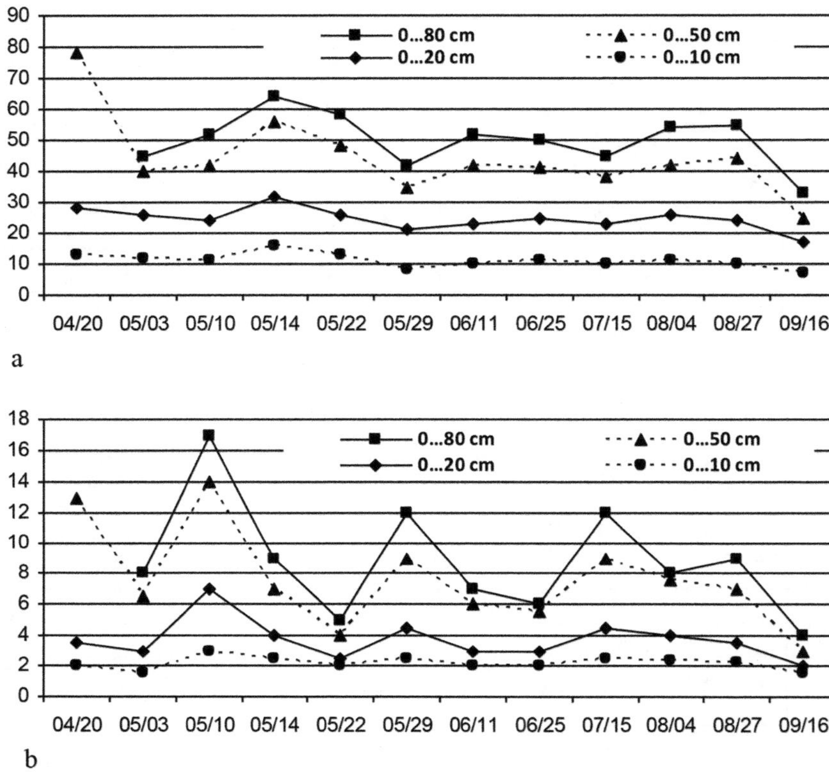


**FIGURE 3.6** Dynamics of reserves of total (a) and labile (b) humus carbon on soil layers, 2010 (t/ha).

ground litter. Really, ground litter is characterized as organogenic horizon in which there is a transformation of the plant's rests and of all organic material arriving on a surface of soil.

Formation of tree waste is a process dynamical in time and differentiated within the forest ecosystems space. Accumulation of organic carbon on a surface of a soil component depends first of all on a relation between size of its annual incoming and size of expense. It is known that incoming of the vegetative rests on a soil surface in forest ecosystems occurs nonuniformly and their specific accessory varies strongly on months.

Tree waste transformation process is influenced by biotic and abiotic factors. Mechanical destruction of tree waste is caused by cycles of freezing – thawing, drying – humidifying that leads to fixing of organic carbon.

During the spring and autumn periods destruction of tree waste at rather weak microbiological activity there are maximum losses and the subsequent migration of various elements out of ground litter.

At tree waste incoming in the generated ground litter it includes in a cycle of transformation of organic matter. Along with incoming tree waste the ground litter transformed also. A consecutive line of transformations is caused by action of some factors: changing and concerning constants. Among the first are character of the vegetative rests and a hydrological mode of soils. It is possible to ascribe to the second next factors: environmental conditions of location (microclimate), granulometric and mineralogical structure of soils, position in system of geochemical landscapes.

The constancy of structure of the incoming vegetative rests leads to stabilization of processes of decomposition and formation of a profile of ground litter with other things being equal [30].

For an estimation of speed and intensity of circulation of organic matter in forest biogeocenoses it is necessary to know reserves of ground litter.

Considering questions of spatial variability of carbon reserves in components of boreal ecosystems, it is possible to notice that under conditions of a southern taiga the data of last years are absent on reserves of organic carbon in ground litter, including annual tree waste.

According to some researches carbon reserves of ground litter horizon of forest soils of a taiga zone can give various values, essentially different from each other. For example, by results of the researches spent in forest biogeocenoses of a sub-band of a southern taiga of the Zvenigorodsky biological research station of the Moscow State University (Moscow Region, Russia), reserves of carbon of ground litter in spruce forests vary within limits 7.1–29.5 t/ha (0.71–2.95 kg/m$^2$) [31]. On researches of the same authors, in 2010 carbon reserves in ground litter of spruce ecosystems of the Kostroma region made 4–19 t/ha (0.4–1.9 kg/m$^2$). In spruce biogeocenoses in Moscow Region carbon reserves have been estimated in 2–71 t/ha (0.2–7.1 kg/m$^2$).

Such essential divergences in an estimation of reserves of an organic matter of a ground litter are obviously connected with use of various approaches and estimation techniques, with different terms of carrying out of researches. At the same time, one of the prospective reasons can be also various features of a forest-vegetation cover, a considerable variation on climatic conditions and zone differences of formation of ecosystems.

Organic matter of forest ground litter transformed during destruction process migrates on a soil profile and is fixed in lower lying horizons. The transformed organic material in a considerable amount is observed in the top horizon of soil.

In horizon of a forest ground litter (0–5 cm) of investigated South taiga biogeocenosis reserves of organic matter on the average for a 2009 season of supervision have made 42.94±2.10 t/ha. The received value is quite co-ordinated with not numerous data of other researchers. By results of the researches spent in birch-spruce young growth in Komi Republic reserves of a ground litter are estimated in 46.1 t/ha, with accumulation for a growth season on 37% [32]. In a mature spruce forest of Komi Republic reserves of a ground litter vary within limits from 26 to 77 t/ha [33].

Analyzing results of studying of dynamics of organic matter reserves in ground litter it is necessary to note the following (Figure 3.7). During the various periods of a growth season significant wavy fluctuations of carbon reserves are observed.

In the beginning of a growth season reserves of carbon of a ground litter made 38.81±11.38 t/ha. To the end of a season of observation reserves have increased by 20.1% and have made 48.68±9.20 t/ha. The maximum value of reserves of organic matter of a ground litter is noted in the second 10-day period of June (55.64±3.04 t/ha). The minimum has been fixed in the end of May and has made 26.47±4.37 t/ha. On the diagram (see Figure 3.7) the tendency of increase in reserves of an organic material from the beginning of the period of observation to its end is well appreciable.

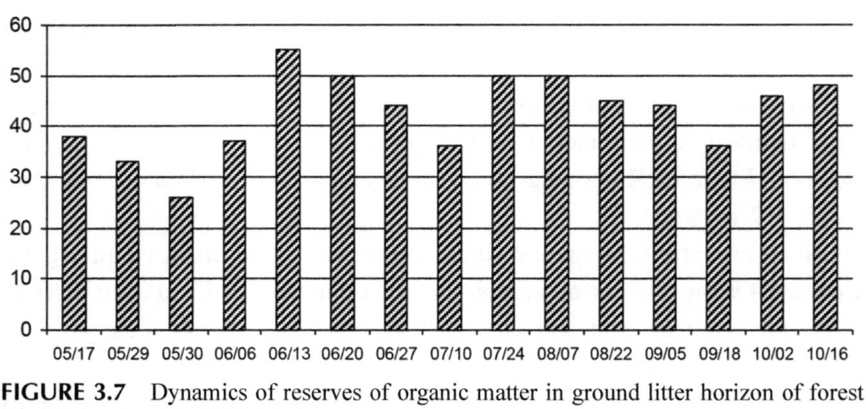


**FIGURE 3.7** Dynamics of reserves of organic matter in ground litter horizon of forest taiga biogeocenosis, 2009 (t/ha).

The possible reason of such variation of parameter of organic material reserves is, on the one hand, increase of destructive activity of soil microorganisms in connection with favorable weather-climatic conditions, as a result of it reserves decrease. But on the other hand – adverse physical factors of environment with an insufficient mode of humidifying which brake destructive activity soil biota that leads to sharp increase in reserves of organics. Despite considerable significant fluctuations of size of reserves during a season, values in the beginning and in the end of a season differ doubtfully that indicates relative resistance and stability of system of soil organic matter.

The second season of observation (2010) was characterized by the abnormal weather phenomena, a combination of a drought and high air temperatures in the middle of summer. Average value of reserves during observation upon forest biogeocenosis of a southern taiga was established on a mark 48.64±3.06 t/ha that is 11.72% higher than in 2009. However, the difference though is significant, but is not so essential.

In the beginning of a 2010 season reserves made 50.25±12.15 t/ha (Figure 3.8). The minimum value of reserves of organic matter of a ground litter makes 33.42±7.12 t/ha. At the same time low values of reserves have been noted in the first and second 10-day periods of May and have made 38.14±11.61 t/ha and 37.24±3.25 t/ha accordingly. The maximum of carbon reserves has been noted in the end of a growth season in the first 10-day period of August which has made 65.07±4.24 t/ha. In the end of a

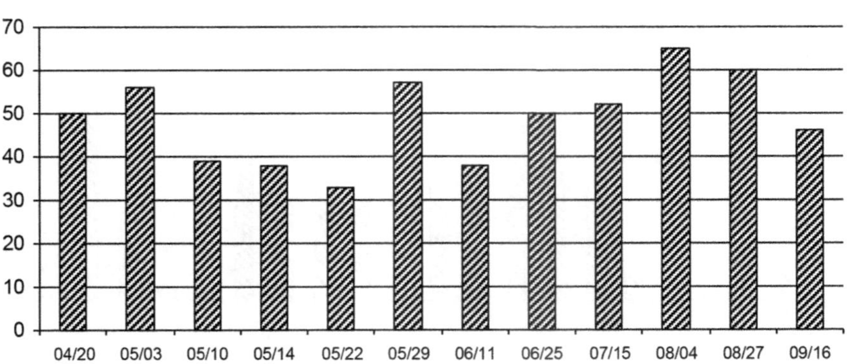


**FIGURE 3.8** Dynamics of reserves of organic matter in ground litter horizon of forest taiga biogeocenosis, 2010 (t/ha).

season of measurements reserves of organic matter of a ground litter have made 45.76±9.02 t/ha.

The received data testify that for a 2010 season of observation reserves of the total organic matter of ground litter horizon have decreased by 8.94%. Considerable deviations both towards increase of reserves of organic matter, and towards fall are observed. It is obvious that abnormal high air temperatures in a combination to unstable mode of humidifying during a 2010 season of researches have affected processes of transformation of organic compounds.

2009 and 2010 growth seasons differ on character of fluctuations of values of carbon reserves. Fluctuations of values of reserves during the investigated periods substantially depended on climatic factors of year of researches.

Weather-climatic conditions and moisture provision influence microbiological activity, and also promote destruction processes, redistribution and accumulation of organic compounds in a soil profile [22].

Destruction and mineralization of an organic material of ground litter, and accordingly organic carbon compounds as a whole on a profile has dynamical character: the maximum is marked during the summer period, and decrease in the content of carbon from spring to the middle of summer and increase by the autumn is observed.

The 2009 season of observation, as a whole, was characterized by values of air temperatures and precipitations close to average long-term values (Figure 3.9). The temperature maximum of the investigated period

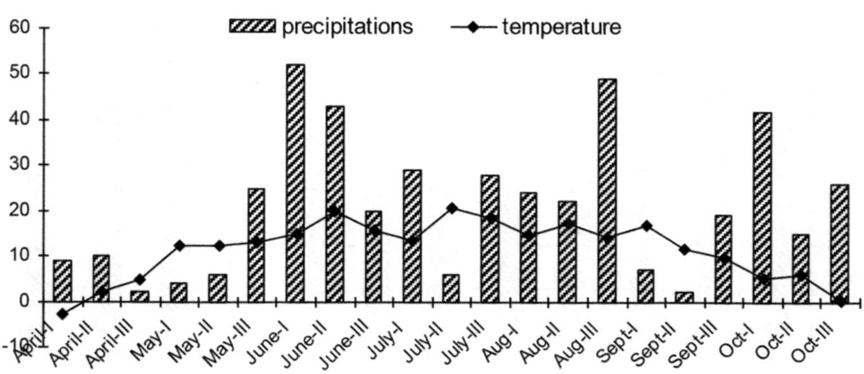


**FIGURE 3.9** Air temperature and amount of precipitations in 2009 growth season on 10-day periods of each month.

has had for July, and a temperature minimum – for the second 10-day period of September. The greatest amount of precipitations has been fixed in the first 10-day periods of June, July, and the third 10-day periods of July and August. The least amount of precipitation has been noted in first two 10-day periods of May (4 and 6 mm accordingly), the second 10-day period of July (6 mm) and first two 10-day periods of September (7 and 2 mm accordingly). There are periods where high airs temperatures are combined with a small amount of precipitations are observed. The droughty period with high temperatures is pointed for the second 10-day period of July and the first 10-day period of September.

Comparing and analyzing diagrams of dynamics of change of organic matter reserves in ground litter in a combination with weather-climatic conditions of growth period accurate dependence of parameters of a carbon reserve on considered factors of environment is traced. Favorable temperatures and an optimum amount of precipitation promoted activation of activity of soil microorganisms; in this connection there was a decrease in reserves of organic matter (see Figure 3.7). And on the contrary, the periods with low temperatures, or too high, and insufficient humidifying promoted decreasing of activity of soil destructors, hence, conducted to increase in reserves of carbon.

However it is necessary to remember that processes of destruction and transformations occur not at once, but are differentiated and shifted in time and space, i.e., after approach of favorable conditions for decomposition of carbon compounds decrease will be observed in its reserves with some delay. In a soil profile favorable conditions for increase in carbon reserves will remain still some time.

Forest ground litter as soil horizon which is on a surface is most sensitive to influence of climatic factors in comparison with underlying horizons [34].

In an abnormal climatic season of 2010 dynamics of organic matter reserves of a litter differ considerably from previous year (Figures 3.8 and 3.10).

The maximum temperatures for a 2010 growth season were observed from third 10-day period of June till the second 10-day period of August and have made 20.9°C and 27.2°C, and precipitations were absent almost completely. For three summer months the temperature maximum had for

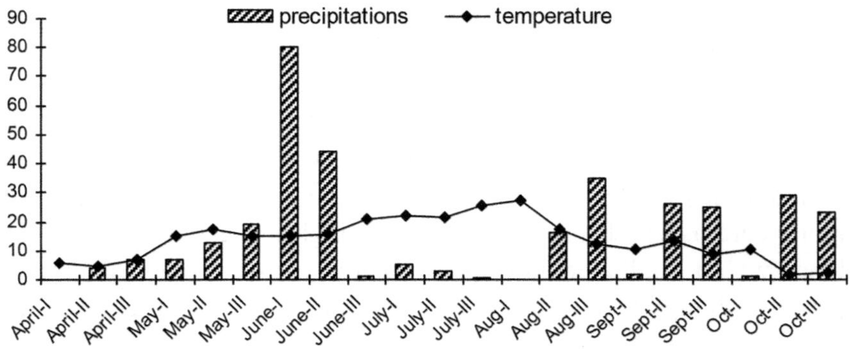


**FIGURE 3.10** Air temperature and amount of precipitations in 2010 growth season on 10-day periods of each month.

July and has made 23.1°C, at a minimum of precipitations – 9 mm for a month. The droughty period of summer months with high air temperatures and a minimum amount of precipitation has made sufficient strong impact on a condition of ecosystems, their productivity, and also on dynamics of organic matter reserves in a soil component of an ecosystem.

Reserves of the total organic matter decrease in the second and third 10-day periods of May, 2010 at rise in temperature and amount of moisture. Activation of destructive activity of soil biota can be one of possible explanations of such phenomenon.

It is known that within the usual temperatures observed on a terrestrial surface, speed of decomposition of organic matter raises with rise in temperature. However, during a summer drought the opposite picture is observed. Parameters of reserves of organic matter in ground litter increase stably and reach the maximum value in the first 10-day period of August and then decrease slowly. Temperature parameters also increase but thus precipitations are absent practically. In the first 10-day period of August there were not precipitations at all (0 mm). It is obvious that in litter horizon, owing to adverse weather factors, the considerable amount of tree waste income, which is not transformed under conditions of sharp deficiency of moisture.

Thus, change of organic matter reserves of soil is a result of interaction of the complex and various factors crossed in biogeocenosis. Therefore, character and the reasons of fluctuation of organic carbon reserves in boreal biogeocenoses demand the further studying.

### 3.3.3 DYNAMICS OF RESERVES OF CARBON IN A SOIL LAYER OF A FOREST TAIGA ECOSYSTEM

At studying of reserves and circulation of organic carbon in ecosystems most usually it is considered only aboveground biomass and organic matter of ground litter. However, underlying horizons of soils contain significant amount of organic matter actively participating in circulation of carbon and other elements.

It is known that reserves of soil carbon increase from equator to poles and reach a maximum in boreal zone. Boreal forests accumulate (in live and dead phytomass) considerable much amount of carbon in a year gain of wood than rainforests. In the conditions of a hot and humid climate tropic ecosystems spend almost 100% of annual primary production on autotrophic and heterotrophic nutrition. In boreal forests essential amounts of carbon collects in a dead biomass (a dead wood, wind fallen trees, soil humus) whereas in tropical forests dead phytomass and soil carbon compounds are quickly processed by microorganisms and carbon is lost with $CO_2$ excretion at respiration [35]. Thus, soils of boreal forests are characterized by the greatest reserves of soil carbon.

In the course of vital activity of forest ecosystems considerable amount of organic carbon which migrates on a profile penetrating into thickness of soil collects in a soil component. Quantitative distribution of organic carbon on soil horizons varies. Non-uniform distribution can be defined as various capacities of horizons, features of structure and properties of horizons, and substantially fluctuation of environment factors.

It is known that forest ground litter accumulates the greatest amount of organic matter in the form of organic rests of different degree of decomposition. However, reserves of humus compounds per se in soil are small in comparison with other horizons (Figure 3.11).

It is connected with low capacity of the given horizon and high mobility of specific humic compounds formed in it in boreal zone. The greatest amount of organic carbon of humic compounds contains in illuvial horizon that is caused by its high capacity.

Studying and estimation of reserves of soil organic carbon becomes complicated by dynamical character of this soil component. In soils of the investigated key sites within two years of observation significant seasonal dynamics of the content of organic matter is revealed.

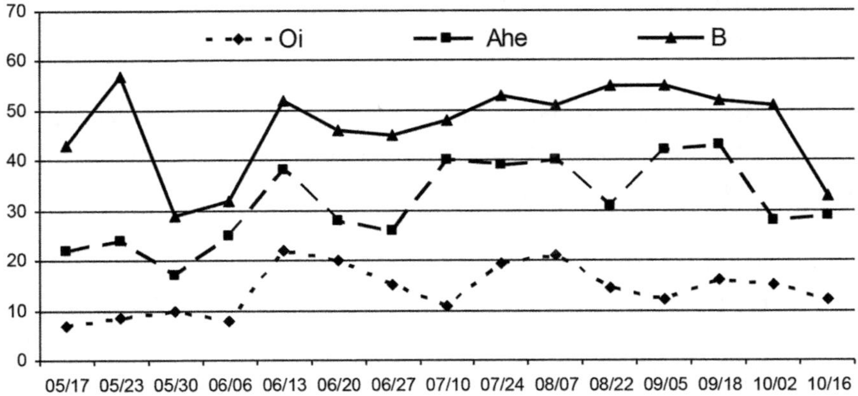


**FIGURE 3.11** Dynamics of reserves of humus total carbon on soil horizons, 2009 (t/ha).

In the beginning of the 2009 growth period the minimum reserves of carbon in horizon $O_i$ – 6.52±1.2 t/ha are fixed. The maximum value of reserves of organic carbon in the given horizon (21.86±0.82 t/ha) has been noted in the second 10-day period of June. In the end of the growth period reserves of organic carbon have made – 12.57±1.08 t/ha that significantly exceeds values in the beginning of a season.

Horizon $A_{he}$ differs by the greatest amplitude of fluctuation of carbon reserves. In the beginning of a season they were 23.03±6.77 t/ha, in the end – 29.41±6.21 t/ha. The maximum value of reserves is fixed in the middle of September, it has made 43.32 t/ha whereas the minimum reserve is noted in the third 10-day period of May and has made 16.64±1.4 t/ha.

In the heaviest horizon B there are the greatest reserves of organic carbon of humic matter. In the beginning of a growth season they made 43.71±6.04 t/ha, in the end – 33.48±2.66 t/ha. Reserves of the total carbon in this horizon in the end of a season were significantly reduced by 23.4%.

The received values of reserves of the total carbon in soil horizons of a South taiga spruce forest have appeared close to results of the researches spent in territory of an middle taiga (Republic of Komi, Russia) where carbon reserves on horizons have made: $O_i$ (0–5 cm) – 14.62±2.69 t/ha, $A_{he}$ (5–20 cm) – 16.36±0.27 t/ha, B (20–80 cm) – 41.71±1.2 t/ha [36].

Recalculation of reserves of carbon by soil horizons has been made for an estimation of reserves of the total carbon and their dynamics in a soil profile as a whole (Figure 3.12).

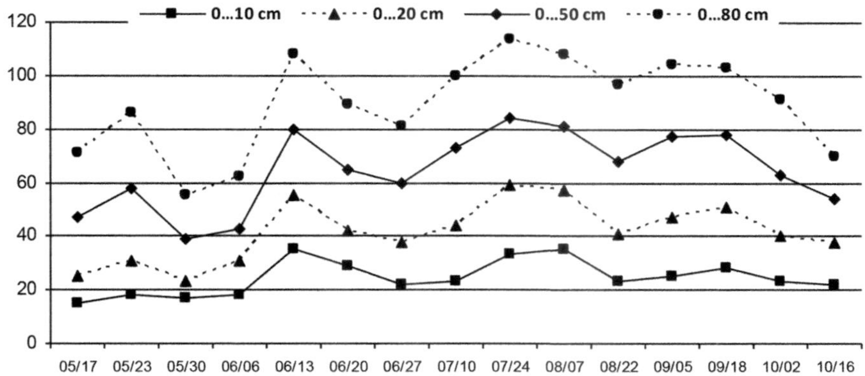


**FIGURE 3.12** Dynamics of reserves of total carbon on soil layers, 2009 (t/ha).

The obtained data is close to non-numerous observations of other researchers. On materials of the State Government of Forest Fund for January, 1st, 1998 average reserves of organic carbon on soil layers 0–30 cm, 0–50 cm, and 0–100 cm are defined as 64±8, 75±9 and 91±10 t/ha accordingly [37]. There is data on carbon deposition in soils of spruce forests in middle taiga zone where its content makes from 84.85±4.30 to 96.5±34.8 t/ha [36].

In some publications [35, 38] the opinion on the admitted statistical errors and low reliability of the data on carbon reserves in soils is expressed. Discrepancy in definition of carbon reserves is connected, obviously, with high dynamism of this parameter, which varies essentially depending on terms of carrying out of observation. Presented figures shown that carbon reserves vary essentially in all soil layers during a season.

Analyzing results, it is possible to note the following. Some soil horizons are characterized by considerable fluctuations of reserves during a season and in this case it is difficult to established the general laws of behavior of system of soil carbon.

Considering reserves on layers it is possible to note some laws. The more considerably thickness of a soil layer the higher scope of fluctuations of reserves of the total carbon during a season. However, it appears that the more considerably a thickness of a layer, the more stably is a system in an annual cycle. Reserves of the total carbon in a layer 0–80 cm in the beginning (70.44±14.31 t/ha) and the end (70.63±12.04 t/ha) of growth season do not differ significantly.

The maximum values of carbon reserves in different soil layers were on the end of July–September and vary within 96.86–111.94 t/ha. The minimum reserves of the total carbon have been fixed in the end of spring and have made 53.00±3.95 t/ha.

If the carbon content in a layer 0–10 cm by the end of season has increased significantly by 36.55% than in a layer 0–80 cm for the same period increase in reserves have made 0.27% only. Such indicators show that the soil component represents self-regulated, stable system where all parameters are in dynamic balance. According to some information, values of deposition of carbon depend on age of forest planting. Young growing plantings actively increase carbon reserves in phytomass; there is a replenishment of pools of a ground litter and soil. In process of aging of a forest stand deposition decreases up to full stabilization of carbon reserves at formation of resistant mature forests [39]. Average age of forest stand in investigated biocenosis makes 60–80 years and many processes, such as an exchange of matter, energy, and the information within system, specific and trophic structure were generated and have plastic and dynamical character that promotes stability of system of soil carbon.

The 2010 growth season was characterized by abnormal high air temperatures and a minimum amount of precipitations than sharply differed from previous year.

The maximum values of carbon reserves in all horizons ($O_i$, $A_{he}$, and B) have been noted in the first 10-day period of August and have made 25.15±2.84, 52.21±1.7, and 60.45±3.67 t/ha accordingly (Figure 3.13).

In ground litter horizon the minimum values of reserves are noted in the beginning of a growth season. In the season beginning in horizon $O_i$ reserves of the total carbon have made 13.15±3.54 t/ha, on its end – 15.35±3.14 t/ha. The increase in reserves in horizon has made 14.32%. In horizons $A_{he}$ and B reserves in the beginning of season have made 29.41±6.13 and 36.27±9.55 t/ha accordingly, in the end of a season – 18.24±8.66 and 31.62±3.87 t/ha accordingly. In horizons $A_{he}$ and B there was a decrease in reserves by 37.98% and 12.82% accordingly. However, in each separately taken horizon change of reserves in the end of season is doubtful in comparison with the beginning of a season.

In a layer 0–80 cm carbon reserves in the season beginning have made 74.10±12.80 t/ha, and in its end – 62.76±10.49 t/ha, carbon reserves have

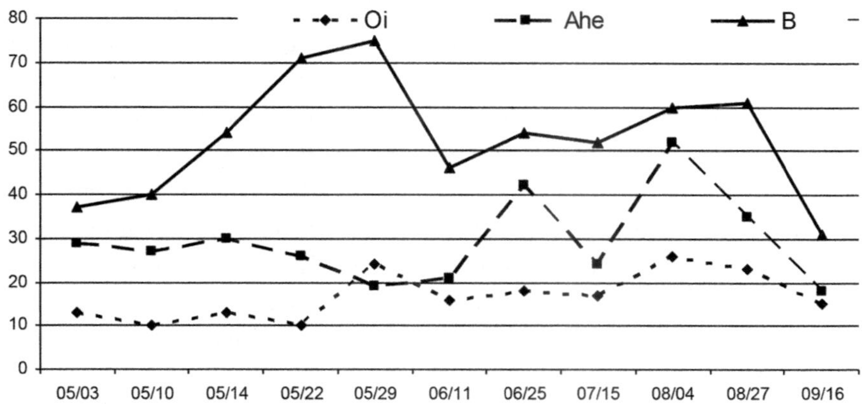


**FIGURE 3.13** Dynamics of reserves of total carbon on soil horizons, 2010 (t/ha).

decreased on 15.3% (Figure 3.14). However, because of the large scope of fluctuations of a parameter decrease is doubtful. The maximum values of carbon reserves had for the first 10-day period of August and have made 129.19±9.63 t/ha. The minimum reserves of the total carbon have been noted in the second 10-day period of September and have made 62.76±10.49 t/ha.

Observable dynamics of reserves of the total carbon in horizons of a soil component during seasons 2009 and 2010 (see Figures 3.11 and 3.13) has shown that in horizon $O_i$ for two seasons of observation carbon

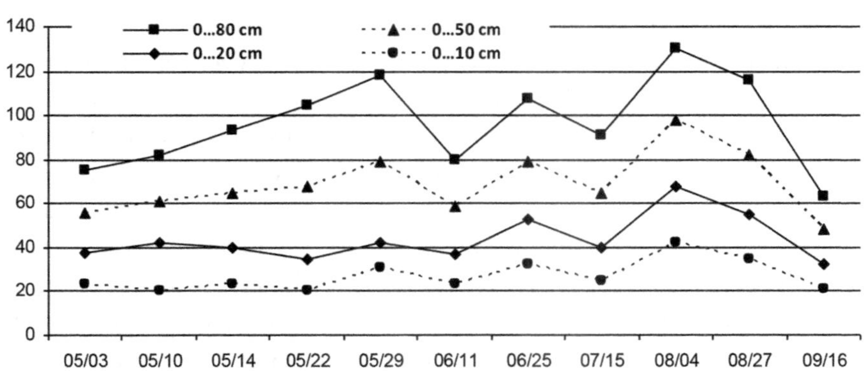


**FIGURE 3.14** Dynamics of reserves of total carbon on soil layers, 2010 (t/ha).

reserves have increased, and in 2009 reserves have increased more than by 2 times. Other situation is observed in horizon $A_{he}$. For a growth season of 2009 reserves of the total carbon increased, as well as in horizon $O_i$, but in 2010 when there was an abnormal high air temperatures in territory of the European part of Russia reserves of carbon of this horizon have decreased practically twice. Character of dynamics of carbon reserves is rather close in both years of research. In the end of the period of observation carbon reserves in soil horizons are restored practically to initial values.

The greatest reserves of the total carbon of humus are contained in horizon B that is caused by its high capacity in comparison with other horizons.

By consideration of carbon reserves in soil layers similar character of their dynamics during growth seasons differ on climatic parameters is revealed. In the beginning and in the end of seasons carbon reserves both on horizons, and as a whole in a soil layer, are close. In the beginning of a growth season carbon reserves in a soil layer 0–80 cm have made 70.44±14.31 and 74.10±12.80 t/ha in 2009 and 2010 according to, and in the end – 70.63±12.04 and 62.76±10.49 t/ha. The greatest amplitude of fluctuation of carbon reserves is noted during the growth period of 2010. Carbon reserves in a layer 0–80 cm vary especially strongly that, certainly, is connected with abnormal hydrothermal conditions and decrease in intensity of processes of photosynthesis depending on them, and microbiological activity. As a result of over-drying humification of organic mass and carbon migration on a profile of soils is complicated. At approach of favorable conditions, processes of microbiological transformation of organic matter that leads to increase of the content of soil carbon start to amplify.

The structure of soil organic matter is complex. Except conservative concerning stable part of soil organic matter there are also labile components.

Abundant actual material is collected till now many-sided characterizing a conservative part of organic matter of soil (the total carbon). It cannot be refer to labile (mobile) carbon – data on it and its reserves in a soil component of an ecosystem are absent practically in the literature.

It is considered that labile components of humus, which are, on the one hand, a source of nutrients accessible to plants and microorganisms, and with another – actively replenish neogenic humic substances are more subject to seasonal dynamics. Character of transformation of humus is thus defined by balance between receipt of organic rests and processes of their humification and mineralization. V.M. Volodin with co-authors [40] considers that character of seasonal dynamics of humus labile forms depends on activity of biocenosis. When it is highest, the content of humus decreases. After biocenosis activity decreases the content of humus raises again. The labile part of humus in this case serves as though as "buffer" between the live population and a steady part of soil humus.

Besides definition of reserves of the total carbon of humus components during researches reserves and dynamics of labile carbon for both seasons of observation (Figures 3.15 and 3.17) have been defined.

For two seasons significant essential fluctuations of reserves of labile carbon both on horizons and on layers of soils are observed.

In the beginning of a 2009 season in horizon $O_i$ reserves of labile carbon have made 1.90±0.40 t/ha. To end of a season reserves of compounds of labile carbon have made 2.92±0.71 t/ha. The maximum value of reserves (6.44±1.13 t/ha) is noted in the third 10-day period of June; the minimum of carbon reserves has made 1.90±0.40 t/ha in the beginning of a season of researches. In horizon $A_{he}$ the least value of reserves of labile carbon has made 3.97±0.8 t/ha in the beginning of the period (the second 10-day period of May). Further the increase in reserves of labile carbon was observed and on end of a season they made 7.75±1.36 t/ha. The maximum reserves have been noted in the second 10-day period of September (11.63±2.06 t/ha).

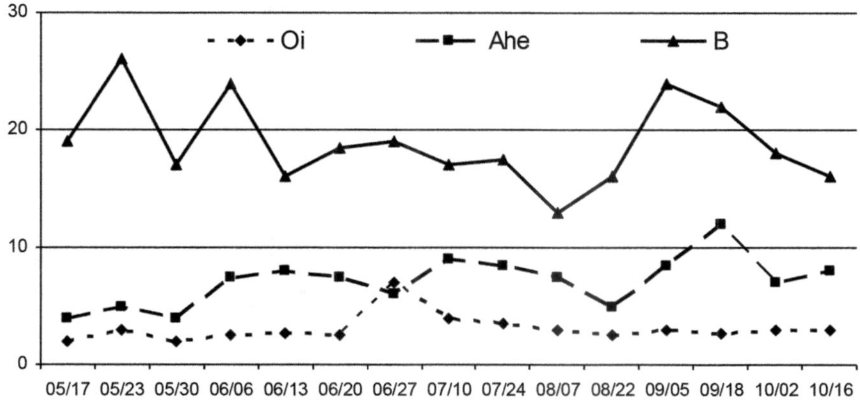


**FIGURE 3.15** Dynamics of reserves of labile carbon on soil horizons, 2009 (t/ha).

If in the beginning of a season in horizon B carbon reserves have made 19.16±2.45 t/ha than in its end – 15.81±1.53 t/ha. The maximum reserve of carbon, which has made 24.55±4.86 t/ha is noted in the third 10-day period of May; minimum – 13.02±5.34 t/ha is noted in the first 10-day period of August.

Reduction of amount of labile humus components in the end of vegetation is caused, obviously, by synthesis of more stable substances. So, for example, in soils of Western Siberia in August the amount of labile humus substances was minimal [41].

Reserves of labile carbon in the end of a season in a layer (0–80 cm) have made 25.64±5.13 t/ha. The maximum values of carbon have been noted in September, minimum in May (Figure 3.16).

Reserves of labile carbon in a layer (0–10 cm) by the season end have increased twice. Also the positive tendency is traced in a layer 0–50 cm where the increase in reserves has made 11.89%. Though in the top horizons reserves have increased in the end of a season in comparison with the beginning, in a layer 0–80 cm the value of reserves remained stable. That is, as a whole the system of labile organic matter of soils is stable enough and, despite seasonal fluctuations, is resistant.

During the 2010 growth season having high air temperatures and a minimum amount of precipitations values of reserves of labile carbon in horizons $O_i$ and $A_{he}$ have appeared minimal. Reserves of labile fractions of

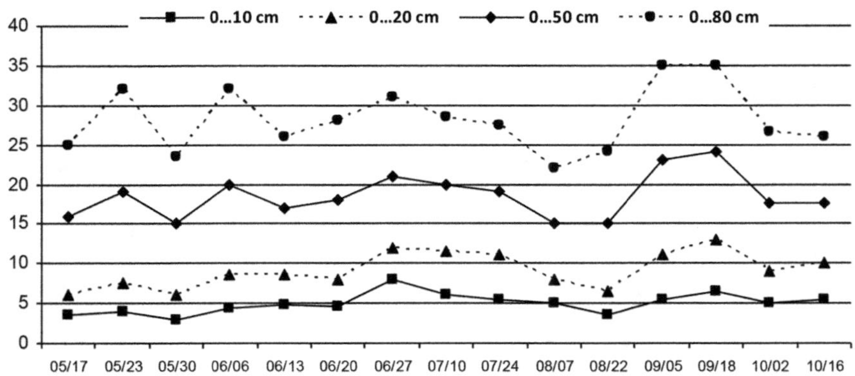


**FIGURE 3.16** Dynamics of reserves of labile carbon on soil layers, 2009 (t/ha).

carbon in horizon $O_i$ in the beginning of the period of researches were up to standard 1.75±0.36 t/ha, in season end have made 1.52±0.32 t/ha. The minimum reserve of labile carbon in horizon $O_i$ has made 1.36±0.14 t/ha in the second 10-day period of May; the maximum value 2.66±0.59 t/ha has been noted in the first 10-day period of May (Figure 3.17).

In horizon $A_{he}$ the maximum value of carbon reserves has made 11.86±2.73 t/ha in the first 10-day period of May; the minimum reserve of labile carbon has had for the second 10-day period of September. Reserves of labile compounds of carbon to end of the growth period have made 2.76±1.42 t/ha. As a whole for a season the amount of organic carbon has decreased by 62.19%.

In horizon B the maximum value of reserves of labile compounds of carbon has made 51.52±4.14 t/ha in second half of July; the minimum value makes 16.74±2.11 t/ha in the end of a season.

Carbon reserves in a layer of a soil profile 0–80 cm in the end of a season have decreased in comparison with the beginning of the growth period (Figure 3.18).

The maximum of carbon reserves is noted in the second 10-day period of July and has made 64.40 t/ha; then there is a sharp decrease to the end of a season. Reserves of labile carbon in a layer 0–80 cm to season end make 21.33±4.98 t/ha whereas in the beginning of a growth season made 26.54±3.27 t/ha. The general decrease in reserves makes

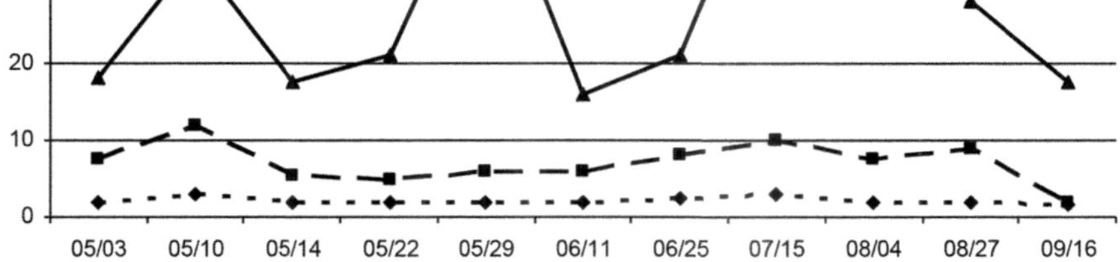


**FIGURE 3.17** Dynamics of reserves of labile carbon on soil horizons, 2010 (t/ha).

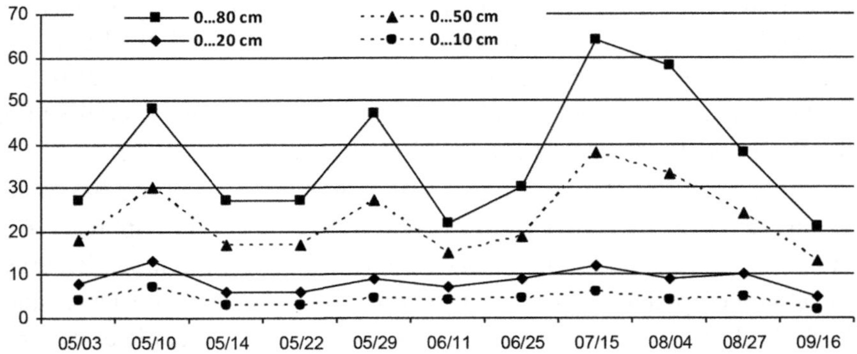


**FIGURE 3.18** Dynamics of reserves of labile carbon on soil layers, 2010 (t/ha).

18.84%. However, decrease in reserves of labile carbon by the season end is insignificant that also indicated relative stability of the system despite adverse climatic factors.

Dynamics of reserves of labile carbon both on horizons and on layers of a soil component distinctly shows considerable influence of environmental factors such as air temperature and amount of precipitation. Fluctuations of reserves in illuvial horizon (horizon B) are especially great.

The analysis of dynamics of reserves of labile carbon during the 2009 growth period has shown that the significant increase in reserves by the season end occurs only in horizons $O_i$ (0–5 cm) and $A_{he}$ (5–20 cm). In horizon of a forest ground litter and eluvial horizon reserves have increased by 1.5 times during 2009 growth season. Insignificant decrease has occurred only in illuvial layer of a soil profile. In 2010 in all horizons there was a decrease in reserves of labile forms of carbon in the end of a season. The most considerable falling was in eluvial horizon more than by 2.5 times.

As a whole the amplitude of fluctuations of reserves of humus organic carbon increases at increase in capacity of a considered layer, however, thus the heavier is soil thickness the more stably is carbon system as a whole. Even during seasons extreme on climatic parameters the soil component of an ecosystem is rather stable that promotes stability of all ecosystem as a whole.

## 3.4 CONCLUSIONS

In arable podzolic soil statistically significant dynamics of the content and reserves of the humus total carbon as well as its labile part is observed. Character of this dynamics is distinguished in different years. The minimum value of reserves in 2009 is noted in first half of growth season, maximum – in second half of growth season. In 2010 character of dynamics is opposite. The greatest amplitudes of fluctuation of the carbon content are characteristic for arable horizon. For carbon of labile compounds the general trend of reduction of the content from the beginning by the end of season is noted. The system of organic matter of arable soil is more dynamical and unstable than in soil of forest biogeocenosis. In different years various values of carbon reserves in the end and in the beginning of seasons are marked.

Forest ground litter is the major component of taiga forest biogeocenoses through which there is a flow of organic carbon from the top layer into soil horizons. Processes of transformation of organic matter of a ground litter influence essentially character and speed of circulation of carbon in forest biogeocenoses. The system of organic matter of a forest ground litter is stable and possesses dynamic balance. Even abnormal hydrothermal conditions of 2010 have not discomposed the system. Distinctions in average reserves of organic matter of a ground litter in 2009 and 2010 were insignificant though character of dynamics of reserves during seasons differs.

In a soil layer of South taiga forest biogeocenoses significant dynamics of reserves of the humus total carbon and its labile part is observed. Character of dynamics of carbon reserves is rather close in both years of researches. The minimum values of reserves of humus organic carbon are characteristic as a rule for beginning of a growth season, maximum – for second half of growth season. Despite considerable fluctuations of carbon reserves during the growth periods, in the beginning and the end of seasons of research parameters did not differ significantly that can be a sign of stability of all system of soil organic matter. Stability of system remains in annual dynamics too.

## KEYWORDS


- **forest ground litter**
- **humus**
- **mobile carbon**
- **seasonal dynamics**


## REFERENCES


1. Kudeyarov, V. N. Role of soil in carbon circle. Pochvovedenie ("Soil Science" in English). 2005, 8, 915–923 (in Russian).
2. Römkens, P. F. A. M., Bouwman, L. A., Boon, G. T. Effect of plant growth on copper solubility and speciation in soil solution samples. Environmental Pollution. 1999, 106, 315–321.
3. Fytianos, K., Katsianis, G., Triantafyllou, P., Zachariadis, G. Accumulation of heavy metals in vegetables grown in an industrial area in relation to soil. Bulletin of Environmental Contamination and Toxicology. 2001, 67, 423–430.
4. Ge, Y., Murray, P., Hendershot, W. H. Trace metal speciation and bioavailability in urban soils. Environmental Pollution. 2000, 107, 137–144.
5. Skáodowski, P., Maciejewska, A., Kwiatkowska, J. The effect of organic matter from brown coal on bioavailability of heavy metals in contaminated soils. Soil and Water Pollution Monitoring, Protection and Remediation, 2006, 3, 299–307.
6. Ashworth, D. J., Alloway, B. J. Influence of Dissolved Organic Matter on the Solubility of Heavy Metals in Sewage-Sludge-Amended Soils. Communications in Soil Science and Plant Analysis. 2008, 39, 538–550.
7. Khan, A. G., Kuek, C., Chandhry, T. M., Khoo, C. S., Hayes, W. J. Role of plants, mycorrhizae and phytochelators in heavy metal contaminated land remediation. Chemosphere. 2000, 41, 197–207.
8. Arnesen, A. K. M., Singh, B. R. Plant uptake and DTPA-extractability of Cd, Cu, Ni and Zn in a Norwegian alum shale soil as affected by previous addition of dairy and pig manures and peat. Canadian Journal of Soil Science. 1999, 78, 531–539.
9. Elliot, H. A., Liberati, M. R. Huang, C. P. Competitive adsorption of heavy metals by soils. Journal of Environment Quality. 1986, 15, 214–219.
10. Ferrand, E., Dumat, C., Leclerc-Cessac, E., Benedetti, M. Phytoavailability of zirconium in relation with its initial speciation and soil characteristics. Plant and Soil. 2006, 287, 313–325.
11. Lo, K. S. L., Yang, W. F., Lin, Y. C. Effects of organic matter on the specific adsorption of heavy metals by soil. Environmental Toxicology and Chemistry. 1992, 34, 139–153.

12. Saar, R. A., Weber, J. H. Fulvic acid: Modifier of metal-ion chemistry. Environmental Science and Technology. 1982, 16, 510–517.
13. Quenea, K., Lamy, I., Winterton, P., Bermond, A., Dumat, C. Interactions between metals and soil organic matter in various particle size fractions of soil contaminated with waste water. Geoderma, Elsevier. 2009, 1, 217–223.
14. Impellitteri, C. A., Lu, Y., Saxe, J. K., Allen, H. E., Peijnenburg, W. J. G. M. Correlation of the partitioning of dissolved organic matter fractions with the desorption of Cd, Cu, Ni, Pb and Zn from 18 Dutch soils. Environment International. 2002, 28, 401–410.
15. Boucher, U., Lamy, I., Cambier, P., Balabane, M. Decomposition of leaves of metallophyte Arabidopsis halleri in soil microcosms: fate of Zn and Cd from plant residues. Environmental Pollution. 2005, 135, 323–332.
16. Senesi, N. Metal-humic substances complexes in the environment. Molecular and mechanistic aspects by multiple spectroscopic approach. In: Adriano, D. C. (ed.): Biogeochemistry of trace metals. Lewis Publ., Boca Raton, 1993, 429–496.
17. Zhang, M., Alva, A. K., Li, C., Calvert, D. V. Chemical association of Cu, Zn, Mn and Pb in selected sandy citrus soils. Soil Science. 1997, 162, 181–187.
18. Santos, A., Bellin, I. C., Corbi, P. P., Cuin, A., Rosa, A. H., de Oliviera Resende, M. O., Rocha, J. C., Melnikov, P. Complexation of metal ions by humic substances and α-amino acids. A comparative study. Proceedings of the 11th Biennial Conference of the International Humic Substances Society and the 6th Humic Substances Seminar, held in Boston, MA from July 21–27, 2002. New York: Taylor & Francis, 2003, 271–273.
19. Shikhova, L. N. Content and dynamics of heavy metals in soils of north-east of European part of Russia. DSc Thesis. Saint-Petersburg State Agrarian University. 2005, 396 p. (in Russian).
20. Shikhova, L. N., Lisitsyn E.M. Dynamics of content and store of humus carbon in arable podzolic soils of south taiga sub-zone in Kirov region. Herald of Udmurt University. Biology. Earth Sciences. 2014, 2, 7–13 (in Russian).
21. Lisitsyn, E. M., Shikhova, L. N., Ovsyankina, A. V. Edaphic stress-factors of north-east of European Russia and problems of plant breeding. Agricultural biology. 2004, 3, 42–60 (in Russian).
22. Orlov, D. S., Biryukova, O. N., Sukhanova, N. I. Organic matter in soils of Russian Federation. Moscow: Nauka ("Science" in English), 1996, 254 p. (in Russian).
23. Shchepachenko, D. G., Mukhortova, L. V., Shvidenko, A. Z., Vedrova, E. F. Reserves of organic carbon in soils of Russia. Pochvovedenie ("Soil Science" in English). 2013, 2, 123–132 (in Russian).
24. Arinushkina, E. V. Handbook of Reference Methods for chemical analysis of soils. Moscow: Moscow State University Publ., 1972. 490 p. (in Russian).
25. Kogut, B. M., Bulkina, L.Yu. Comparative estimation of repeatability of methods for determining of labile forms of chernozem humus. Pochvovedenie ("Soil Science" in English). 1987, 4, 143–145 (in Russian).
26. Alexandrova, L. N. Soil organic matter and processes of its transformation. Leningrad: Nauka ("Science" in English). 1980, 287 p. (in Russian).
27. Drichko, V. F., Bakina, L. G., Orlova, N. E. Stable and labile parts of humus in sod-podzolic soil. Pochvovedenie ("Soil Science" in English). 2013, 1, 41–47 (in Russian).

28. Orlov, D. S. Humic acids in soil and general theory of humification. Moscow: Moscow State University Publishing House. 1990, 325 p. (in Russian).
29. Kudryavtsev, V. A. Forming of forest ground litter in vegetation conditions of spruce oxalis forests. Lesnoe khozyaistvo ("Forest Industry" in English). 2009, 3, 16–17 (in Russian).
30. Bogatirev, L. G., Demin, V. V., Matyshak, G. V., Sapozhnikov, V. A. About some theoretic aspects of investigation of ground litter. Lesovedenie ("Forest science" in English). 2004, 4, 17–28 (in Russian).
31. Podvezennaya, M. A., Ryzhova, I. M. Relationships between the variability of soil carbon reserves and the spatial structure of plant cover in forest biogeocenoses. Herald of Moscow State University. Serial 17. Soil Science. 2010, 4, 3–9 (in Russian).
32. Pristova, T. A. Characteristics of wood abscission and reserves of forest ground litter in leaved plantations in middle taiga. Herald of Biology Institute of Komi Scientific Center of Ural Branch of Russian Academy of Sciences. 2011, 9, 7–9 (in Russian).
33. Kuznetsov, M. A., Osipov, A. F. Vegetative abscission as a component of biological circle of carbon in waterlogged coniferous associations of middle taiga. Herald of Biology Institute of Komi Scientific Center of Ural Branch of Russian Academy of Sciences. 2011, 9, 10–12 (in Russian).
34. Shikhova, L. N., Zubkova, O.A., Russkhih, E. A., Koryakina, E. V. Dynamics of carbon reserve in a soil layer of taiga forest ecosystem. Herald of Udmurt University. Biology. Earth Sciences. 2011, 4, 31–39 (in Russian).
35. Moiseev, B. N., Filipchuk, A. N. IPCC's (Intergovernmental Panel on Climate Change) method for calculation of annual deposition of carbon and estimation of its application for Russian forests. Lesnoe khozyaistvo ("Forest Industry" in English). 2009, 4, 11–13 (in Russian).
36. Pristova, T. A. Components of carbon circle in leaved-coniferous plantation of middle taiga. Herald of Biology Institute of Komi Scientific Center of Ural Branch of Russian Academy of Sciences. 2009, 8, 12–16 (in Russian).
37. Chesnykh, O. V., Zamolodchikov, D. G., Utkin, A. I. Total reserves of biological carbon and nitrogen in soils of forest fund of Russia. Lesovedenie ("Forest Science" in English). 2004, 4, 30–42 (in Russian).
38. Treifeld, R. F. Contribution of forest management into realization of international obligations of Russia on problem of global warming of climate. Lesnoe khozyaistvo ("Forest Industry" in English). 2007, 3, 7–8 (in Russian).
39. Zamolodchikov, D. G., Karelin, D. V. Absorption of carbon by forests of Valdai national park and perspectives of its realization within climatic agreements. Natural, culture-historical and touristic potential of Valdai Hills, its protection and use. Saint-Petersburg: Publishing-Printing Center of Saint-Petersburg University of Technology and Design, 2010, 69–17 (in Russian).
40. Volodin, V. M., Masyutenko, N. P., Velyukhanova, O. V. Dynamics of organic matter in soil at agricultural use of chernozem. Materials of Scientific-Practical Conference "Agriculture in XXI Century. Problems and Ways of Its Decision," 25–27 October 2000. Kursk: Research Institute of Farming and Soil Erosion Protection. 2001, 206–210 (in Russian).
41. Tjurin, I. V. Soil Organic Matter and Its Role in Fertility. Moscow: Nauka ("Science" in English), 1965, 317 p. (in Russian).

## CHAPTER 4

# HEAVY METALS IN THE SYSTEM SOIL–PLANT WHEN USING SEWAGE SLUDGE FOR FERTILIZER

GENRIETTA YE. MERZLAYA,[1] RAFAIL A. AFANAS'EV,[2] OLGA A. VLASOVA,[3] and MICHAIL O. SMIRNOV[4]

*D.N. Pryanishnikov All-Russian Scientific Research Institute of Agrochemistry, 31A, Pryanishnikov St., Moscow, 127550, Russia; [1]E-mail: lab.organic@mail.ru; [2]E-mail: rafail-afanasev@mail.ru; [3]E-mail: cool.vlasova2013@yandex.ru; [4]E-mail: User53530@yandex.ru*

## CONTENTS



### ABSTRACT

Studies have shown that municipal wastewater sludge can, with appropriate training be used as fertilizer. In order to avoid environmental and production pollution with heavy metals found in many types of precipitation you must comply with the relevant regulations of their use for agricultural crops, including rates and frequency of the application in soil.

## 4.1 INTRODUCTION

Urbanization of the population is accompanied by the accumulation of huge masses of organic wastes in the form of sewage sludge, which because of the possible content of pollutants, especially heavy metals, are not widely applied in agricultural production and thereby practically excluded from the biological cycle of substances in agriculture [1–3]. At the same time, municipal wastewater sludge contains more than 50% organic matter, macro- and microelements, which are useful for plants. The problem of bioutilization of sewage sludge can be successfully solved by the creation on their basis of valuable fertilizing means, such as composts, organic-mineral fertilizers, including pelleted ones. During the processing pathogenic microflora and other pathogenic organisms in them are destroyed, and by reducing humidity and granulation one can transport them over long distances.

## 4.2 MATERIALS AND METHODOLOGY

In the experience laid out in the climatic conditions of the North-West region of Russia (Vologda region), we studied the effect and aftereffect of fertilizers on the basis of sewage sludge on heavy metal content in the system soil-plant. The research was conducted in the crop rotation link: fiber flax (cultivar Zaryanka), potato (cultivar Elizabeth). In the scheme of our experiment we included a control without fertilization (variant 1) and six variants with fertilizers, including variants 2–4 – organic fertilization system with increasing doses of compost, variant 5 – with mineral fertilization system, variant 6 – with organic-mineral fertilization system, version 7 – with pelleted organic-mineral fertilizer (Table 4.1). Variants 5 and 6 had nutrient content equivalent to variant 3. Experiment had three replications. Siting of variants was systematic. The area of the experimental plots was 10 $m^2$ (4 x 2 $m^2$). Experiment was laid out on three fields of crop rotations, entered sequentially by year (2010–2012).

**TABLE 4.1** The Chemical Composition of Organic Fertilizers, on Average, Over Years of Research

| Fertilizer | Ash content, % | pH | Mass fraction, % of dry matter | | | |
|---|---|---|---|---|---|---|
| | | | Organic matter | N | $P_2O_5$ | $K_2O$ |
| Compost | 33.2 | 6.3 | 66.8 | 1.95 | 0.8 | 0.3 |
| Pelleted organic-mineral fertilizer | 77.5 | 7.5 | 22.5 | 2.8 | 3.1 | 2.5 |
| The standard of Russia [4] | Not rated | 5.5–8.5 | No less than 20 | No less than 0.6 | No less than 1.5 | Not rated |

The soil of the experimental plot was sod-low podzol medium loamy, by agrochemical properties – medium reclaimed. In the layer 0–20 cm it had the high content of humus – 3.9% and mobile phosphorus ($P_2O_5$) – 230 mg/kg, the medium content of potassium ($K_2O$) – 113 mg/kg, slightly acidic reaction of medium – $pH_{KCl}$ 5.3. The content of total forms of heavy metals and arsenic before laying experience was (mg/kg): Cu – 5, Zn – 25, Mn – 277, Co – 6, Pb – 9, Cd – 0.5, Ni – 10, Cr – 11, Hg – 0.02, As – 1.

The compost was made from peat and sewage sludge of the city Vologda. Pelleted organic-mineral fertilizer was produced by the firm CJSC (Closed Joint Stock Company) "Twin Trading Company" from dewatered sewage sludge with the addition of mineral nitrogen and potassium to 5% by weight, with the size of granules 14x20 $mm^2$. The fertilizer was applied into the soil before sowing of fiber flax in the years of laying-out. The aftereffect of fertilizers was studied in the cultivation of the second culture of field crop rotation link – potato. The chemical composition of the applied fertilizers (see Table 4.1) indicates their high value. The content of heavy metals and arsenic in fertilizers (Table 4.2) below the standards adopted in Russia.

Applied fertilizers were no danger in sanitary-bacteriological respect, as there were not detected pathogenic enterobacteria, and the group index of Escherichia coli was within acceptable values [4]. Effective specific activity of natural radionuclides in fertilizers amounted to no more than 10 Bq/kg at an acceptable value 300 Bq/kg [5].

All field and laboratory research experiences were conducted by standard methods [6]. Statistical processing of experimental data was

**TABLE 4.2** The Content of Toxic Elements (Total Forms) in Fertilizers, in Average Years of the Research

| Fertilizer | Content, mg/kg of dry matter | | | | | | | | | |
|---|---|---|---|---|---|---|---|---|---|---|
| | Cu | Zn | Pb | Cd | Ni | Cr | Mn | Co | Hg | As |
| Compost | 45 | 140 | 14 | 1.1 | 13 | 12 | 217 | 4.3 | 0.11 | 1.2 |
| Pelleted organic-mineral fertilizer | 406 | 1584 | 70 | 14.6 | 34 | 236 | 701 | 6.7 | 1.40 | 2.6 |
| The standard of Russia [4], no less than | 750 | 1750 | 250 | 15 | 200 | 500 | - | - | 7.5 | 10 |

implemented by the dispersion method according to Dospekhov [7] using program STRAZ.

## 4.3 RESULTS AND DISCUSSION

According to the obtained results, applied fertilizers had a positive impact on the yield increase of the studied cultures of the field crop rotation link. When applying compost in increasing doses from 2 to 6 t/ha we observed consistently increase of flax straw yields (Figure 4.1), but its greatest values in the 3-year average were obtained in the variant applying 4 t/ha of organic fertilizer (29.8 C/ha). The highest seed yield of fiber flax (3.3 C/ha) was also achieved in the variant of granulated organic-mineral fertilizer, and the dose of compost 6 t/ha (Figure 4.2). Fertilizers from sewage sludge and pelleted organic-mineral fertilizer positive influence on morphological signs of flax straw, increasing technical flax length by 1.8–6.7 cm compared to control.

Fertilizers from sewage sludge has not had any significant impact on the content of heavy metals and arsenic in fiber flax (Tables 4.3 and 4.4), although applying compost at a dose of 6 t/ha resulted in a slight increase in the content of lead and arsenic.

The existing variation of values was due to the diversity of soil fertility, and unaccounted random factors and, as a rule, did not go beyond the permissible values, provided by the SanReg (Sanitary regulations and norms) 2.3.2.1078–01 [9] and Maximum Allowable Concentrations (MAC) [10].

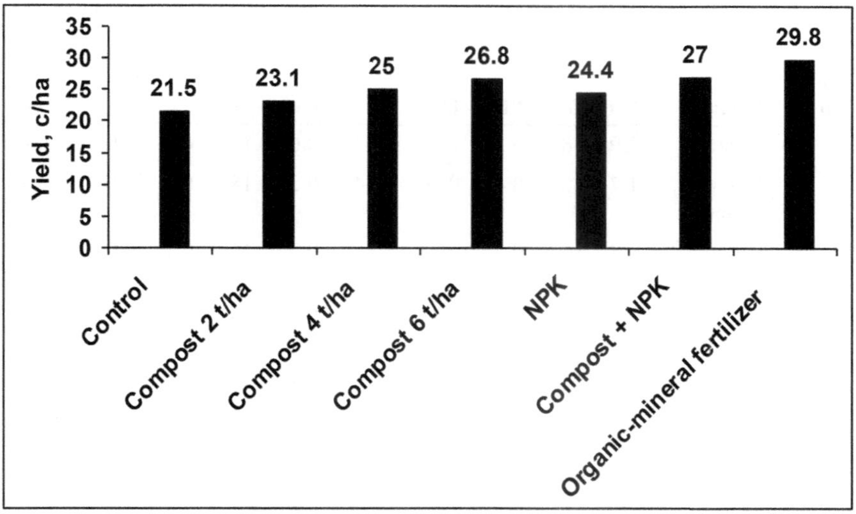


**FIGURE 4.1** The flax straw yield in the 3-year average (LSD0.5 = 1.5).

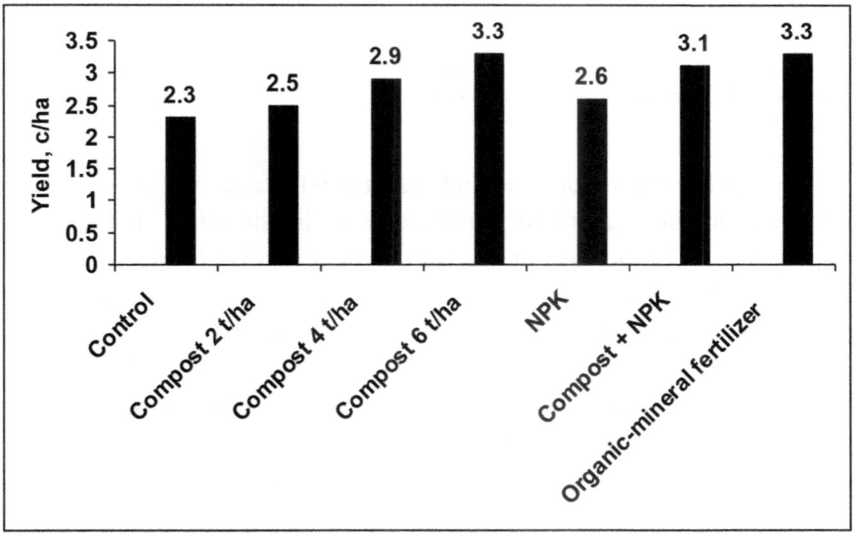


**FIGURE 4.2** The flax seed yield in the 3-year average (LSD0.5 = 0.2).

**TABLE 4.3** Influence of Fertilizers on the Content of Heavy Metals and Arsenic in the Straw of Flax (Average for the Years of Research), mg/kg Dry Weight

| Variant | Fertilizer | Cu | Zn | Pb | Cd | Ni | Cr | Mn | Co | Hg | As |
|---|---|---|---|---|---|---|---|---|---|---|---|
| 1 | Control | 1.9 | 18 | 0.5 | 0.28 | 0.5 | 0.6 | 19 | 0.1 | 0.01 | 0.08 |
| 2 | Compost, 2 t/ha | 1.7 | 18 | 0.6 | 0.26 | 0.4 | 0.5 | 18 | 0.1 | 0.01 | 0.13 |
| 3 | Compost, 4 t/ha | 1.8 | 16 | 0.5 | 0.23 | 0.5 | 0.5 | 17 | 0.1 | 0.01 | 0.07 |
| 4 | Compost, 6 t/ha | 1.7 | 15 | 0.7 | 0.24 | 0.6 | 0.6 | 16 | 0.1 | 0.01 | 0.13 |
| 5 | NPK | 1.7 | 15 | 0.6 | 0.25 | 0.4 | 0.5 | 17 | 0.1 | 0.01 | 0.08 |
| 6 | Compost+ NPK | 1.5 | 13 | 0.6 | 0.20 | 0.4 | 0.7 | 16 | 0.1 | 0.01 | 0.11 |
| 7 | Pelleted organic-mineral fertilizer, 4 t/ha | 1.7 | 14 | 0.5 | 0.25 | 0.4 | 0.8 | 18 | 0.2 | 0.01 | 0.06 |
| Normal content [8] | | 2–12 | 15–150 | 0.1–5.0 | 0.05–0.2 | 0.4–3.0 | 0.2–1.0 | | | 0.005–0.01 | 0.1–1.0 |

*Note*: Variant 5: NPK equivalent to variant 3.

The application of organic and mineral fertilizers had no significant influence on the content of macroelements in the straw and seeds of flax. Some of the increase in nitrogen content relative to the control was observed when analyzing the straw only in the variant of mineral fertilizers and applying 6 t/ha of compost. The content of phosphorus and potassium in straw and seeds of flax, for different versions varied slightly.

The results of the potato tuber yield depending on the effectiveness of fertilizers are presented in Figure 4.3.

On the control (without fertilizers) yield of the tubers in the 3-year average amounted to 187 C/ha. Higher yields are obtained in variants with the introduction of pelleted organic-mineral fertilizer (224 C/ha) and compost at a dose of 6 t/ha (213 C/ha) and also in the case of application of organic-mineral system (201 C/ha), where we found the resulting significant increase in the crop yield, formed respectively 19.8; 13.9 and 7.5% in relation to the control.

**TABLE 4.4** Influence of Fertilizers on the Content of Heavy Metals and Arsenic in the Seeds of Flax (Average for the Years of Research), mg/kg Dry Weight

| Variant | Fertilizer | Cu | Zn | Pb | Cd | Ni | Cr | Mn | Co | Hg | As |
|---|---|---|---|---|---|---|---|---|---|---|---|
| 1 | Control | 9.2 | 42 | 0.9 | 0.13 | 0.8 | 0.4 | 15 | 0.2 | 0.01 | 0.05 |
| 2 | Compost, 2 t/ha | 6.4 | 36 | 0.7 | 0.09 | 0.6 | 0.3 | 13 | 0.1 | 0.01 | 0.07 |
| 3 | Compost, 4 t/ha | 10.6 | 44 | 0,9 | 0.11 | 0.7 | 0.3 | 14 | 0.2 | 0.01 | 0.06 |
| 4 | Compost, 6 t/ha | 10.0 | 48 | 1.1 | 0,13 | 0.9 | 0.4 | 21 | 0.1 | 0.01 | 0.05 |
| 5 | NPK | 7.1 | 48 | 0.9 | 0.14 | 1.0 | 0.4 | 17 | 0.1 | 0.01 | 0.1 |
| 6 | Compost + NPK | 5.6 | 35 | 0.7 | 0.10 | 0.8 | 0.2 | 18 | 0.2 | 0.01 | 0.05 |
| 7 | Pelleted organic-mineral fertilizer, 4 t/ha | 8.9 | 42 | 0.7 | 0.13 | 1.1 | 0.3 | 15 | 0.1 | 0.01 | 0.04 |
| SanReg 2.3.2.1078–01 (seeds) | | | | 0.5 | 0.1 | | | | | 0.02 | 0.2 |

*Note*: Variant 5: NPK equivalent to variant 3.

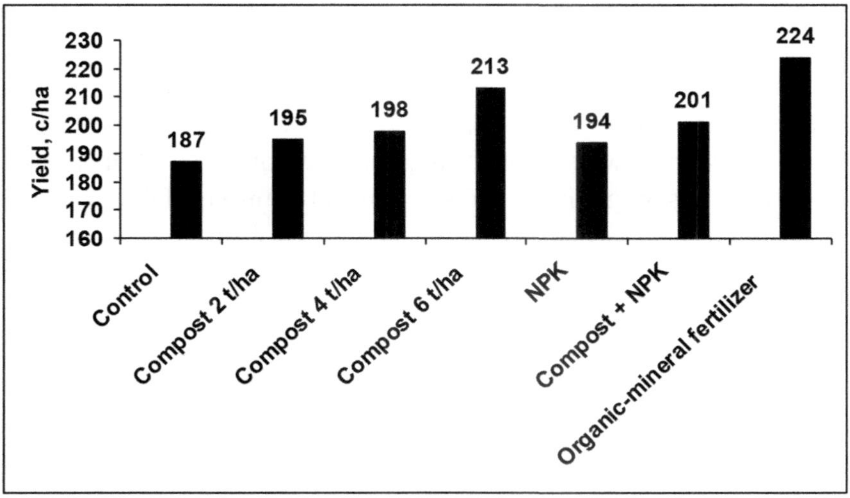


**FIGURE 4.3** The yield of potato tubers, 3-year average (LSD05= 8.3 C/ha).

The content of heavy metals and arsenic in potato tubers varied for different variants, but clear patterns of their accumulation in relation to unfertilized control were observed (Table 4.5).

When comparing the content of heavy metals and arsenic in potato tubers with the allowable concentrations adopted in Russia by SanReg and MAC exceeding the latter was not installed.

The study of the chemical composition of potato tubers depending on the applied fertilizers showed that the nitrogen content, phosphorus and potassium by variants have changed slightly. However, in all variants of fertilizers starch content in the tubers increased. Higher starch content obtained in the variants with compost, where this figure was more the control for 2.3–2.9%.

According to the analysis of productivity of crop rotation link flax – potato, it has increased under the influence of applied fertilizers. The most productive link – 41.8 and 38.8 kg/ha of fodder units is obtained at application of pelleted organic-mineral fertilizer and 6 t/ha of compost. It was 22 and 13% higher than the control without fertilizers. Thus, according to the research results it can be concluded that compost based on sewage sludge of Vologda treatment facilities with peat and pelleted organic-mineral fertilizer was provided in the years of their introduction (2010, 2011 and 2012) a significant increase in the yield of production of flax and in the years of aftereffect (2011, 2012 and 2013) – potato tubers. At the same time, the use of mineral system was effective only in the year of fertilizers application.

For objective environmental assessment of agrocenoses it is important determining the heavy metal balance. When calculating the balance in an article of receipt we were taken into account intake of nutrients with compost on the basis of sewage sludge and with mineral fertilizers (Tables 4.6–4.8).

Among the studied crops the biggest removal of elements with the harvest, especially zinc and manganese, demonstrated potato. Moreover, the higher removal of elements we observed when applying compost at a dose of 6 t/ha and pelleted organic-mineral fertilizer at a dose of 4 t/ha. Removal of elements with the harvest of flax had the same tendencies.

The calculations showed that during a rotation of crop rotation there was a positive economic balance of heavy metals: copper, cadmium, chromium, manganese, mercury in variants with application of compost at doses of 4 and 6 t/ha and pelleted organic fertilizer; about lead, nickel, cobalt and arsenic –

**TABLE 4.5** The Content of Heavy Metals and Arsenic in Potato Tubers (Average for the Years of Study)

| No. of variant | Fertilizer | Cu | Zn | Pb | Cd | Ni | Cr | Mn | Co | Hg | As |
|---|---|---|---|---|---|---|---|---|---|---|---|
| 1 | Control | 1.4 | 7.1 | 0.4 | 0.027 | 0.2 | 0.15 | 4.2 | 0.08 | 0.01 | 0.02 |
| 2 | Compost, 2 t/ha | 1.2 | 7.5 | 0.3 | 0.027 | 0.2 | 0.12 | 4.4 | 0.05 | 0.01 | 0.03 |
| 3 | Compost, 4 t/ha | 1.2 | 7.5 | 0.3 | 0.025 | 0.2 | 0.14 | 5.0 | 0.04 | 0.01 | 0.02 |
| 4 | Compost, 6 t/ha | 1.2 | 5.8 | 0.3 | 0.020 | 0.1 | 0.16 | 4.6 | 0.04 | 0.01 | 0.02 |
| 5 | NPK | 1.4 | 8.2 | 0.3 | 0.021 | 0.2 | 0.18 | 6.5 | 0.03 | 0.01 | 0.02 |
| 6 | Compost + NPK | 1.1 | 6.6 | 0.4 | 0.022 | 0.2 | 0.17 | 5.5 | 0.05 | 0.01 | 0.02 |
| 7 | Pelleted organic- mineral fertilizer, 4 t/ha | 1.2 | 8.4 | 0.3 | 0.030 | 0.2 | 0.13 | 4.5 | 0.05 | 0.01 | 0.02 |
| SanReg 2.3.2.1078–01 (seeds) | | — | — | 0.5 | 0.03 | — | — | — | — | 0.02 | 0.2 |
| MAC | | 30 | 50 | 5.0 | 0.3 | 3.0 | 0.5 | — | 1.0 | 0.05 | 0.5 |

Note: Variant 5: NPK equivalent to variant 3.

**TABLE 4.6** The Application of Heavy Metals and Arsenic with Organic and Mineral Fertilizers, g/ha

| No. of variant | Fertilizer | Cu | Zn | Pb | Cd | Ni | Cr | Mn | Co | Hg | As |
|---|---|---|---|---|---|---|---|---|---|---|---|
| 1 | Control | — | — | — | — | — | — | — | — | — | — |
| 2 | Compost, 2t/ha | 90 | 280 | 29 | 2.1 | 27 | 24 | 434 | 9 | 0.2 | 2.4 |
| 3 | Compost, 4t/ha | 180 | 560 | 57 | 4.2 | 54 | 48 | 868 | 17 | 0.4 | 4.8 |
| 4 | Compost, 6t/ha | 270 | 840 | 86 | 6.4 | 80 | 72 | 1302 | 26 | 0.7 | 7.2 |
| 5 | NPK | 1.4 | 5.7 | 7.5 | 0.4 | 4.4 | — | — | — | — | — |
| 6 | Compost + NPK | 91 | 283 | 32 | 2.3 | 29 | 24 | 434 | 9 | 0.2 | 2.4 |
| 7 | Pelleted organic-mineral fertilizer, 4 t/ha | 1624 | 6336 | 279 | 58.4 | 136 | 944 | 2804 | 27 | 5.6 | 10.4 |

*Note*: Variant 5: NPK equivalent to variant 3.

in variants with application of increasing doses of compost from 2 to 4 t/ha and pelleted organic fertilizer; as for zinc – only variant of granulated organic fertilizer. However, many elements are necessary for the growth and development of plants, thus, normalized application of fertilizers on the basis of the sludge improves the nutrient regime of the soil.

Balance calculations showed that the danger of soil contamination with heavy metals through the application of compost on the basis of sewage sludge was not observed.

To characterize the overall potential contamination of soil by pollutants one can use such factor as total soil pollution – $Zc$, defined by the formula:

$$Zc = \Sigma\, Kc - (n - 1), \tag{1}$$

where $Kc$ is the ratio of the concentration of the ingredient in the contaminated soil to background concentrations; $n$ is the number of defined ingredients.

When the amount of the total pollution $Zc$ less than 16, soil belongs to category 1 (the weakest of pollution), 16–32 – to 2 category, 32–128 – to

**TABLE 4.7** Influence of Organic and Mineral Fertilizers on the Economic Removal of Heavy Metals and Arsenic Through Years of Research, g/ha

| No. of variant | Fertilizer | Cu | Zn | Pb | Cd | Ni | Cr | Mn | Co | Hg | As |
|---|---|---|---|---|---|---|---|---|---|---|---|
| ***Fiber-flax*** | | | | | | | | | | | |
| 1 | Control | 6.2 | 49.2 | 1.3 | 0.6 | 13 | 1.3 | 45 | 0.3 | 0.01 | 0.2 |
| 2 | Compost, 2 t/ha | 5.6 | 50.1 | 1.5 | 0.6 | 1.0 | 1.3 | 45 | 0.3 | 0.02 | 0.3 |
| 3 | Compost, 4 t/ha | 7.6 | 53.9 | 1.5 | 0.6 | 1.4 | 1.4 | 46 | 0.2 | 0.02 | 0.2 |
| 4 | Compost, 6 t/ha | 8.0 | 55.5 | 2.1 | 0.7 | 1.8 | 1.6 | 49 | 0.3 | 0.02 | 0.4 |
| 5 | NPK | 6.1 | 49.2 | 1.6 | 0.6 | 1.3 | 1.3 | 46 | 0.3 | 0.02 | 0.2 |
| 6 | Compost + NPK | 5.9 | 46.6 | 1.7 | 0.6 | 1.3 | 1.9 | 48 | 0.2 | 0.02 | 0.3 |
| 7 | Pelleted organic-mineral fertilizer, 4 t/ha | 8.0 | 55.5 | 1.8 | 0.8 | 1.5 | 2.4 | 60 | 0.8 | 0.02 | 0.2 |
| ***Potato*** | | | | | | | | | | | |
| 1 | Control | 85 | 640 | 27 | 2.5 | 26 | 30 | 720 | 4.3 | 0.3 | 1.1 |
| 2 | Compost, 2 t/ha | 104 | 636 | 27 | 2.8 | 28 | 30 | 820 | 5.3 | 0.3 | 1.8 |
| 3 | Compost, 4 t/ha | 92 | 771 | 27 | 2.2 | 31 | 32 | 452 | 4.1 | 0.4 | 1.1 |
| 4 | Compost, 6 t/ha | 109 | 951 | 30 | 2.8 | 41 | 31 | 859 | 5.1 | 0.4 | 1.5 |
| 5 | NPK equivalent to variant 3 | 141 | 637 | 24 | 7.4 | 43 | 30 | 416 | 3.1 | 0.3 | 1.6 |
| 6 | Compost, NPK | 98 | 1073 | 32 | 7.2 | 42 | 37 | 878 | 4.7 | 0.4 | 1.4 |
| 7 | Pelleted organic-mineral fertilizer, 4 t/ha | 173 | 987 | 37 | 10.8 | 49 | 41 | 668 | 6.0 | 0.5 | 2.1 |

*Note*: Variant 5: NPK equivalent to variant 3.

**TABLE 4.8** The Balance of Heavy Metals and Arsenic in Sod-Podzol Soil, t/ha

| No. of variant | Fertilizer | Cu | Zn | Pb | Cd | Ni | Cr | Mn | Co | Hg | As |
|---|---|---|---|---|---|---|---|---|---|---|---|
| 1 | Control | –92 | –689 | –29 | –3 | –5 | –32 | –765 | –5 | 0 | –1 |
| 2 | Compost, 2 t/ha | –18 | –397 | 0.4 | –1.1 | 22.8 | –7.6 | –423 | 3.0 | –0.1 | 0.4 |
| 3 | Compost, 4 t/ha | 82 | –255 | 28.8 | 1.6 | 48.1 | 15.4 | 378 | 12.9 | 0.1 | 3.5 |
| 4 | Compost, 6 t/ha | 154 | –156 | 53.6 | 3.0 | 77.0 | 39.7 | 403 | 20.5 | 0.2 | 5.4 |
| 5 | NPK | –144 | –672 | –17.4 | –7.6 | –0.2 | – | – | – | – | – |
| 6 | Compost + NPK | –12 | –829 | –1.0 | –5.4 | 23.9 | 14.3 | –484 | 3.7 | –0.2 | 0.8 |
| 7 | Pelleted organic- mineral fertilizer, 4 t/ha | 1444 | 5303 | 241.2 | 47.0 | 130 | 901 | 2087 | 20.2 | 5.1 | 8.2 |

*Note*: Variant 5: NPK equivalent to variant 3.

**TABLE 4.9** The Total Elemental Contamination (Zc) of Sod-Podzol Medium Loamy Soil

| | The concentration coefficient, Kc | | | | | | | | | | |
|---|---|---|---|---|---|---|---|---|---|---|---|
| Fertilizer | Cu | Zn | Pb | Cd | Ni | Cr | Mn | Co | Hg | As | Zc |
| Compost, 2 t/ha | 0.91 | 0.95 | 0.86 | 1.06 | 0.92 | 1.01 | 1.10 | 0.97 | 0.67 | 0.95 | 0.40 |
| Compost, 4 t/ha | 0.93 | 0.95 | 0.93 | 1.00 | 0.93 | 0.94 | 1.00 | 0.95 | 0.67 | 1.14 | 0.45 |
| Compost, 6 t/ha | 0.93 | 0.95 | 0.86 | 1.09 | 0.89 | 0.96 | 1.00 | 1.05 | 0.67 | 0.95 | 0.34 |
| NPK | 1.09 | 1.04 | 0.97 | 1.23 | 0.90 | 0.93 | 0.81 | 0.94 | 1.00 | 1.00 | 0.90 |
| Compost, 6 t/ha | 1.05 | 1.00 | 1.00 | 1.11 | 0.97 | 0.96 | 0.81 | 0.94 | 1.00 | 0.95 | 0.81 |
| Pelleted organic-mineral fertilizer, 4 t/ha | 1.07 | 1.00 | 0.86 | 1.20 | 0.86 | 0.92 | 0.84 | 0.92 | 1.00 | 0.90 | 0.58 |

3 category, more than 128 – to 4 category (the most severe pollution) [11]. In the conditions of ongoing field experience *Zc*, calculated with respect to the content of heavy metals and arsenic in the control variant (background concentration), ranged from 0.34 to 0.90 (Table 4.9), that means insignificant level of pollution.

Thus, the normalized fertilizer application on the basis of sewage sludge did not significantly increase content of heavy metals and arsenic in the production of flax and potatoes. Increasing doses of compost on the basis of sewage sludge did not reveal clear patterns in the change of the content of toxic elements in soil and plant products. The content of heavy metals and arsenic in the soil and production of flax and potatoes when fertilizing compost met hygienic standards adopted in Russia.

## 4.4 CONCLUSIONS

1. The application of sewage sludge as pelleted organic fertilizer or composts is an effective method of increasing the yield of agricultural crops in the conditions of sod-podzol soils.
2. The application of such fertilizers in rates of 4–6 t/ha (calculated on the dry matter) provided a significant increase in the yield of straw and seed fiber flax, seed potato tubers.
3. The resulting vegetable production was consistent with the basic technological requirements and the content of heavy metals and arsenic met acceptable health standards.
4. We did not observe a significant contamination with heavy metals and arsenic sod-podzol medium loamy soil fertilized with compost and pelleted organic-mineral fertilizer.

## KEYWORDS

- **bioutilization**
- **compost**
- **fiber flax**
- **municipal wastewater**
- **potato**

## REFERENCES

1. The strategy of using sewage sludge and compost in farming [Ed. N. Z. Milatshenko]. Moscow. Agroconsult. 2002, 140 p. (In Russian).
2. Resources of organic fertilizers in agriculture of Russia (Information and analytical reference book) [Ed. Eskov, A. I.]. Vladimir: Russian Research Institute of Organic Fertilizers and Peat (In Russian), 2006, 200 p. (In Russian).
3. Pahnenko, E. N. Sewage sludge and other non-traditional organic fertilizers. Moscow: Laboratory of Sciences, 2007, 311 p. (In Russian).
4. State Standard of the Russian Federation R 17.4.3.07–2001. The nature conservancy. The soils. Requirements to the properties of sewage sludge when used as fertilizer. Moscow: Information-publishing center of the Russian Ministry of Health. 2001 (In Russian).
5. State Standard of the Russian Federation R 54651–2011. The organic fertilizers on the basis of sewage sludge. (In Russian).
6. Afanasev, R. A., Merzlaya, G. Ye. Methodical recommendations on studying the effectiveness of non-traditional organic and organic-mineral fertilizers. Moscow: Russian Academy of agricultural Sciences. 1999, 40 p. (In Russian).
7. Dospekhov, B. A. Methodology of field experiment. Moscow: Kolos (Ear In Russian), 1979, 416 p. (In Russian).
8. Mineev, V. G. Agrochemistry and biosphere. Moscow: Kolos ("Ear" In Eng.), 1984, 245 p. (In Russian).
9. Sanitary regulations and norms. SanReg 2.3.2.1078–01. Hygienic requirements for safety and nutrition value of food products. (In Russian).
10. Temporary maximum allowable concentrations of certain chemical elements in feed for farm animals. Department of veterinary of the State agroindustrial complex of the USSR. 1987 (In Russian).
11. Milatshenko, N. Z., Litwak, Sh. I. (Eds.). Methodological and organizational basis for implementation of agro-environmental monitoring in intensive farming (on the basis of Geographical network of experiments). Moscow: All-Russian Scientific research Institute of Agrochemistry. 1991, 356 p. (In Russian).

## CHAPTER 5

# DYNAMICS OF EXCHANGE POTASSIUM IN THE SOILS DURING PROLONGED TRIALS WHEN FERTILIZER APPLYING

RAFAIL A. AFANAS'EV,[1] GENRIETTA YE. MERZLAYA,[2] and MIKHAIL O. SMIRNOV[3]

*Pryanishnikov All-Russian Scientific Research Institute of Agrochemistry, 31A, Pryanishnikova St., Moscow, 127550, Russia, E-mail: [1]rafail-afanasev@mail.ru, [2]lab.organic@mail.ru, [3]User53530@yandex.ru*

## CONTENTS



## ABSTRACT

In long-term field experiments the use of potassic fertilizers in the first rotations of crop rotations increases the content of exchange potassium in soils with its subsequent stabilization or even decrease due to the transformation of mobile forms of the element in less mobile forms. Due to the inverse transition of non-exchangeable forms of soil potassium content of its exchangeable forms can be maintained at a relatively stable level even when there is negative economic balance of this element.

## 5.1 INTRODUCTION

Potassium is one of the most important indicators of soil fertility, as along with nitrogen and phosphorus it refers to the basic elements of plant nutrition [1–3]. In 70–80 years of the last century, i.e., in the period of intensive chemicalization of agriculture, the content of exchange potassium in the soils of Russia was gradually increased. In recent decades, a dramatic reduction in the use of mineral fertilizers up to 5–10 kg/ha and organic to 0.4–0.5 t/ha leads to a noticeable depletion of exchange potassium in soils [4]. The productivity of agricultural crops, in particular grain harvest, in recent decades, despite a negative balance of major nutrients in soils, remains almost at the same level [5]. So for example in 1986–1990, grain yield from the entire harvested area in Russia was 17.4 c/ha, whereas in 2001–2004 – 18.0 c/ha [6, 7]. This may be due to the fact that, firstly, in recent decades from use in agriculture was eliminated marginal lands, and, secondly, agricultural plants at the present time are provided with nutrients, including potassium, which come from soil reserves, developed in years of intensive chemicalization. Thus in some regions (Lipetsk, Tambov, Moscow regions) is still not revealed a certain regularity in the decrease of exchange potassium in the arable layer of soils [8]. Features of dynamics of exchange potassium in different soils more clearly expressed in a controlled environment of long-term field experiments that are subject to review in this article.

## 5.2 MATERIALS AND METHODOLOGY

Study of the dynamics of exchange potassium was conducted in long-term field experiments on soils significantly differing in their genetic characteristics: sod-gleyic heavy loamy drained (Lithuanian Scientific Research

Institute of Agriculture, years of research: 1961–1974) [9], sod-podzolic heavy loamy (Central Experimental Station of All-Russian Scientific Research Institute of Agrochemistry, Moscow region, years of research: 1964–1992) [10], sod-podzolic light loamy soils (branch of All-Russian Scientific Research Institute of Agrochemistry, Smolensk region, years of research: 1979–2008), sod-podzolic loamy sand (Grodno Experimental Station, Belarus, years of research: 1964–1979) [11] and common chernozem (Stavropol Scientific Research Institute of Agriculture, years of research: 1962–1992) [12]. The content of the exchange forms of the element was determined by the method of Egner–Rim in Lithuania, the method of Maslova or Kirsanov in the nonchernozem zone, the method of Machigin – in the Stavropol region [13]. The purpose of this article is to justify the trends of the potassium dynamics in different soils.

## 5.3 RESULTS AND DISCUSSION

The study of the dynamics of exchange potassium showed that if there is the positive economic balance its content in the soil generally increases, but not proportionally to the intensity of balance. This, in particular, is confirmed by the data obtained in long-term experience of Central Experimental Station of All-Russian Scientific Research Institute of Agrochemistry conducted on loamy sod-podzolic soil (Moscow region). On average over the seven rotations of field rotation in the variant with the annual application of $N_{175}P_{103}K_{247}$ (natrium-phosphorus-kalium) + 122.5 t/ha of manure by increasing the content of exchange potassium (method of Maslova) in the topsoil at 76 mg/kg (milligram/kilogram) to improve the content of exchange potassium at 10 mg/kg was spent 474 kg/ha of fertilizer deposited in excess removal, while it would have required, according to the calculations, applying about 30 kg/ha. Unrecorded amounts of potassium over 3 t/ha are moved to an inactive form and may partially migrate to subsurface horizons.

From Figure 5.1 it also follows that the content of exchange potassium in the topsoil increased only in the first three rotations of field rotation, but by the end of the seventh rotation the content of exchange potassium decreased to 227 mg/kg due to a more intensive transition of exchangeable forms of an element in non-exchangeable. The tendency of transition of

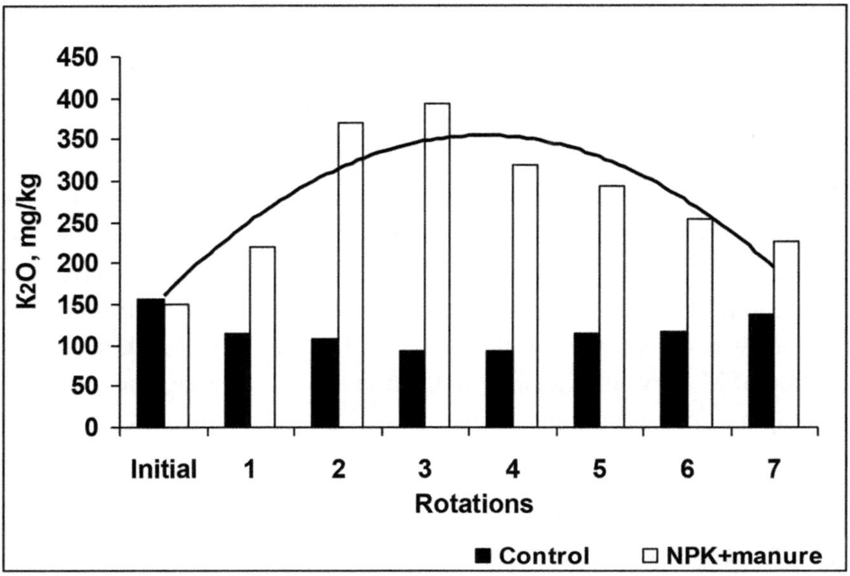


**FIGURE 5.1** Dynamics of exchange potassium content in the sod-podzolic heavy loamy soil (Moscow region).

exchange potassium in non-exchangeable form is confirmed and analyzed in other experiments.

In a field experiment on sod-podzolic light loamy soil (Smolensk region) in the variant of organic-mineral system of fertilizers ($N_{75}P_{75}K_{75}$ + manure of 9.6 t/ha) with a positive balance of the element over 30 years in the amount of 1280 kg/ha, the exchange potassium content (Kirsanov's method) in the topsoil not only not increased but even decreased from 146 to 115 mg/kg (Figure 5.2).

In the subsurface soil (20–100 cm) exchange potassium content in the end of the last (fourth) rotation compared with the first rotation also decreased to 76 kg/ha. Therefore, in one-meter soil layer researchers marked an intense transition of fertilizer potassium deposited in excess of the removal in non-exchangeable state. However, during the three aftereffect years in this variant, the exchange potassium content in the topsoil increased from 115 to 181 mg/kg due to reverse change of soil potassium from non-exchangeable forms in exchangeable. Average yield of the crops of the field crop rotation for the first 30 years from the start of

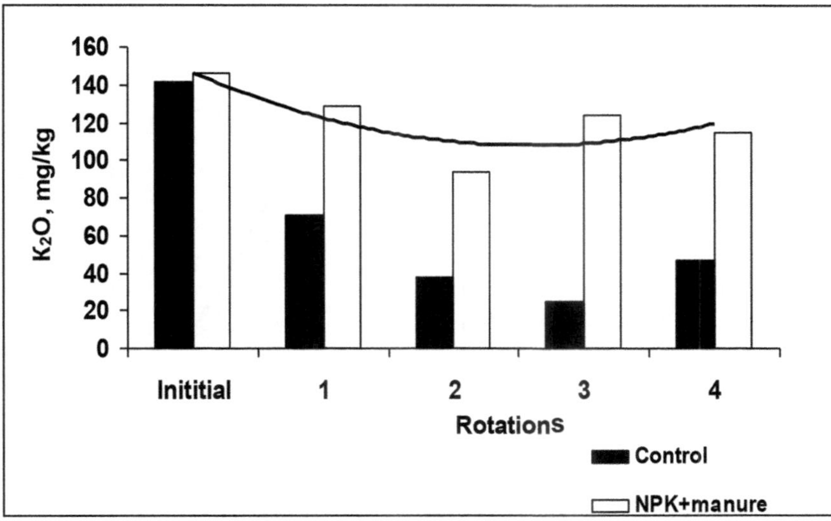


**FIGURE 5.2** Dynamics of exchange potassium content in the sod-podzolic light loamy soil (Smolensk region).

the experiment when applying fertilizers in organic-mineral variant was 3 t g.u./ha (grain unit/hectare). In the subsequent 3-year period of aftereffect it was 3.1 t g.u./ha, i.e., remained almost at the same level.

In the experience with using modern chernozem (Stavropol region) the exchange potassium content (according to the method of Machigin) in the topsoil in the variant $N_{120}P_{120}K_{150}$ with positive balance of potassium in the first three rotations of the 6-field crop rotation increased. At a negative balance, when in the fourth and fifth rotations one applied only nitrogen-phosphorus fertilizer, its content is almost unchanged (Figure 5.3). So when applying potassic fertilizers in excess of the removal in the amount of 896 kg/ha for 18 years (the first three rotations), the exchange potassium content increased by 87 mg/kg, and increase of the exchange potassium content by 10 mg/kg required beyond the removal 103 kg/ha a.s. (active substance) of potassic fertilizers. With a deficit balance of potassium in the last 12 years (4th and 5th rotations) in the amount of 789 kg/ha, the exchange potassium content in the topsoil was within the measurement accuracy. This means that the removal of potassium by crops of crop rotation for the last two rotations was compensated for soil reserves of non-exchangeable forms of the element. At the same time the exchange potassium storage in the

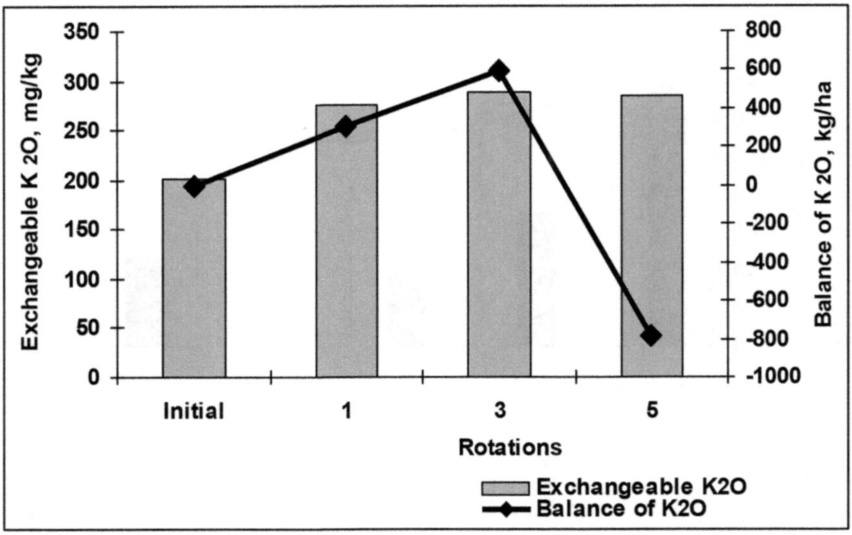


**FIGURE 5.3** Balance and exchange potassium content in the common heavy loamy chernozem soil in the variant NPK (Stavropol region).

subsurface soil (20–100 cm) not only not decreased, but even increased by 260 kg/ha. The productivity of field rotation for the first three rotations amounted to 4.1 t g.u./ha, and for the last two rotations – 3.6 t g.u./ha. In other words, the productivity of crops without the use of potassic fertilizers in this variant decreased by only 14%.

In sod-gleyic drained clay loam (Lithuania), medium provided with exchange potassium (according to Egner-Rim), its content in the topsoil when applying $N_{225}P_{324}K_{350}$ for rotation increased in the second rotation compared with the first even if the total deficit of potassium was more than 500 kg/ha (Figure 5.4). In the control variant, where fertilizer was not applied, there was a regular decline in the availability of soil with the element.

In Belarus when using sod-medium podzol loamy sand soil with an annual negative balance of this element exchange potassium content (according to Kirsanov) in the first three rotations of field rotation slightly increased also due to the non-exchangeable forms, but then by the end of the fourth rotation, with the depletion of their soil storage, a clear tendency has developed to the decline of the exchangeable form, due to the low availability of loamy sand soils with potassium (Figure 5.5).

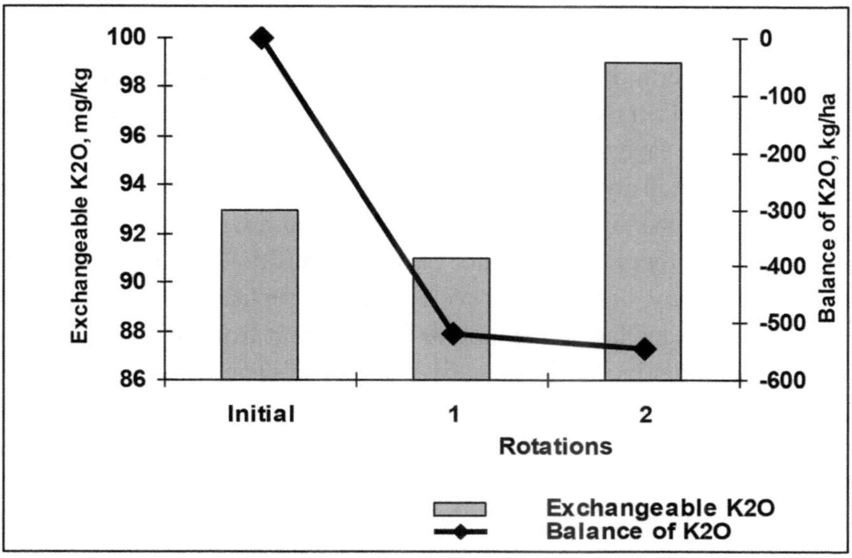


**FIGURE 5.4** Balance and content of exchange potassium in the sod-gleyic heavy loamy drained soil in the variant NPK (Lithuania).

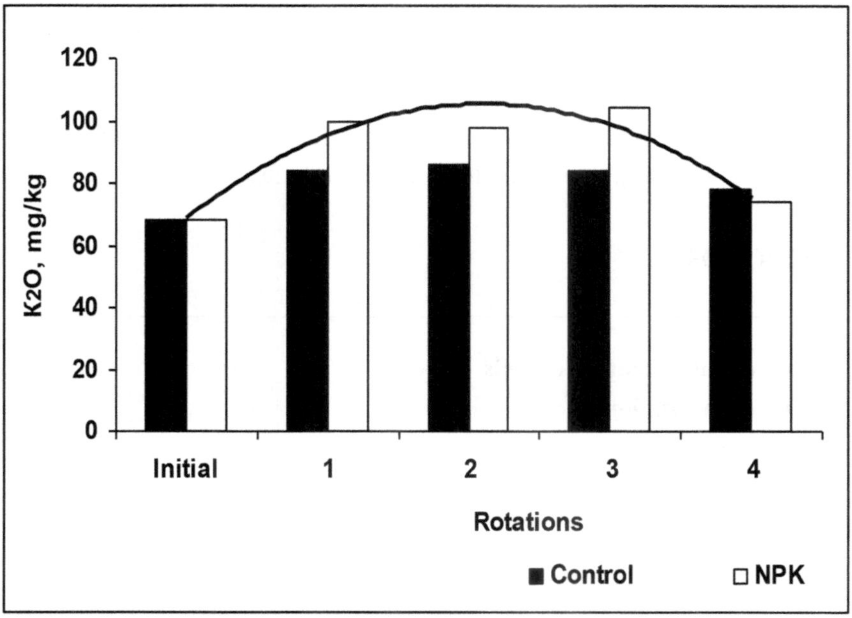


**FIGURE 5.5** Dynamics of exchange potassium content in the sod-podzolic loamy sand soil (Belarus).

## 5.4 CONCLUSIONS

1. Long-term dynamics of potassium mobile forms in prolonged use of fertilizers show its transformation as in plow horizons and subsurface layers of different soils.
2. The intensity of the transformation processes usually increases during the time of interaction of potassium with soil.
3. The authors observed not only the transition of mobile forms in sedentary, but also the reverse, i.e., the replenishing in topsoil and subsurface layers the mobile form content from soil storage, especially distinct with a negative balance of elements.
4. The above process provides a certain aftereffect of fertilizers, expressed in comparative stability of potassium regime of the soil and maintaining crop productivity of crop rotations.
5. The results of long-term field experiments are generally consistent with the processes inherent in arable soils; that explains the relative stability of crop yields.
6. The transformation of mobile potassium in different soils can serve as a basis to predict the provision of soils and plants with this element and the expediency of potassic fertilizer application, which based on the identified patterns.
7. However, according to other studies, the excess potassium fertilizers can have a negative impact on soil and plants, and their effects can be compared to the action of pollutants.

## KEYWORDS

- **chemicalization of soils**
- **crop rotation**
- **mineral fertilizers**

## REFERENCES


1. Pchelkin, V. U. Soil potassium and potassic fertilizers. Moscow: "Ear" ("Kolos" in Russian), 1966, 336 p. (in Russian).
2. Sobachkin, A. A., Zabavskaya, K. M., Cheban, V. M. et al. Research on the efficiency of potassic fertilizers at crop rotation saturation with intensive cultures (field crop rotation). The results of long-termed researches with fertilizers in regions of the country. Proceedings of Pryanishnikov All-Russian Scientific Research Institute of Agrochemistry. 1978, 100–120 (in Russian).
3. Sychev, V. G., Mineev, V. G. Role of Pryanishnikov All-Russian Scientific Research Institute of Agrochemistry in solving complex problems of agriculture chemicalization. "Fertility" ("Plodorodie" in Russian), 2011, 3, 2–4 (in Russian).
4. Romanenko, G. A. (Ed.). The concept of development of agrochemistry and agrochemical service of agriculture of the Russian Federation for the period till 2010, Moscow: Pryanishnikov All-Russian Scientific Research Institute of Agrochemistry, 2005, 80 p. (in Russian).
5. Chekmarev, P. A. The results of the agricultural year in the field of crop production and implementation of the state program in 2010, Problems of Agrochemistry and Ecology, 2012, 2, 63–64 (in Russian).
6. Mineev, V. G. The History and Status of Agricultural Chemistry at the Turn of the twenty-First Century. Moscow: Publishing House of Moscow State University, 2006, 795 p. (in Russian).
7. Sychev, V. G., Aristarhov, A. N. The Main Stages and the Strategy for the Development of Agrochemical Service of Agricultural Production in the Russian Federation. Fertility, 2004, 5, 2–6 (in Russian).
8. Sychev, V. G. (Ed.). Advisory agrochemical service in the Russian Federation. Results and prospects (40 years of Agrochemical service). Moscow: Pryanishnikov All-Russian Scientific Research Institute of Agrochemistry, 2005, 569 p. (in Russian).
9. Plesyavichyus, K. I. Comparison of the fertilizer system used on heavy soils of different mechanical composition. The results of long-termed researches with fertilizers in regions of the country. Proceedings of Pryanishnikov All-Russian Scientific Research Institute of Agrochemistry. 1982, 12, 4–82 (in Russian).
10. Efremov, V. F. Study of the role of manure organic matter in improvement of sod-podzol soil fertility. The results of long-termed researches in the system of geographical series of experiments with fertilizers in Russian Federation. Moscow: Pryanishnikov All-Russian Scientific Research Institute of Agrochemistry, 2011, 47–71 (in Russian).
11. Shuglya, Z. M. Fertilizer system in crop rotation. The Results of Long-Termed Researches with Fertilizers in Regions of the Country. Proceedings of Pryanishnikov All-Russian Scientific Research Institute of Agrochemistry. 1982, II, 94–118 (in Russian).
12. Shustikova, E. P., Shapovalova, N. N. Productivity of ordinary chernozems after prolonged mineral fertilization. The results of Long-Termed Researches in the System of Geographical Series of Experiments with Fertilizers in Russian Federation. Moscow: Pryanishnikov All-Russian Scientific Research Institute of Agrochemistry, 2011, 1, 331–351 (in Russian).
13. Sokolov, A. V. (Ed.). Agrochemical Methods of Soil Research. Moscow: "Science," 5th edition. 1975. 656 p. (in Russian).

# CHAPTER 6

# HEAVY METALS IN SOILS AND PLANTS OF URBAN ECOSYSTEMS (ON THE EXAMPLE OF THE CITY OF YOSHKAR-OLA)

OLGA L. VOSKRESENSKAYA, ELENA A. ALYABYSHEVA, ELENA V. SARBAYEVA, and VLADIMIR S. VOSKRESENSKIY

*Mari State University, 1 Lenin Square, Yoshkar-Ola, Republic of Mari El, 424000, Russia; E-mail: voskres2006@rambler.ru*

## CONTENTS



## ABSTRACT

The chapter describes and analyzes the problem of soil pollution by heavy metals in urban areas (evidence from Yoshkar-Ola). Accumulation of the most common and hazardous heavy metals – lead, copper, zinc and cadmium in soils of different functional zones of the city is examined in the work. It has been revealed that in the city of Yoshkar-Ola soils are contaminated by heavy metals in comparison with background concentrations; relatively high concentrations of heavy metals are found in the industrial zone of the city and in some streets of the residential zone. The excess of the maximum permissible concentration of copper and in most areas – of zinc is recorded in soils of the studied areas. This affected woody plants growing in the urban environment; accumulation of heavy metals in their assimilation apparatus is higher than in plants growing in the background territory. However, the maximum permissible concentration in the leaves (needles) of woody plants has not been identified, it indicates the barrier function of the root system of the test plants.

## 6.1 INTRODUCTION

Human impact on the environment is becoming a major force of the evolution of urban ecosystems, especially through the impact on soil and plants. Plant censes is able to maintain the ecological potential of cities to prevent rapid degradation of their ecosystems if they have stable composition balance and high potential productivity [1] The anthropogenic impact on the components of urban ecosystems is manifested both directly (air pollution by emissions of motor vehicles, industries and enterprises; soil contamination by consumer, construction and industrial wastes, heavy metals, ash from thermal power plants, deterioration of soil fertility due to its contamination by toxic substances, salinization, compaction, the use of acidic soil mixtures and waterlogging of the area; surface and ground water pollution by wastes of domestic character; the deterioration of habitat due to the destruction of system links by pathogens and pests) and indirectly (changes in the composition and the conditions of existence of natural phytocenoses) [2, 3].

Currently, one of the most important environmental problems is pollution by heavy metals [4]. They enter the atmosphere with gaseous compounds, as well as technological dust. With rainwater they enter the soil which holds positively charged metal ions due to its good absorption capacity. Their systematic entry into the soil even in small amounts can lead to permanent accumulation of metals in soil [5, 6].

In soil, metals are firmly bonded to the humic substances forming a poorly soluble compound, part of the absorbed bases, clay minerals, and migrate in the soil solution in the profile. The concentration of heavy metals in the soil solution is the most important environmental characteristics of the soil, as it determines the migration of heavy metals in the profile and their absorption by plants. Mobile forms of heavy metals are concentrated mainly in the upper layers of soil, which contain a lot of organic matter and where biochemical processes are active [7].

Although many heavy metals are not necessary for normal functioning of plants, they can be actively absorbed by them and preserve toxic properties, affecting negatively the plant. Their direct impact on plant starts with the contact and sorption by above-ground organs, mainly by leaves. In the urban environment, woody plants can act as an important barrier to the spread of heavy metals. Their assimilation organs (leaves) with well-developed exchange surface absorb and precipitate the highest number of air pollutants, but they are affected much more than other organs [2, 8].

The objectives of the study were to examine the content of heavy metals (copper, lead, cadmium, zinc) in the soil of recreational, residential and industrial zones of Yoshkar-Ola, as well as to reveal the peculiarities of accumulation and transformation of heavy metal ions in the tissues of some woody plant species growing in various functional zones of the city.

The study was conducted in Yoshkar-Ola, the capital city of the Republic of Mari El. The city is situated on the Mari lowlands, in the eastern part of the East European Plain, 50 kilometers north of the Volga River, on its left tributary – the River Malaya Kokshaga.

The soil cover of the Republic of Mari El refers to taiga forest soils. The main background of the soil cover is soddy-podzolic loamy, sandy soils, which are characterized by the development of the humus horizon connected with the sod-forming process. The soil cover of the Mari lowlands is mosaic. The extensive development of deposits of sand and sandy loam, weak dissected relief enhance the effect of meso- and microforms of relief and poorly mineralized ground water on the soil formation. The main background of the soil cover of interfluvial plains is soddy, heavy and medium podzolic soils of zone type. They have low fertility, high acidity, a small amount of humus, nitrogen and other nutrient elements for plants.

In the city, soil is under the influence of the same factors of soil formation as natural undisturbed soil, but in the city anthropogenic factors of formation of soil prevail over natural factors, and so-called "cultural layer" is formed which is the upper layer of the ground with the imprints of human activity. The accumulation of this layer is due to landslides in the process of earthworks, soil bedding, and accumulation of various wastes. Thickness of cultural layer varies from a few centimeters in towns to several tens of meters in cities [9].

## 6.2 MATERIALS AND METHODOLOGY

To assess the degree of contamination of the urban environment by heavy metals, twelve test areas within Yoshkar-Ola were selected (Figure 6.1). Two of them are in the recreational zone of the city (Forest park "Pine Wood" and Park of Culture and Leisure named after 30th anniversary of the Komsomol). Five of them are in the residential zone (Nekrasov, Podolskikh Kursantov Street, Pushkin, Ryabinin, Eshpaj Streets), located in the central part of the city where there are main areas of residential buildings. Five of them are in the industrial zone (Lermontov, University, Solovyov, Karl Marx, Stroiteley Streets), where such big industrial enterprises as joint-stock companies "Marbiopharm," "Contact," "Chromatec" and some other engineering, metalworking, building materials, chemical and food manufacturing companies are concentrated. In the city of Yoshkar-Ola road transport causes significant pollution of the environment. In the industrial zone, traffic flow is almost 2 times higher than in the residential zone [10].

Soil sampling for analysis was carried out in accordance with the State Standard of Russia – 17.4.4.02–82 [11]. Spot soil samples were taken in layers from depths of 0–20 cm with the envelope method; then five spot soil samples from one test area were mixed.

The objects of the study were middle-aged woody plants of three species *Betula pendula* Roth., *Tilia cordata* Mill., *Pinus sylvestris* L. These plants grew in the same functional zones where soils were studied. The Forest park "Pine Wood" was studied as a recreational zone of the city,

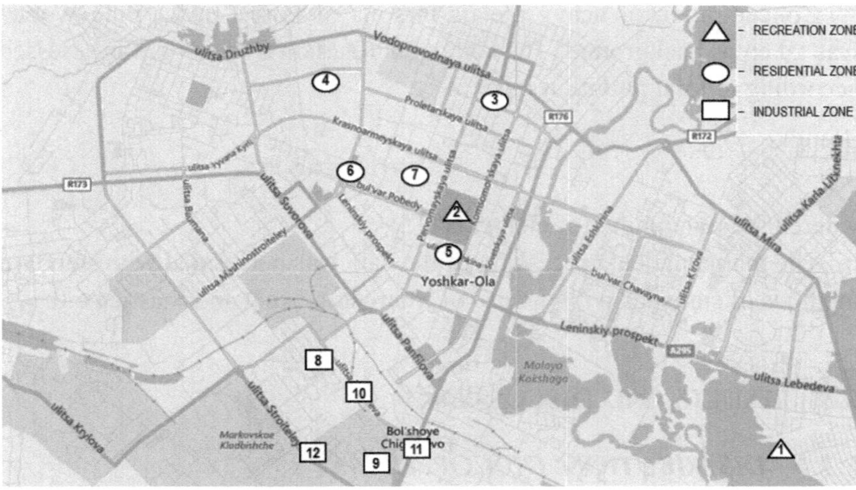


**FIGURE 6.1** The studied areas in the city of Yoshkar-Ola. Note. 1 – Forest park "Pine Wood," 2 – Park of Culture and Leisure named after 30th anniversary of the Komsomol, 3 – Nekrasov Street, 4 – Podolskikh Kursantov Street, 5 – Pushkin Street, 6 – Ryabinin Street, 7 – Eshpaj Street; 8 – Lermontov Street, 9 – University Street, 10 – Solovyov Street, 11 – Karl Marx Street, 12 – Stroiteley street.

Ryabinin Street – as a residential zone, Stroiteley Street – as an industrial zone. Studied leaves (needles) were collected in July 2014. Sampling and preparation for ultimate analysis were carried out with standard methods [12]. To analyze the content of elements in soils and plants, selected samples were air-dried, then the content of mobile forms of heavy metals in soil and plant samples was determined with atomic absorption method in the accredited laboratory of the Chemistry Department of Volga State Technological University [13, 14].

Standard list of chemical soil research included determination of the content of heavy metals of Hazard classes 1 and 2. Earlier studies showed us the content of only copper, lead, zinc and cadmium in the soils of Yoshkar-Ola. Assessment of chemical contamination of soils was carried out according to parameters developed during joint geochemical and hygienic studies of the urban environment. The concentration coefficient of the chemical was calculated, which is determined by the allocation of its real content in the soil (C) to the background content ($C_b$):

$$Kc = C/C_b.$$

Concentration of heavy metals in soils of Forest park "Pine Wood" was taken as background. The soil contamination by metals was assessed according to total pollution index $Z_C$:

$$Zc = \sum_{i}^{n} Kc - (n - 1),$$

where *n* – total number of elements [15].

The mathematical processing of the data was performed with standard methods of statistics with the use of Microsoft Excel and Stastica 6.0.

## 6.3 RESULTS AND DISCUSSION

### *6.3.1 DISTRIBUTION OF COPPER*

According to the results of our study (Figure 6.2), the content of mobile forms of copper in all soil samples was high enough, even in the soil samples from the recreational zone of the city, which is characterized by minimal anthropogenic impact. The lowest concentration of this heavy metal – about 2.8 mg/kg was found in the soil of the Forest park "Pine Wood," only in this area of research copper concentration did not exceed maximum permissible concentration (3 mg/kg), in the soil of the Park of Culture and Leisure named after 30th anniversary of the Komsomol, the concentration of copper ions was 1.6 times higher and reached 4.5 mg/kg. In this case, a high background content of mobile forms of copper may be due to slight soil texture of soil horizons, the action of the soil water regime, as well as increasing amount of absorbed bases and base saturation, usually seen in soddy-podzolic soils [5, 16].

In the residential zone of Yoshkar-Ola mobile forms of copper concentration in soil varied from 7.6–7.7 mg/kg in Nekrasov and Podolskikh Kursantov Streets to 14.1 mg/kg in Ryabinin Street. Soils of the residential zone of the city were characterized by the excess of MPC by 2.5–4.7 times. In the industrial zone of the city all the studied soil samples were characterized by the excess of MPC of copper concentration, the maximum content of this heavy metal was in Karl Marx and Stroiteley Streets (where the copper content exceeded MPC by 8.7 and 11.4 times, respectively).

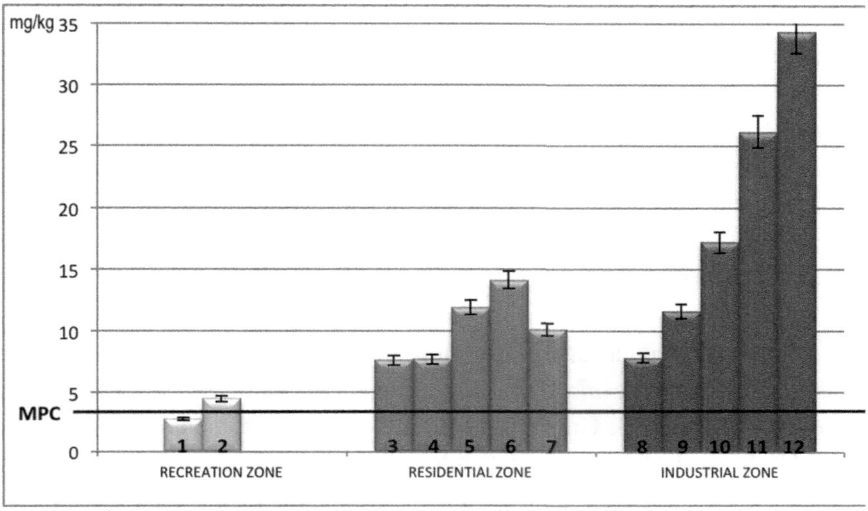


**FIGURE 6.2** The content of mobile forms of copper in soils of the city of Yoshkar-Ola. Note. 1 – Forest park "Pine Wood," 2 – Park of Culture and Leisure named after 30th anniversary of the Komsomol, 3 – Nekrasov Street, 4 – Podolskikh Kursantov Street, 5 – Pushkin Street, 6 – Ryabinin Street, 7 – Eshpaj Street; 8 – Lermontov Street, 9 – University Street, 10 – Solovyov Street, 11 – Karl Marx Street, 12 – Stroiteley Street, MPC – maximum permissible concentration.

In the urban environment, soil contamination by copper compounds is the result of the work of industrial enterprises and the use of copper-containing substances: municipal waste, solutions for spraying plants from pests, fertilizers, etc. Some local anomalies in soils can be the result of corrosion of construction materials containing copper alloys (e.g., electrical wires, pipes). Abrasion of the contact wires of trams and trolley buses also causes pollution by copper in urban areas. The most important factor in the soil pollution by copper is the capability of the surface layer of soil to accumulate copper, therefore the content of the metals in the soil can be very high in some cases [6, 7].

### 6.3.2 THE LEAD CONTENT

Lead is one the most common metals in the surface layer of soil. Lead content is usually examined to estimate the impact of transport on soil.

According to the results of our study (Figure 6.3), the lead content in the soil of the recreational zone of the city was low and reached 2.5–3 mg/kg. In soil samples collected in the residential zone the lead content varied from 2.6 (Nekrasov, Podolskikh Kursantov, Pushkin Streets) to 8.1 mg/kg (Eshpaj Street), where MPC was exceeded by 1.3 times (6 mg/kg).

In soils of the industrial zone the lead content exceeded MPC in almost all areas of the study, but the maximum accumulation of the element was recorded in soils in Karl Marx and Stroiteley Streets (MPC was exceeded by 3.7–4 times).

Lead enters the urban environment in different ways: most of the lead (70%) goes into the air with the exhaust gases of vehicles, which use leaded petrol as fuel. Currently, leaded petrol is not used as a fuel in the city. High level of lead content is found in the sediments of urban waste water; Pb is found in consumer and industrial wastes, which include car batteries, cans with lead solder, paint and other lead-containing materials.

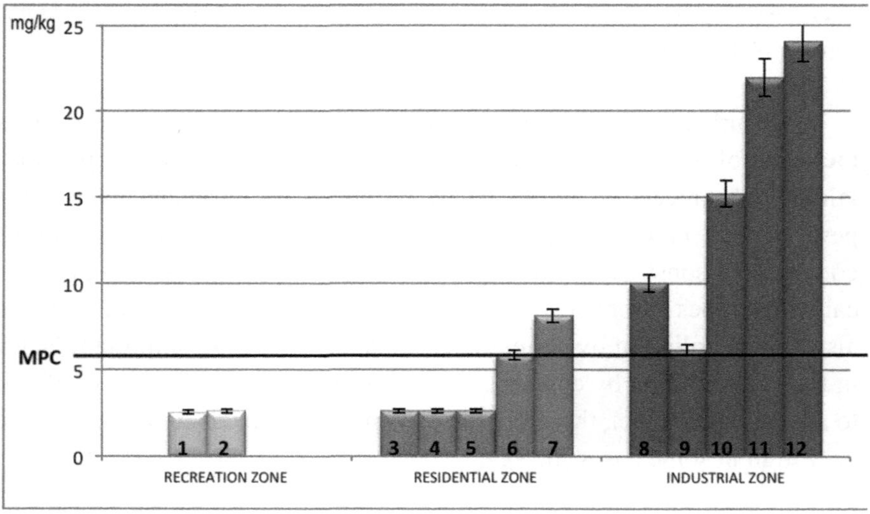


**FIGURE 6.3** The content of mobile forms of lead in the soil of Yoshkar-Ola. Note. 1 – Forest park "Pine Wood," 2 – Park of Culture and Leisure named after 30th anniversary of the Komsomol, 3 – Nekrasov Street, 4 – Podolskikh Kursantov Street, 5 – Pushkin Street, 6 – Ryabinin Street, 7 – Eshpaj Street; 8 – Lermontov Street, 9 – University Street, 10 – Solovyov Street, 11 – Karl Marx Street, 12 – Stroiteley Street; MPC – maximum permissible concentration.

Another way of lead entry into the environment is its use in construction and for decorative purposes due to high resistance of its oxides to corrosion in the atmosphere; lead compounds are widely used in a coating of all surfaces, however, these coatings are easily decomposed under the effect of acid rains [6, 9]. Progressive increase in the percentage abundance of Pb in soils is due to technogenic lead [17].

### 6.3.3 THE ZINC CONTENT

Zinc is not only a nutrient element for plants, but also a soil pollutant. As shown in Figure 6.4, the zinc content in the soil of the recreational zone of the city (both in the Forest park "Pine Wood" and in the Park of Culture and Leisure named after 30th anniversary of the Komsomol) was low – about 10 mg/kg, which is significantly lower than MPC.

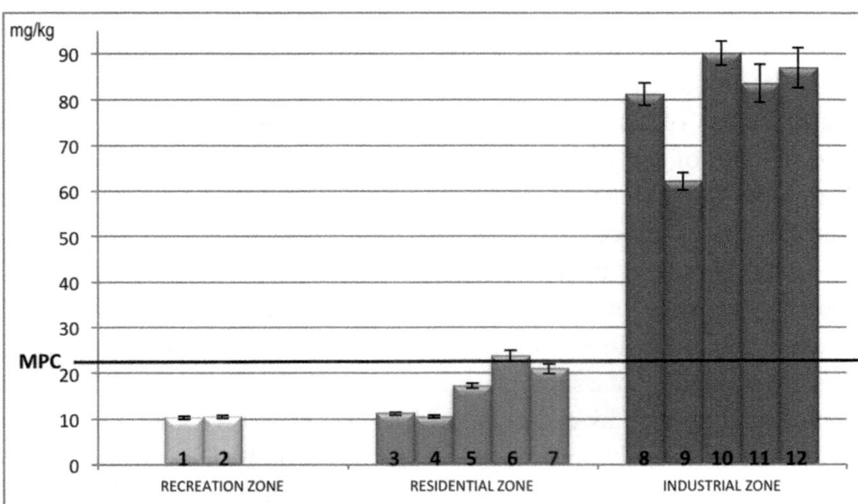


**FIGURE 6.4** The content of mobile forms of zinc in soils of Yoshkar-Ola. Note. 1 – Forest park "Pine Wood," 2 – Park of Culture and Leisure named after 30th anniversary of the Komsomol, 3 – Nekrasov Street, 4 – Podolskikh Kursantov Street, 5 – Pushkin Street, 6 – Ryabinin Street, 7 – Eshpaj Street; 8 – Lermontov Street, 9 – University Street, 10 – Solovyov Street, 11 – Karl Marx Street, 12 – Stroiteley Street; MPC – maximum permissible concentration.

In the residential zone of Yoshkar-Ola, zinc content slightly exceeded MPC only in the soil in Ryabinin Street (23.6 mg/kg). In all soil samples of the industrial zone of the city zinc content exceeded MPC by 2.7–3.9 times, maximum excess was in the soils of Lomonosov and Stroiteley Streets.

In town zinc enters the roadside area due to abrasion of different components, erosion of galvanized surfaces, the tire wear, the use of additives in oils containing this metal. Anthropogenic sources of zinc include nonferrous metal companies, agrotechnical activities. Observations show that soil contamination by zinc in some areas has led to an extremely high accumulation of it in the surface soil now. Increase in the concentration of zinc in the soil cover can occur as a result of continuous use of sewage sludge of cities as organic matter, as well as the combustion of rubber, which contains this element [18].

### *6.3.4 CADMIUM CONTENT*

Cadmium is generally found in soil in relatively low concentrations, but soil contamination by this metal can pose a serious risk to health. In the studied areas of Yoshkar-Ola mobile forms of cadmium were found in two samples of soil of the industrial zone – Stroiteley Street (1.2 mg/kg), Karl Marx Street (0.6 mg/kg). The concentration of the heavy metal in soil samples did not exceed MPC.

Cadmium enters urban ecosystems due to some reasons. It goes into the atmosphere from motor vehicles, as rubber tires and lubricating oils contain cadmium. In some cases, cadmium can be found in fertilizers (superphosphate, potassium phosphate, etc.). However, the level of environmental pollution by cadmium in Yoshkar-Ola is very low.

To assess intensity and hazard of soil pollution by heavy metals in the studied areas, the coefficient of concentration of elements ($K_C$) was calculated as it is an indicator that is used to identify local technogenic anomalies connected with gas and dust emissions of vehicles and industries, their accumulation in urban areas (Table 6.1).

According to the degree of soil contamination by mobile forms of copper, studied areas in Yoshkar-Ola form the following increasing

**TABLE 6.1** The Coefficient of Concentration of Heavy Metals and Total Pollution Index (Zc) of Soil of the City of Yoshkar-Ola

| Studied areas | $K_C$ | | | | Zc |
|---|---|---|---|---|---|
| | **Cu** | **Pb** | **Zn** | **Cd** | |
| Recreational zone | | | | | |
| Forest park "Pine Wood" | 1.00 | 1.00 | 1.00 | – | 0 |
| Park of Culture and Leisure named after 30th anniversary of the Komsomol | 1.5 | 1.04 | 1.67 | – | 0 |
| Residential zone | | | | | |
| Nekrasov Street | 2.53 | 1.04 | 1.15 | – | 1.72 |
| Podolskikh Kursantov Street | 2.57 | 1.04 | 1.02 | – | 1.63 |
| Pushkin Street | 3.97 | 1.04 | 1.68 | – | 3.69 |
| Ryabinin Street | 4.7 | 2.32 | 2.25 | – | 6.27 |
| Eshpaj Street | 3.37 | 3.24 | 2.05 | – | 5.66 |
| Industrial zone | | | | | |
| Lermontov Street | 2.6 | 4.00 | 8.45 | – | 12.05 |
| University Street | 3.87 | 2.44 | 6.09 | – | 9.40 |
| Solovyov Street | 5.73 | 6.08 | 8.82 | – | 17.63 |
| Karl Marx Street | 8.73 | 8.76 | 7.96 | 0.6 | 22.45 |
| Stroiteley Street | 11.4 | 9.60 | 8.32 | 1.2 | 26.32 |

sequence: Forest park "Pine Wood" < Park of Culture and Leisure named after 30th anniversary of the Komsomol < Nekrasov Street < Podolskikh Kursantov Street < Lermontov Street < Eshpaj Street < University Street < Pushkin Street < Ryabinin Street < Solovyov Street < Karl Marx Street < Stroiteley Street.

According to the degree of soil contamination by mobile forms of lead, studied areas form the following increasing sequence: Forest park "Pine Wood" < Park of Culture and Leisure named after 30th anniversary of the Komsomol = Nekrasov Street = Podolskikh Kursantov Street = Pushkin Street < Ryabinin Street < Lomonosov Street < Eshpaj Street < Lermontov Street < Solovyov Street < Karl Marx Street < Stroiteley Street.

According to the degree of soil contamination by mobile forms of zinc, studied areas form the following increasing sequence: Forest park "Pine Wood" < Podolskikh Kursantov Street < Nekrasov Street < Park of Culture and Leisure named after 30th anniversary of the Komsomol < Pushkin

Street < Eshpaj Street < Ryabinin Street < Lomonosov Street <Karl Marx Street < Stroiteley Street < Lermontov Street < Solovyov Street.

"The Guidelines for assessing the degree of soil pollution by chemicals" [15] gives the assessment scale of the danger level of soil contamination by the total pollution index (Table 6.2). It was found that the degree of soil pollution in the recreational and residential zones of Yoshkar-Ola is permissible. In the industrial zone the total index was considerably higher, but it indicates "moderately hazardous" degree of soil contamination only in two areas – in Karl Marx and Stroiteley Streets. Such living conditions can affect human health and cause increase in the total incidence of people.

Due to the wide spread of heavy metals, control activity of their accumulation in ecosystems has increased significantly. Research on peculiarities of heavy metal accumulation by woody plants is connected with the need to estimate biosphere and environment stabilizing functions of plants which act as a phytofilter to the spread of pollutants in the environment. Heavy metals can be dangerous not only to humans but also to plants, causing premature aging and reducing efficiency of their health-related functions [19].

Our study was aimed at identifying the relationship between the content of heavy metals in soil and their accumulation in assimilation tissues of plants.

**TABLE 6.2** Assessment Scale of the Danger Level of Soil Contamination by the Total Pollution Index (Zc)

| Level of pollution | Zc | Changes in public health indicators in polluted areas |
|---|---|---|
| Permissible | less than 16 | The lowest level of morbidity in children and minimal frequency of functional abnormalities |
| Moderately hazardous | 16–32 | The increase in the total incidence of people |
| Hazardous | 32–128 | The increase in the total incidence of people, frequently and chronically ill children, children with cardiovascular system disorders |
| Extremely hazardous | more than 128 | The increase in the incidence of children, women's reproductive function disorder (increase in toxicosis of pregnancy, premature births, mortinatality, small for gestational age babies) |

When studying the *copper* content in leaves (needles) of woody plants, it was found (Table 6.3) that copper content in *T. cordata* and *B. pendula* was almost twice as much as in *P. sylvestris*. However, the copper content in the needles of *P. sylvestris* was minimal and varied from 1.1 to 2.5 mg/kg. Maximum concentration of copper was in all studied plants growing in the industrial zone of Yoshkar-Ola; it was 6.2 mg/kg in *B. pendula* and 4.9 mg/kg in *T. cordata*. The results of the analysis of variance showed that plants growing in the industrial zone had statistically significant differences in the accumulation of copper in comparison with the plants growing in residential and recreational zones ($p = 0.0017$). Significant differences in the accumulation of copper by different plant species were also revealed ($p < 10^{-4}$).

Despite the fact that copper in plants is actively involved in photosynthesis, respiration, recovery, nitrogen fixation, biochemical processes as part of enzymes, which perform the substrate oxidation by molecular oxygen, copper in high concentrations can have significant toxic effects [20]. Normal metal content in plants is up to 20 mg/kg. In high concentration the element affects negatively, causing growth inhibition and slow metabolic processes.

**TABLE 6.3** Heavy Metal Content in the Leaves of Woody Plants (mg/kg)

| Species | Heavy metals | Recreational zone | Residential zone | Industrial zone |
|---|---|---|---|---|
| Scots pine *(P. sylvestris)* | Cu | 1.10 ± 0.100 | 2.40 ± 0.102 | 2.50 ± 0.102 |
| | Pb | 0.23 ± 0.012 | 0.43 ± 0.020 | 0.49 ± 0.024 |
| | Zn | 6.60 ± 0.640 | 10.60 ± 0.401 | 12.90 ± 0.602 |
| | Cd | 0.02 ± 0.002 | 0.04 ± 0.003 | 0.05 ± 0.001 |
| Weeping birch *(B. pendula)* | Cu | 3.90 ± 0.120 | 4.80 ± 0.210 | 6.20 ± 0.303 |
| | Pb | 0.90 ± 0.021 | 1.10 ± 0.061 | 1.30 ± 0.091 |
| | Zn | 12.10 ± 0.660 | 17.40 ± 0.792 | 22.20 ± 1.104 |
| | Cd | 0.06 ± 0.002 | 0.08 ± 0.005 | 0.09 ± 0.002 |
| Small-leaved lime *(T. cordata)* | Cu | 2.70 ± 0.110 | 4.90 ± 0.211 | 4.90 ± 0.202 |
| | Pb | 0.64 ± 0.041 | 0.60 ± 0.040 | 1.20 ± 0.051 |
| | Zn | 11.30 ± 0.060 | 14.10 ± 0.603 | 16.90 ± 0.611 |
| | Cd | 0.09 ± 0.005 | 0.11 ± 0.005 | 0.14 ± 0.002 |

Thus, revealed concentrations of heavy metals are within the physiological norm for the studied plants, in spite of high copper content in the soil of studied habitats. Although the soil affects the element composition of plants, plants are capable of selective absorption of elements from the nutrient depending on the internal needs of the organism.

While studying the accumulation of lead by plants it was found that its content in the needles of *P. sylvestris* was low, and depending on the location it varied from 0.23 mg/kg (recreational zone) to 0.49 mg/kg (industrial zone). The lead content in the leaves of *B. pendula* in the industrial zone of Yoshkar-Ola reached 1.4 mg/kg. The lead content in the leaves of *T. cordata* was slightly lower than in the leaves of *B. pendula* and varied from 0.64 mg/kg (Forest park "Pine Wood") to 1.2 mg/kg (industrial zone). Statistically significant differences between the content of lead in this species in different habitats have not been identified, despite the significant differences in the content of the metal in the soil. Apparently, the accumulation of lead in leaves of woody plants is not only due to its intake from the soil, air pollution by lead also makes a certain contribution to it. The root system of the plants can perform the role of a barrier, accumulating a larger part of the absorbed lead and limiting its intake in aboveground organs, therefore the significant accumulation of this heavy metal in the assimilation organs of the studied species was not observed.

The toxic effect of lead on plants is mainly due to the disfunction of fundamental biological processes (photosynthesis, respiration, etc.) However, it is believed that lead is less dangerous to plants than other heavy metals. Lead ions in soil lose their mobility quickly as a result of the formation of slightly soluble compounds, due to the absorption by organic and mineral colloids. Low phytotoxicity of lead is also due to the presence of well-functioning system of inactivation in plants, but very high concentrations of this element in the soil can significantly inhibit the growth and development of plants [1, 21, 22].

When estimating the *zinc* content in the studied woody plants, it was found that the concentration of zinc in the needles of *P. sylvestris* (Table 6.3) varied from 6.6 to 12.9 mg/kg. Considerably more zinc was accumulated in the leaves of *T. cordata* – the concentration of zinc in the leaves of the species in the recreational zone was 11.30 mg/kg, which was less than in plants growing in the residential zone (14.1 mg/kg). The maximum zinc content was in *T. cordata*, which grew in the industrial zone of the city (16.9 mg/ kg).

Of all studied woody plants the highest zinc content was in *B. pendula* in all functional zones of the city, and varied from 12.1 to 22.2 mg/kg. The zinc content in the leaves of *B. pendula* growing in the industrial zone of Yoshkar-Ola was 1.8 times higher than that of plants in recreational zone. According to the results of analysis of variance, it was revealed that plants of all species studied in the industrial and residential zones had statistically significant differences in comparison with plants in the recreational zone.

The results of the research showed that the zinc content in woody plants growing in all studied zones of the city is within the physiological norm. However, our results showed that the concentration of zinc in the studied woody plants was growing with the increasing technological environmental impact. The accumulation of this element by the plants was significantly higher in areas with high zinc content in the soil. Comparing the concentration of heavy metals in soils of different areas of Yoshkar-Ola and their accumulation in plants, it should be noted that the excess of MPC in content of mobile forms of zinc in the soil of residential and industrial zones was recorded. In these zones zinc content in leaves of the studied plants is higher than in plants in the recreational zone.

Environmental pollution by zinc significantly affects the concentration of this element in the plant. In ecosystems, where zinc is a component of air pollution, above-ground parts of the plants accumulate most of it. On the other hand, plants growing on soils contaminated with zinc accumulate most of this element in the root system. Excessive accumulation of metals in plants in some contaminated areas can be a real danger to human health [18].

The toxicity of *cadmium* ions is 2–20 times more than the toxicity of other mentioned heavy metals. The accumulation of cadmium is usually seen only in areas with intense technological environmental impact near large enterprises. If a plant has a high concentration of cadmium, more often it leads to disruption of growth [6, 23].

As can be seen in Table 6.3, the cadmium content in the needles of *P. sylvestris* was 0.02–0.05 mg/kg, in the leaves of *B. pendula* – 0.06–0.09 mg/kg and *P. sylvestris* – 0.09–0.24 mg/kg. At the same time there was an increase in the content of this metal in all test plants growing in residential and industrial zones of Yoshkar-Ola. However, the heavy metal content was lower than the MPC values and did not have statistically significant differences. The studied soils contained cadmium in very low concentration, which could not lead to the accumulation of the metal in leaves, therefore it can be assumed that the accumulation of this element in plants can be the result of air pollution, when cadmium enters the leaves through cuticles.

## 6.4 CONCLUSIONS

Thus, it has been found that the soils of Yoshkar-Ola experience technological impact, resulting in the accumulation of heavy metals, mainly copper, lead and zinc as compared with natural (background) soils. This is especially characteristic of the soils in residential and industrial zones of the city. In the soil of industrial zone, the copper content was 2.6–11.6 times more than MPC, the lead content – 1.2–4 times more, zinc content – 2.7–4 times more. The scheme of metal accumulation in the studied soils is as follows: Cd < Pb < Cu < Zn.

According to the total index, soil pollution in recreational and residential zones of Yoshkar-Ola has a permissible level of pollution. The soils of the two areas in the industrial zone – Karl Marx and Stroiteley Streets (where the main industrial companies of the city are situated and there is a high flow of vehicles) are moderately dangerous.

Metal phytoaccumulation in the city is characterized by species specificity and depends on growth conditions, especially on polymetallic soil pollution and air pollution. Heavy metal content in woody plants growing in the industrial zone was higher than in other functional zones of the city; the content of heavy metals in woody plants changed in the same way as in the soil: Cd < Pb < Cu < Zn. The amount of heavy metals in plants does not exceed the physiological norm despite high concentrations of these elements in the soil, indicating the selective absorption of these metals by plants. The distribution pattern of heavy metals has species specificity, their content decreases as follows: *Betula pendula* > *Tilia cordata* > *Pinus sylvestris.*

## ACKNOWLEDGEMENT

We thank the staff of the Laboratory of the Chemistry Department of Volga State Technological University for providing equipment (Analyst 400 Atomic Absorption Spectrometer) for assessment of heavy metal concentrations.

## KEYWORDS

- **cadmium**
- **copper**
- **lead**
- **soil pollution**
- **woody plants**
- **zinc**

## REFERENCES

1. Alekseev, Y. V. Heavy Metals in Soils and Plants, Leningrad, Agropromizdat, 1987, 142p (in Russian).
2. Ilyin, V. B. Heavy Metals in Soil-Plant System, Novosibirsk, Science, 1991, 151p (in Russian).
3. Fellenberg, G. Environmental Pollution. Introduction to Environmental Chemistry, Moscow, Peace, 1997, 232p (in Russian).
4. Izrael, Y. A. Ecology and Control of the Environment, Moscow, Gidrometeoizdat, 1984, 560p (in Russian).
5. Bowen, H. J. M. Environmental Chemistry of the Elements, London, Academic Press, 1979, 317p.
6. Voskresenskiy, V. S., Voskresenskaya, O. L. Impact of the Factors of the Urban Environment on the Functional State of Woody Plants: A Monograph, Yoshkar-Ola, Mari State University Publishing House, 2011, 194p (in Russian).
7. Vasiliev, A. A., Chashchin, A. N. Heavy Metals in the Soils of the City of Chusovoy: Assessment and Diagnosis of Contamination, Perm, Perm State Agricultural Academy, 12011, 97p (in Russian).
8. Kulagin, A. A., Shagieva, Y. A. Woody Plants and Biological Preservation of Industrial Pollutants, Moscow, Science, 2005, 190p (in Russian).

9. Ecology of the City of Yoshkar-Ola: Scientific Publication. Yoshkar-Ola, Mari State University Publishing House, 2007, 300p (in Russian).
10. Alyabysheva, E. A., Sarbayeva, E. V., Voskresenskaya, O. L., Voskresenskiy, V. S. Ecological Assessment of Urban Environment, Yoshkar-Ola, Mari State University Publishing House, 2013, 96p (in Russian).
11. GOST 17.4.4.02–82 "Soils. Methods of Sampling and Sample Preparation for Chemical, Biological and Helminthological Analysis." Moscow, 1982, 11p (in Russian).
12. Zyrin, N. G., Obukhov, A. I. Spectral Analysis of the Soil, Plants and Other Biological Objects, Moscow, Moscow State University Publishing House, 1977, 334p (in Russian).
13. Guidelines for Conducting Field and Laboratory Research of Soils and Plants in the Control of Metal Pollution of the Environment. Moscow, 1981, Gidrometeoizdat (in Russian).
14. Methods for Measurement of the Mass Fraction of Acid-soluble Forms of Metals (Copper, Lead, Zinc, Nickel, Cadmium) in Soil Samples with Atomic Absorption Analysis. The USSR State Committee for Hydrometeorology, 1990, 32p (in Russian).
15. Guidelines for Assessing the Risk of Soil Contamination by Chemicals. Moscow, Ministry of Healthcare of the USSR, 1987, 25p (in Russian).
16. Hygienic Standards GN 2.1.7.2041–06 "Maximum Permissible Concentration (MPC) of Chemicals in the Soil." Moscow, 2006, 7p (in Russian).
17. Vodyanitsky, Y. N. Heavy and Extra Heavy Metals and Non-metals in Contaminated Soils. 2009, Moscow, Soil Institute Named after V. V. Dokuchaev of Russian Academy of Agricultural Sciences, 95p (in Russian).
18. Robson, A. D. Zinc in Soil and Plants, Australia, Klumer Academic Publishing, 1993, 320p.
19. Voskresenskaya, O. L., Voskresenskiy, V. S., Skochilova, E. A., Kopylova, T. I., Alyabysheva, E. A., Sarbayeva, E. V. Organism and Environment: Factorial Ecology, Yoshkar-Ola, 2005, 175p (in Russian).
20. Kodom, K. Heavy Metal Pollution in Soils from Anthropogenic Activities, LAP LAMBERT Academic Publishing, 2011, 120p.
21. Kabata-Pendias, A., Kabata-Pendias, H. Trace Elements in Soils and Plants, Moscow, Peace, 1989, 439p (in Russian).
22. Egoshina, T. L., Shikhova, L. N. Lead in Soils and Plants of the Northeast of the European Part of Russia, Orenburg, Orenburg State University Bulletin, 2008, *10(92)*, 135–141 (in Russian).
23. Alyabysheva, E. A., Sarbayeva, E. V., Kopylova, T. I., Voskresenskaya, O. L. Industrial Ecology, Yoshkar-Ola, 2010, 110p (in Russian).

# CHAPTER 7

# BIOREMEDIATION OF OIL-CONTAMINATED SOILS

SARRA A. BEKUZAROVA,[1] LARISSA I. WEISFELD,[2] and EUGENE N. ALEXANDROV[2]

[1]*Gorsky State Agrarian University, 37, Kirov St., Vladikavkaz, Republic of North Ossetia – Alania, 362040, Russia, Tel. +79188257323, E-mail: bekos37@mail.ru*

[2]*Emanuel Institute of Biochemical Chemistry, Russian Academy of Sciences, 4, Kosygin St., Moscow, 119334, Russia, E-mail: liv11@yandex.ru*

## CONTENTS



### ABSTRACT

The cleaning from oil contamination is nowadays a global problem. The application of chemical substances for the neutralization of hydrocarbons brought more harm than good. Soil is a living ecosystem, because it is the habitat of microbes, mushrooms, worms, and smallest insects, plants root systems. The soil is respiring. Hydrocarbons disturb gas exchange, air temperature, kill living world, disturb ecology not only in the soil, but also in surrounding atmosphere. The paper includes a brief overview of the current state of pollution and attempts of soil restoration, avoidance of chemicals application, and soil environment enrichment by biological methods using plants, biological preparations, and mineral fossils. The concrete experiment of living systems application is presented.

## 7.1 INTRODUCTION

Under extensive pollution of soils and sediments by oil hydrocarbons in the conditions of oil and gas industry development it is necessary to develop urgent measures of soil detoxication and preservation of soil organisms with the aim of agricultural use. The soil contamination by hydrocarbons occurs in the regions of oil harvesting and under oil products transportation. This contamination happens because of evaporation of light oil fractions, pipe breaks on industrial pipelines, unpredictable flowing and accidents at oil wells, leaks under repair works [1]. Oil and oil products remain in the soil in the form of chemical compounds; liquid fractions are flushed with water, volatile fractions escapes to the atmosphere. Under oil pollution the following factors are interacting: (a) unique and permanently changing complex oil composition; (b) ecosystem heterogeneity; (c) diversity and variability of external natural factors, both physical (e.g., oscillations of environment temperature, soil structure, the state of its aeration, atmospheric humidity), as well as biological (state and modification of microbiota) [1–5].

Oil contamination suppresses plant development [6, 7]. Soils contaminated by oil are 5–6° warmer, which is important for soil microorganisms, especially under general climate warming [8].

The ecosystem can clean itself from oil, but *self-cleaning* is a long process. It is necessary to undertake measures aimed to decrease the harmful effect of oil products on microbiological and soil-plants complex.

In XX century scientists found that the agriculture with intensive application of chemicals and deep ploughing of topsoil negatively affects soil microorganisms, which leads to the decrease of yields. In Russia the trend of avoiding of chemicals application both in plant cultivation as well as in the cleaning of contaminated soils, in particular under oil contamination, is increasing. It is possible to renew the soil, restore its fertile layer (humus) using "living systems" technologies. The microbiological methods are based on the decomposition of oil products by microorganisms – oil destructors, which are initially presented in the soil, as well as introduced during the cleaning process [5].

Bioremediation of soil pollutants represents a set of technologies of cleaning and restoration of contaminated ecosystems with the application of "living systems," namely plants, microorganisms, animals etc. [9].

Complex preparations, in particular the preparation containing microbes of genres *Pseudomonas, Azotobacter, Bacillus*, were successfully applied for bioremediation of oil-contaminated soils [10–12].

In the aerated soil the hydrocarbons quicker evaporate or are exposed to microbiological processes [8].

The ways of soil cleaning are studied in different regions and economic zones of Russia. In the permafrost zone the natural suppression of soil flora occurs. The Republic of Yakutia is one of the most contaminated territories. The collective of authors from the Institute of oil and gaze problems of Siberian branch of Russian Academy of Sciences [13–15] performed the biological cleaning of soils using microbiological oil destructors basing on indigenous hydrocarbons oxidizing microorganisms in combination with perennial plants with addition of local zeolites. The biological destruction of oil pollution was achieved. The composition of samples became practically identical to background one.

Plants are used as indicators of oil pollution [16, 17]. Biodegradation of soil because of oil hydrocarbons occurs as a result of the disturbance of metabolism of living soil organisms and plant root system, although it was shown that the microdoses of hydrocarbons could even stimulate the growth of cereals roots. The toxicity of oil-contaminated soils is determined by morphological parameters of plants, namely length of roots and shoots of cereal germs. The ratio of these indicators in control was 1:1 and under oil pollution it reached 4:1.

Bome and Nazyrov [16] (Tyumen State University) found a specific reaction of perennial cereal herbs (*Bromus inermis*, *Festuca rubra*) and red clover (*Trifolium pratense*) on the effect of oil hydrocarbons at different orthogenesis stages in laboratory and field conditions. In laboratory experiment under seeds treatment by oil (in concentrations of 0.3; 0.6; 0.9%) *Bromus inermis* germs were least sensitive. The soil contamination by oil (in concentration 9%) in vegetation vessels led to suppression of growth of *Trifolium pratense* shoots and to the activation of *Bromus inermis* growth.

*Minerals* play an important role in the restoration of disturbed soils. North Ossetian scientists suggest applying local natural zeolite-containing clays in the restoration of contaminated soils. Deposits of clays of local origin are situated in tributaries of Terek River, in their upper and lower parts. They have different chemical composition [18–21].

Marzoyev and Aborov [16] applied for the restoration of contaminated soils natural clays irlits and lexinit from Northern Caucasus. The following chemical composition these types of clay was detected (from and to in %%): silicon 53.7–48.9; aluminum 16.4–15.2; iron 3.94–2.3; calcium 2.5–25.0; potassium 1.75–1.86; manganese 0.1–1.14; magnesium 1.82–2.53; sodium 1.1–1.6; phosphorus 0.2–2.5; sulfur 0.9–1.1; copper 3.94–2.93; cobalt 1.0–1.9; molybdenum 0.8–0.6; and iodine 0–0.2; pH 3.0–8.46.

Zherukov et al. [17] and Alborove et al. [18] also used local minerals, namely dialbekulit and irlits. The dialbekulit clay (deposit of Terek river) contains (in %%): silicon 46.5; iron 7.1; calcium 37; cobalt 0.1; zinc 1.1; nickel 1.7, and phosphorus 1.7. The reaction of environment is alcaline (pH 9.1). Irlit 7 clay differs by its composition from dialbekulit and contains (in %%): silicon 54, aluminum 16, iron 4, sulfur 2.5, potassium 2 and vitally important, elements (copper, cobalt, molybdenum, zinc, selenium) within the range of 0.1–0.9%. The reaction of environment is acid (pH 3.8). Irlit 1 clay contains (in %%) silicon 54, aluminum 28, iron 7, sulfur 2, calcium 3, magnesium 1.7, manganese 1.7, potassium 2.1, sodium 1.1, sulfur 2.0, selenium 0.8. The reaction of environment is neutral. The dialbekulit clay harvested in other deposit contains (in %): silicon 46.5; calcium 37; zinc, iron, copper, nickel within the range 1.2–2; phosphorus 0.2, and potassium 0.9; pH 9.16. It also contains in microdoses Co, Mg and other microelements. In all cases no heavy metals and radioactive nuclides detected. The alanit clay was also harvested in the Terek basin (it will be described later). All clays have high absorption capacity because of high silicon content (46–54%) they are able to absorb water and other substances and they content necessary chemical macro- and microelements. The high amount of elements in the natural clays is also necessary for the life of biological objects brought into the soil.

The biotechnological method of soil cleaning from oil products, which we propose, contributes to the restoration of natural soil fertility and is less costly than commonly applied chemical fertilizers or complex technologies of soil treatment, which are now proposed. The method has a principally new basement related with the application of natural clays, microbiological preparations (Berkon, Baykal EM-1) and phenotypic activator (para-aminobenzoic acid).

We propose the following scheme of soil restoring measures:

1. To perform the ploughing of contaminated areas by the method contributing the highest soil aeration, i.e., without turnover of topsoil. Plowing helps aeration, plowing without turning the reservoir enrichment soil by root residues and promotes to the growth of soil organisms, microorganisms introduced into the soil from outside.
2. To develop the methods of soil regeneration by means of plant crops, which provide the highest soil enrichment by useful substances and able to absorb hydrocarbons. The method includes the sowing of annual crop mixed with zeolite-containing clay in the first year, its skewing and ploughing of plants into the soil at the beginning of the seeds ripening phase. The soil fertilization by effective microbiological preparations Barkon and Baykal EM-1 is also included.
3. To apply the biologically active substance, namely para-aminobenzoic acid (PABA).
4. It is possible to mix zeolite-containing clays with waste products of sugar production as molasses, corn starch and natural mineral water.

The method of soil bio-remediation is described below [26].

## 7.2 MATERIALS AND METHODOLOGY

The soil plot of 10 m$^2$ was polluted by oil. We applied biologically active substances: para-aminobenzoic acid (PABA), microbiological preparation

Baykal EM-1, clay sediments of local origin, namely alanit, and stevia leaves.

The first year the plot contaminated by oil was irrigated by 0.1–0.2% water solution of PABA. Previously PABA was dissolved in the hot water.

The *working solution* of PABA was prepared in the proportion of 100 g of powder PABA for 10 L of water with the temperature 80–85°C. Then the stevia (*Stevia rebaudiana* Bertoni) leaves were added to the hot solution in the proportion of 0.2%. If fresh leaves were disabling, the pharmaceutical preparation was applied. In this case 20 mL of stevia solution were mixed with 10 L of liquid.

At the same time the solution of biological preparation Baykal EM-1 was prepared in the proportion of 1 portion of the preparation to 100 portions of non-chlorinated water not containing any bactericide preparations, i.e., 100 g of the biological preparation for 10 L of water. After the cooling of PABA solution down to 20–25°C both solutions were mixed. The total solution volume amounted 20 liters. This amount of solution was applied for irrigation of the oil-contaminated plot.

After 2–3 weeks later seeds of amaranth (*Amaranthus* L.) mixed with alanit clay is sown at the plot treated by para-aminobenzoic acid. After the plants reach the phase of branching, the plot is repeatedly irrigated by the working solution.

The mature amaranth plants were ploughed at the beginning of seeds ripening. The next year after amaranth ploughing, legumes and cereal herbs were sown.

## 7.3 RESULTS AND DISCUSSION

The total results of the experiment are presented in the Table 7.1.

The results of experiment show considerable decrease of hydrocarbons and oil in the contaminated soil, this means that the method showed to be effective. The content of oil and its hydrocarbons decreased in all variants of the experiment. The amaranth seeds, which had time to ripen, will germ after ploughing and produce new plants. The next year, after amaranth ploughing, legume and cereal herbs are sown, namely coronilla, sainfoin, clover, alfaalfa, and melilot. Perennial herbs, in particular legumes, enrich

**TABLE 7.1** The Decrease of Toxicity of Oil-Contaminated Soils by the Biotechnological Method [21]

| Treatment of contaminated plot | Oil concentration, g/kg | Decrease of oil content, % | Content of hydrocarbons, mg/kg |
|---|---|---|---|
| Control (without irrigation) | 66.4 | - | 2680 |
| Plot irrigation by PABA water solution, 0.1–0.2% | 57.0 | 14.2 | 1812 |
| Plot irrigation by water solution of Baykal EM-1 (without stevia) | 48.2 | 27.5 | 1620 |
| Plot irrigation by working solution | 36.8 | 44.1 | 860 |
| Irrigation of amaranth crop by working solution in the phase of branching | 42.5 | 36.0 | 1180 |
| Plot irrigation by working solution: PABA (0.1–0.2%) + stevia (0.2%) + Baykal EM-1 + amaranth crop with irrigation in the phase of branch | 24.8 | 62.7 | 362 |

the soil by nitrogen-fixing bacteria, which are able to absorb heavy metals from soil, radioactive nuclides, pesticides, and oil hydrocarbons. The next year it is possible to continue the procedure until the complete cleaning of the plot.

### 7.3.1 CHARACTERISTIC COMPONENTS USED FOR THE PURIFICATION OF SOIL

The *amaranth* has the following important peculiarities: high drought resistance, high reaction to land treatment, adaptability to different soil and climatic conditions, resistance to diseases and pests, high seed productivity and extremely high coefficient of multiplication. By the moment of the beginning of seeds ripening amaranth accumulates high volume of green mass: 50–60 tones per hectare. Amaranth is characterized by high seed productivity, above 3 tones of seeds per 1 ha [22–24] with high concentration of proteins and biologically active substances [24].

*Silicon* is a mandatory member of plants and microorganisms tissues. Its amount in the plants depends on its content in the habitat, as well as on the total plant biomass. At the international Nobel symposium in Stockholm in 1977 silicon was recognized as «element of life». Silicon is characterized by high absorbing capacity, it can bind toxic elements, it accelerates metabolism of plants [26]. The silicon concentration on the natural clays of Northern Ossetia is the highest (15–18% in comparison with the content of other elements) [20, 21]. During one season silicon is able to clean the soil considerably from oil hydrocarbons, radionuclides,, and heavy metals.

Because of the high biomass amaranth applied to the soil the highest amount of silicon, especially its green parts (leaves and stem). The silicon content in dry mass is 3–5%. Most of the silicon (above 50%) is related to the organic components of plant tissue (proteins, lipides, and cellulose).

Besides of silicon, the ploughing of amaranth on the contaminated soil at the beginning of seeds ripening enriches the soil by substances useful for soil microorganisms, which contributes to the soil cleaning from hydrocarbons. Amaranth seeds, which had time to ripen, will germ after ploughing and give new plants.

The practice showed that *amaranth* can grow at any soil, but it hardly withstands acid soil. That is why in our experiments in the first year of sowing we added to the seeds the zeolite-containing clay – *crushed alanit* [24], which has alkaline reaction (pH 8.6). Additionally, clay sediments of alanit are reach by macro- and microelements.

As amaranth seeds are small (the mass of 1000 seeds is 0.5–0.6 g), their sowing in the first year mixed with crushed alanit improves seeds flowability, provides homogenous distribution of mixture over the plot and fertility of the environment of the seeds bed. Under higher alanit concentrations the process of sowing becomes more complicated.

*Alanit* [19, 20] belongs to the natural zeolite-containing clays from Northern Ossetia highlands deposits. It contains the following elements (in %%): silicon – 51–53; aluminum – 16–17; iron – 5–6; calcium – 30–33; potassium – 0.07, phosphorus – 0.38, manganese – 0.04, sulfur – 0.98, magnesium – 1.6%. Also it contains small amounts of zinc, copper, cobalt and other microelements in the forms available for plants. The reaction of the environment is alkaline (pH 8.64) because of the high calcium content. In contrast with basalts, alanit has rather week water loss (about 3%), high heat capacity (the coefficient is 0/34) and absorption capacity; nitrogen content (totally 0.36% is quite sufficient for plant nutrition. Consequently, the synergism of amaranth and alanit decreases the amount of hydrocarbons 4–5 times in the first year.

*Effective biological preparation Baikal EM-1* contains a stable community of physiologically compatible and complementary useful microorganisms. Originally the preparation was obtained from ecologically clean soil of Transbaikalia and was developed in Russia by Shablin [28]. The preparation Baykal EM-1 represents a water solution containing above 80 species of photosynthesing, lactate, nitrogen fixing bacteria, yeasts, actinomycetes, fermenting fungi and products of their metabolism [30].

The microorganisms of the preparation use the soil organic matter within the process of their multiplication and at the same time they enrich its microflora. This fertilizer affects microbiological equilibrium of the soil, establishes new effective symbiosis of plants and organisms in the soil by absorbing light and synthesizing biologically active substances, it improves soil structure because of its chemical and physical sorption activity, neutralizes harmful effect of heavy metals salts. Its developers created new direction named humus biotechnology [29]. The principle difference of Baykal-EM-1 from other microbiological preparations consists in that it is multi-component and not harmful for humans, animals and soil microflora.

As it is impossible to mix microbiological preparation with any *bactericide* substances including honey, we added to the working solution sweet leaves of stevia in order to provide the nutrition of microorganisms.

*Stevia* is a natural sweetener because its leaves content stevoside and other diterpene glycosides [30]. Fresh leave of stevia is slightly sweeter than sugar, dry leave is 30–40 times sweeter, extract is 40–50 sweeter and the concentrated product is 80–120 sweeter. Sweet substances of stevia consist from diterpene glycosides: stevoside, rebaudioside A, rebaudioside C, dilkoside A; they include also nicotinic acid, flavonoids, aminoacids, pectins, volatile oils, mineral elements, and vitamins.

*Para-aminobenzoic acid* (PABA) [31-37] 4-amino-2-oxibezoic acid ($C_7H_7NO_2$) with molecular mass of 137.1 is soluble in the hot water (80–90°*C*), it is well soluble in benzol, ethanol, acetic acid, it is contained in yeast, legume germs, manure, it is the antagonist of novocain, sulfonamide. PABA represents fine crystalline powder of white or slightly cream colour, it belongs to the group of vitamins B (vitamin H), it is a part of folic acid. PABA is widely synthesized by plants, yeasts and microorganisms and possesses reparagenic and antimutagene properties.

Rapoport [31] discovered the property of PABA to be a non-hereditable regulator of growth, directly connected to pherments. Rapoport [37, 38] called PABA "phenotypic activator." The non-hereditable character of PABA effect was studied on winter wheat.

Rapoport [34] found that PABA positively affects the development of Drosophyls, as well as plants, it contributes to the increase of agricultural crops productivity. He proposed the following scheme of relations enzymes and phenotype: "genes → their heterocatalize (RNA molecules are substrate) → iRNA → iRNA catalize (aminoacids are substrate) → enzymes → their catalize (different substrates) → phenotype." In contrast to chemical mutagens, PABA does not enter in valency relations with enzymes. It is a reparagene to enzyme systems.

In the small concentration (0.03%) PABA has a stimulating effect on plant development [35, 38]. PABA positively affects germinating capacity of seeds, it increases plants resistance to unfavorable environmental factors and to some phytopatogenes, in particular it increases hardiness of winter wheat, improves other characters determining productivity.

The combined effect of salt solutions and PABA on seeds development of barley cultivars with low resistance to salinization. Bekuzarova et al. [38] showed that PABA increases hardiness of winter wheat when sowing without thrashing.

Thus, we substantiated the applied method of biological purification of oil contamination. Microbiological preparations restore and enrich soil. Amaranth seeds, which had time to ripen, will germ after ploughing and will give new plants. Alanit introduced into the soil mixed with seeds of amaranth, adds macro- and microelements, which are necessary supplements for microorganisms and plant root system development. The next year, after amaranth ploughing, legumes and cereal herbs are sowed. The roots of perennial herbs, in particular legumes, enrich soil ecosystem by nitrogen fixing bacteria.

## 7.4 CONCLUSIONS

1. The biological preparations containing wide set of microorganisms were applied for the restoration of oil contaminated soils.

2. The optimal conditions for contaminated soils restoration occur under the application of Baykal EM-1 biological preparation.
3. The biological preparation Baykal EM-1 was dissolved in 0.1–0.2% water solution of para-aminobenzoic acid (PABA).
4. We immersed in PABA 200 g of stevia leaves instead of sugar or other sweeteners and consequently we obtained 2% solution of stevia in PABA. If fresh leaves were disable, the pharmaceutical preparation was applied: 20 mL of stevia solution were mixed with 10 liters of liquid.
5. Repeated irrigation of contaminated soil in the phase of amaranth branching led to hydrocarbons content decrease from 2680 to 1812 mg/kg.
6. The irrigation of oil contaminated soil by the mixture of Baykal EM-1 biological preparation and PABA with stevia leaves conditioned the decrease of oil content by 44.1%.
7. The maximum effect is reached after preliminary sowing of amaranth plants with further sowing of perennial legume herbs. When sowing amaranth, seeds were mixed with alanit cley reach by elements and having alkaline reaction. Alanit contributes to amaranth seeds sowing.
8. Repeated soil irrigation by this working solution during the growth period of absorbing amaranth culture, ploughing of its aboveground biomass as green manure allowed to decrease hydrocarbons content (362 mg/kg) down to maximum permissible concentration of oil by 62.7%.

## KEYWORDS

- **alanit**
- **amaranth**
- **Baykal EM-1**
- **dialbekulit**
- **para-aminobenzoic acid**
- **stevia**

## REFERENCES

1. Pikovsky, Y. I., Puzanova, T. (2012). Environmental issues of oil production in Russia. Heat-energy Company. Russia, 1, 34–37. (In Russian).
2. Pikovsky, Y. I. (2003). Problems of diagnosis and regulation of soil contamination oil and petroleum products. Y. I. Pikovsky, A. N. Gennadiev, S. S. Chernyansky, G. N. Sakharov. Journal of Soil Science, 9,1132–1140. (In Russian).
3. Pikovsky, Y. I. (1993). Natural and man-made flow of hydrocarbons into the environment. Y. I. Pikovsky. Moscow, Publisher Moscow State University, 280p. (In Russian).
4. Pikovsky, Y. I. (1988). Technological transformation of oil flows in soil ecosystems. Y. I. Pikovsky. Restoration of contaminated soil ecosystems, Moscow, Nauka, 7–12. (In Russian).
5. Glyaznetsova, Y. S. (2010). The oil pollution of soils and bottom sediments in the territory of Yakutia. Composition, distribution, transformation. Y. S. Glyaznetsova, I. N. Zuyeva, O. N. Chalaya, & S. Kh. Livshits. Yakutsk. Akhaan. 160p. (In Russian).
6. Nazarov, A. V. (2007). The effect of soil oil pollution on plants. Bulletin of Perm State University, 5(10), 134–139. (In Russian).
7. Kireeva, N. A. (2009). The effect of soil contamination with oil on physiological parameters of plants and rhizospheric microbiota. N. A. Kireeva, E. I. Novoselova, & A. S. Grigoriadi. The effect of soil contamination with oil on physiological parameters of plants and rhizospheric microbiota. Agrocemistry, 7, 71–80. (In Russian).
8. Karalov, A. M. (1989). Regulation of the thermal regime of oil-contaminated lands in terms of their biological recultivation. 8-th all-Union Congress of Soil Science. Abstracts. Book 1, 37.
9. Bioremediation Science Abroad. (2013)._A monthly Review No. 25, August–September, Institute problem science abroad of RAS. URL: www.issras.ru/global_science_review_(In Russian).
10. Kuritsyn, A. V. (2011). Bioremediation of oil contaminated soils on technological platforms. A. V. Kuritsyn, T. V. Kuritsyna, & I. V. Kataeva. Proceedings of the Samara Scientific Center RAS, 1(5), 1271–1273. (In Russian).
11. Kireeva, N. A. (2004). Integrated bioremediation of oil-contaminated soils to reduce its toxicity. N. A. Kireeva, E. M., Tarasenko, T. S. Onegova, et al. Biotechnology, 6, 63–70. (In Russian).
12. Ilarionov, S. A. (2006). You can restore the soil biocenosis subjected to the oil pollution. S. A. Ilarionov, S. Yu. Ilarionova, A. V. Nazarov, & I. G. Kalachnikova. International Scientific-technical Journal: Scientific Technical Centre "TATA," 1, 56–59. (In Russian).
13. Glyaznetsova, Y. S. (2015). An evaluation of the biological treatment effectiveness of oil polluted soils for the Yakutian region. Y. S. Glyaznetsova, I. N. Zueva, O. N. Chalaya, & S. Kh. Lifshits. Biological Systems, Biodiversity, and Stability of Plant Communities. Canada, USA, Apple Academic Press, Inc. 505–516. (In Russian).
14. Lifshits, S. Kh. (2014). Patent RF # 2535746 "Method of recovery of oil-contaminated soils by applying of microbal-plant communities"/ S. H. Lifshits,Y.S. Glyaznetsova, O. N. Chalaya, I. N. Zueva, & L. A. Erofeevskaya. Date publication: 20.12.2014. Bull. # 35. (In Russian).

15. Erofeevskaya, L. A. (2014). Patent RF # 201311150: "Method of purification of permafrost-affected soils and grounds from pollution by oil and oil products." L. A. Erofeevskaya, Yu. S. Glyaznetsova, P. G. Novgorodov, O. N. Chalaya, I. N. Zueva, S. Kh. Lifshits, S. E. Efimov, & A. R. Alexandrov. Date publication: 20.11.2015. Bull. # 16. (In Russian).
16. Bome, N. A. (2016). Plant Response to Oil Contamination in Simulated Conditions. Temperate Crop Science and Breeding. Ecological and Genetic Studies. N. A. Bome, & R. A. Nazyrov. Apple Academic Press, 371–384.
17. Bekuzarova, S. A. (2015). Patent RF #2552057 "Method of Assessment of Contaminated Lands." S. A. Bekuzarova, N. A. Bome, E. A. Goncharova et al. Date publication: 20.11.2015. Bull # 16. (In Russian).
18. Sokaev, K. E. (2004). Natural and agro-climatic resources of the Republic of North Ossetia – Alania. K. E. Sokaev. Agrochemical Herald, 6, 4–8. (In Russian).
19. Sokaev, K. E. (2008). Training manual. K. E. Sokaev & K. H. Byasov. Vladikavkaz, Publishing of Kosta Hetagurov North-Caucasus State University, 53p. (In Russian).
20. Sokaev, K. E. (2010). Ecological-geochemical assessment of soils of the foothills of the Central Caucasus in their long-term agricultural use and application of fertilizers. Sokaev, K. E. Thesis DSc of agriculture. Vladikavkaz, Gorsky State University, 61p. (In Russian).
16. Marzoev, M. B. (2014). Environmentally sound technologies of land recultivation which were violated at open cast mining. M. B. Marzoev, & G. M. Alborov. Proceedings of young scientists, Ecology, 121–127. (In Russian).
17. Zherukov, B. H. (2013). Patent #2486736 "Method of increasing soil fertility." B. H. Zherukov, I. M. Khanieva, Khaniev, M. H., Bekuzarova, S. A., et al. Date publication: 10.07.2013. Bull # 19. (In Russian).
18. Alborov, I. D. (2013). Patent # 2496820 "Sorbent-Ameliorant for Cleaning Oil-Polluted Lands." I. D. Alborov, H. E. Taimaskhanova, S. A. Bekuzarova et al. Date publication 27.10. 2013. Bull. № 30. (In Russian).
19. Zaalishvili, V. B. (2010). Patent # 2396133 "A Method of Rehabilitation of Oil-Contaminated Lands." V. B. Zaalishvili, S. A. Bekuzarova, D-H. S. Bataev et al. Date publication 10.08.2010. Bull. № 22. (In Russian).
20. Zaalishvili, V. B. (2013). Patent # 2481162 "The Method of Reclamation of Oil-Contaminated Land." V. B. Zaalishvili, S. A. Bekuzarova, H. N. Mazhiev et al. Date publication 03.05.2013. Bull. № 13. (In Russian).
21. Bekuzarova, S. A. (2015). Patent RF # 2555595 "The Method of Reproduction of Oil-Contaminated Lands." S. A. Bekusarova, E. N. Alexandrov, L. I. Weisfeld et al. Date publication 10.07.2015. Bull. № 19. (In Russian).
22. Kononkov, P. F. (1998). Amarant-Promising Culture of the XXI Century. P. F. Kononkov, V. K. Gins, & M. S. Gins. Moscow. Publishing House, E. Fedorov. 1998. 310 p. (In Russian).
23. Gulshina, V. A. (2008). Biology of development and feature of biochemical structure of varieties of amaranth (Amarantthus, L.) in the Central-Chernozem Region of Russia. Thesis PhD. Michurinsk. 144 p. (In Russian).
24. Zelenkov, V. N. (2011). Amaranh: Biochemical and Chemical Portrait in Ontogenesis. V. N. Zelenkov, V. A. Gulshina, & A. A. Lapin. Moscow. The Publication of the Russian Academy of Natural Sciences. 104p. (In Russian).

25. Gins, M. S. (2002). Biologically Active Substances of Amaranth. Amaranthine: Properties, Mechanisms of Action and Practical Use. Moscow. Russian Peoples' Friendship University. 183p. (In Russian).
26. Podobed, L. I. (2013). The Bioavailability of Silicon—A New Stage in the Development of Agriculture. Technology Development Center URL: http://nabikat.com/docs/Podobed_biokremniy_dlya_selskogo_hozyaystva.pdf.
27. Bekuzarova, S. A. (2007). Patent RF # 2294094 "The Substrate for Growing Plants." Bekuzarova, S. A., Yudashev, M. A. Date publication 27.02.2007. Bull. № 6. (In Russian).
28. Shablin, P. A. EM-technology. Ulan-Ude. URL: http://baykal-em.ru/proizvoditeli.html.
29. EM-Technology—Biotechnology of the XXI Century (2006). The collection of materials on the practical application of the drug, "Baikal EM-1." Compiled by SA Suhamera. Almaty. Kazahstky National Agrarian University. 76p. (In Russian).
30. Sitnichuk, I. Y. (2002). Development of an effective method of allocating the amount of diterpene glycosides from Stevia rebaudiana Bertoni. E. N. Strizheva, A. A. Efremov, & G. G. Pervyshin. Chemistry of Plant Raw Materials, 3, 73–75. (In Russian).
31. Rapoport, I. A. (1948). Phenogenetically analysis of the independent and dependent of differentiation. Proceedings of the Institute of Cytology, Histology, Embryology, USSR Academy of Sciences, 2, 1, 3–135. (In Russian).
32. Eiges, N. S. (2012). Some aspects of the non-hereditary variability induced on crops using antioxidant para-amino benzoic acid. Autochthonous and introduced plants. Collection of scientific papers. N. S. Eiges, L. I. Weisfeld, et al. The National dendrological park of Ukraine "Sofiyivka" NAS of Ukraine. 8, 71–78 (in Ukrainian).
33. Eiges, N. S. (2011). Modification (non-hereditary) the effect of para-amino benzoic acid on crops. Recent and new directions of agricultural science. N. S. Eiges, L. I. Weisfeld et al. , Vladikavkaz, Gorsky State universitet, 35–37. (In Russian).
34. Rapoport, I. A., (1989). Action PABA in connection with the genetic structure. Chemical mutagens and para-aminobenzoic acid in improving the productivity of agricultural plants. Moscow, Nauka, 3–37. (In Russian).
35. Burakov, A. E. (2012). Patent 2463779 "The method of caring for ornamental plants," A. E. Burakov, S. A. Bekuzarova, L. I. Weisfeld, et al. Date publication 20.08.2012. Bull. 23. (In Russian).
36. Burakov, A. E. (2013). Path for Plant Conservation and the ways to care for them in the winter garden. A. E. Burakov, S. A. Bekuzarova, L. I. Weisfeld, et al. Selection and Genetic Science and Education, dedicated to the anniversary of, F. M. Paria. (19 marth 2013). Uman. Umantsky National University of Horticulture. Uman. Umantsky National University of Horticulture, 20–21. (In Russian).
37. Bome, N. A. (1998). Efficacy of para-aminobenzoic Acid to the Ontogeny of Plants under Stress. N. A. Bome, & A. A. Govorukhina Tyumen, Publishing House Tyumen State University, 2, 176–182. (In Russian).
38. Bekuzarova, S. A. (2013). Activating by para-aminobenzoic acid of sowing properties of seed and of winter grain crops and forage cereals. S. A. Bekuzarova, N. A. Bome, L. I. Weisfeld, F. T. Tsomartova, & G. V. Luschenko. Polymers Research Journal. Nova Science Publishers, Inc., 7, 1, 1–8.

# PART II

# IMPACT OF HEAVY METALS ON VEGETATION

## CHAPTER 8

# PLANTS – BIOINDICATORS OF SOIL CONTAMINATION BY HEAVY METALS

SARRA A. BEKUZAROVA,[1] IRINA A. SHABANOVA,[1] and ALAN D. BEKMURSOV[2]

[1]*Gorsky State Agrarian University, d. 37, Kirov St., Vladikavkaz, Republic of North Ossetia – Alania, 362040, Russia, Tel. +79188257323, E-mail: bekos37@mail.ru*

[2]*K.L. Khetagurov North Ossetian State University, d. 44–46, Vatutin St., Vladikavkaz, RNO-Alania, 362025, Russia, E-mail: 3210813@mail.ru*

## CONTENTS



### ABSTRACT

The soil toxicity was determined by means of the evaluation of the level of their contamination using plants, capable to absorb heavy metals. These plants we name bioindicators. The conclusion about soil toxicity was done and the measures aimed to decrease it were suggested in dependence on the accumulation of lead, cadmium, copper, zinc, nickel etc. in different phases of plants development in doses exceeding maximum permissible concentrations.

## 8.1 INTRODUCTION

The increase of anthropogenic effect on the biosphere led to the pollution of the atmosphere, water and soil resources by industrial emissions and heavy metals (HM), which are deposited in the soil [1].

Presently the total power of sources of anthropogenic pollution exceeds in many cases the power of natural ones, e.g., natural sources of nitrogen oxide produce annually 30 million tones of nitrogen, whereas anthropogenic ones produce 30–50 million tones. Natural and anthropogenic sources produce 30 and above 150 million tones of sulfur dioxide, respectively [2, 3]. About 10 times more lead fall in the biosphere from human activity than from natural pollution. The contamination of the environment by HM harbors a high danger. Lead, cadmium, mercury, copper, nickel, zinc, chromium, vanadium became practically permanent components of the air in industrial cities. The problem of air contamination by lead, which then falls into soil and plants, is especially acute [4].

Soil is a powerful filter, cleaning biosphere from man-caused substances, and its chemical bar, which usually strongly fixes pollutants. Soil can become the source of secondary pollution as a result of their redistribution and migration by groundwater and surface water streams [4].

In order to detect soil contamination in time, biological objects with strong reaction on the state of environment are used, by means of the observation of the changes in their metabolism during their vital functions [5].

Bioindicators are such organisms that the changes in their development are the indicators of the processes of their habitat change. Their indicative importance is determined by the ecological tolerance of the biological system.

An organism is able to maintain its homeostasis within the limits of possible adaptation. A biological system reacts to the environmental influence as whole, not only to particular factors, and the amplitude of physiological tolerance variations is modified by the internal state of the system, namely by nutritional conditions, age, peculiarities of the genotype [5, 6]. Biological objects allow to evaluate the measure of the toxicity of any substances synthesized by man and thus they allow to control their effect with sufficient reliability [6]. A whole series of microorganisms, plants and animals was found to be bioindicators [7–9].

It is known that the ability of different plants to absorb HM from contaminated soils is different, because root systems of different species. In some species the coming of HM in aboveground organs is limited [10].

In some cases lead, chromium, mercury, absorbed by roots, are exposed to strong fixation and are not available to further reallocation in the aboveground parts of plant. Cadmium, zinc, copper and nickel are relatively rapidly translocated from soil to aboveground parts of plants and remarkably change physiological processes. When reaching high concentrations in the soil they evoke worsening of agricultural crops growth, which is expressed in chlorosis, necrosis, and plant stop development, or the agricultural crops yield considerably decreases with the display of these characters.

The HM accumulation and allocation in plant organs has distinctly expressed acropetal character (roots>stems>leaves>fruits) indicating the presence of protective mechanism in plants. This prevents the coming of toxins from roots to aboveground organs. This trend is less expressed on the soils with normal HM concentration than on the soils with elevated one.

HM evoke also other unfavorable changes of soil properties, which are expressed in disturbances of soil microorganism's activity, slowing of humification of plant debris, worsening of soil structure, impoverishment of species composition of microorganisms in contaminated soil [11].

Negative effect of HM is based on the activation of ferments, leading to the changes in the different steps of metabolism. The influence of HM at sub-cellular level is expressed in the disturbance of cell membranes functions and ion transport, as well as in the destruction of mitochondria and chloroplasts [12].

Phytotoxic effect of HM alters in dependence on soil state. There, where the conditions contributing to metals transformation in the mobile state occur, the harmful effect of HM on plants is stronger than on the soils with high sorption capacity relatively to these metals. The analysis of actual state of agricultural production in natural zone of Northern Caucasus, of the basic agrochemical soil properties variation dynamics, shows a stable trend of soil fertility worsening, increasing of degradation processes, worsening of general ecological situation, which conditions the decrease of production and economic indicators, increase of amount of unprofitable branches and companies in the regional agroindustrial complex [4, 13].

Under conditions of ecological stress in North Caucasian region the interest to the organic nitrogen increases, in particular to legumes and their capacity to decrease soil toxicity and restore the fertility because of their unique ability to accumulate organic matter, to absorb heavy metals, to improve soil structure [14].

The application in our research of different legume herbs (*Medicago hybridum, Onobrychis viciifolia* or *O. sativa, Coronilla varia, Melilotus officinalis* Desr. or *M. yellow*) in the combination with new sorbent, namely zeolite containing clay, allowed to detect a series of new regularities concerning the decrease of HM in soil and plants and restoration of soil fertility. The unique ability of legume herbs to absorb heavy metals and radioactive nuclides from soil determined the main aim of our study: to decrease soil toxicity by detecting maximum accumulation of toxicants and their disutility in dependence on the phases of crops development and their position in the crop rotation system.

## 8.2 MATERIAL AND METHODODOLOGY

The experiments were performed at the leached chernozem of Prigorodny district of the Republic North Ossetia-Alania with humus content of 5–6.3% and the following concentrations of nutrients: 0.4% of nitrogen, 0.2–0.3% of phosphorus and 1.62–1.9% of potassium. The reaction of soil (pH) was 5.48–6.92. Top-soil is represented by leached chernozem on the bedrock of pebbles at the depth of 25–80 cm. Initially the research was performed on five cultivated cultivars of red clover (Vladikavkazsky, Daryal, Nart, Farn, Alan), and three wild species of natural mountainous phytocenoses. Corn, mangel-wurzel, oats and clover were precursors in crop rotation (clover was studied as monoculture). The experiments on the toxicity decreasing included four variants: 1 – control (without sideration), 2 – plowing of green mass in the phase of flourishing, 3 – plowing of green mass with natural mud alanit five days after green mass harvesting, 4 – plowing of green mass 7–10 days after hay-crop. Soil humidity, temperature regime and HM concentration were studied in the experiments.

The concentration of heavy metals (copper, zinc, cobalt, manganese, iron, lead, cadmium) was determined in ashes solutions extracted by hydrochloric acid (1:3) by means of atomic absorptive spectroscopy according to the procedure of [12].

Alanit is zeolite containing clay of local origin (steppe zone of Mozdok district of the Republic North Ossetia-Alania) containing 52.7 of silicon ($Si0_2$); 16.6% of aluminum ($Al_2O_3$); 32.7% of calcium (CaO); 6.17% of iron ($Fe_2O_3$); manganese, sulfur, phosphorus, potassium, copper, zinc (within the range of 0.1–0.9%) and other microelements.

## 8.3 RESULTS AND DISCUSSION

The study showed that Vladikavkazsky clover cultivar accumulates mobile cadmium forms after precursor culture of mangel-wurzel more than other cultivars (0.188 mg/kg). The minimum amount was found at the plot with annual clover sawing at the same place (monoculture – 0.122 mg/kg). At the second year of plant life the amount of mobile form of lead increase in Daryal and Vladikavkazsky cultivars up to 2.090 и 1.490 mg/kg, respectively which exceeds maximum permissible concentration (MPC) 6.97 and 4.97 times, respectively. Lower concentrations of this element were found in Nart cultivar in the first (up to 0.215 mg/kg) and second (up to 0.260 mg/kg) years of life. After precursor cultures of mangel-wurzel and corn lead mobility for two years of life of Vladikavkazsky clover cultivar exceeded MPC 1.28–3.34 times and 2.75–4.13 times, respectively. In Farn and Alan cultivars mobile lead exceeded MPC 2.7–4.07 and 1.88–2.08 times, respectively (Table 8.1).

HM accumulation by development phases. *Astragalus* species widely spread in Northern Caucasus, also have high absorptive properties. Recently their bioindicative abilities and at the same time protective

**TABLE 8.1** Content of Mobile Forms of Heavy Metals (mg/kg) in the Soil (0–20 cm) After Different Precursors of Red Clover in the 2nd Year of Life

| Clover cultivar | Precursor | Cd | Pb | Zn | Co | Cu | Mn | Fe |
|---|---|---|---|---|---|---|---|---|
| Daryal | *Trifolium* monoculture | 0.185 | 2.090 | 3.860 | 0.598 | 0.440 | 69.10 | 35.00 |
| Vladikavkazsky | *Trifolium* monoculture | 0.149 | 1.490 | 6.395 | 0.535 | 0.415 | 94.40 | 19.10 |
| Vladikavkazsky | Mangel culture | 0.234 | 1.002 | 3.955 | 0.425 | 0.170 | 93.25 | 1.55 |
| Vladikavkazsky | Corn | 0.228 | 1.240 | 4.870 | 0.475 | 0.415 | 98.50 | 12.90 |
| Farn | *Trifolium* monoculture | 0.145 | 1.220 | 4.564 | 0.386 | 0.298 | 80.15 | 10.50 |
| Nart | *Trifolium* monoculture | 0.178 | 0.260 | 2.791 | 0.370 | 0.256 | 80.90 | 8.96 |
| Alan | *Trifolium* monoculture | 0.183 | 0.625 | 3.843 | 0.395 | 0.342 | 97.25 | 5.98 |
| HCP | — | 0.001 | 0.009 | 0.10 | 0.009 | 0.005 | 0.75 | 0.1 |
| Experimental error S % | — | 0.65 | 1.02 | 2.54 | 2.43 | 2.63 | 1.43 | 1.48 |

functions related with the selenium accumulation in tissues were detected [6, 7].

Table 8.2 shows that aboveground biomass of *Coronilla varia* absorbed the highest amount of copper – from 74.0 to 92.0 mg/kg; *Melilotus officinalis* accumulated the highest amount of copper – up to 96.2 mg/kg; according to iron content *Galega orientalis* showed the highest accumulation capacity (up to 1764.0 mg/kg).

The obtained data shows that the zinc concentration 2.17 times exceeded maximum permissible concentration in sainfoin, 2.18 times in *Medicago sativa* 2.74 times in *Coronilla varia*, 1.63 times in *Galega*, 1.43 times in *Asparagus*.

For copper MPC was exceeded 6 times in *Melilotus officinalis*, 7.38 times in lucerne, 5.38 times in *Coronilla varia.*

For iron MPC in green mass was exceeded 7.5 times in *Melilotus officinalis*, 6.95 times in lucerne, 6.18 times in *Coronilla varia*, 7.05 times in *Galega orientalis*, 10.14 times in *Astragalus* (Table 8.2).

The obtained data from Table 8.2 shows that zinc content exceeded maximum permissible concentration 2.17 times in *Melilotus officinalis*,

**TABLE 8.2** Heavy Metal Content (mg/kg) in Green Mass of Legume Herbs Under Study

| Culture | Zn | Cu | Fe |
|---|---|---|---|
| *Astragalus galegiformis* Lam., introducent | 74.8–81.8 | 9.8–16.6 | 58.4–93.6 |
| Coronilla varia | 61.6–71.6 | 74.0–92.0 | 724.0–1546.0 |
| *Galega orientalis* Lam. | 39.6–42.6 | 68.0–70.0 | 1718.0–1764.0 |
| *Medicago sativa* L. | 53.0–57.0 | 74.0–96.2 | 1638.0–1736.0 |
| Onobrychis arenaria | 53.0–56.8 | 60.0–78.0 | 1550.0–1762.0 |

2.18 times in lucerne, 2.74 times in *Coronilla varia,* 1.63 times in *Galega,* 1.43 times in *Astragalus.*

Our earlier study detected high HM absorptive capacity in *Amaranthus cruentus* L., which contains rather high amount of silicon (50 kg/t), phosphorus (164 kg/t), potassium (156 kg/t), calcium (58 kg/t), magnesium (77 kg/t), as well as many other microelements [10].

Taking into account biological peculiarity of plants absorbing toxicants, we developed the method of culture placement on toxic soils. The cultures of *Coronilla varia* and *Amaranthus* were sown by isolated strips at the width of sowing-machine sweep (machine for cereals and herbs of type 3.6 A). Such way of cultures distribution, when sowing on the green fertilizer, is explained by their low competitiveness. Favanoides, alkaloids and a series of acids, which are contained in *Amaranthus* suppress nitrogen fixing capacity of legume culture *Coronilla varia.* Taking into account biological peculiarities of each species in mixture, *Coronilla varia* as more cold resistant was sown in early spring. 2–3 weeks later *Amarа́nthus* was sown in the free strips. Sowing rates of *Coronilla varia* and *Amaranthus* were 15–20 and 0.5–1 kg/ha, respectively [10]. *Amaranthus* plants accumulate considerable amount of macro- and microelements (vanadium, manganese, molybdenum, cobalt, copper and other elements). Combined sowing of *Coronilla varia* and *Amaranthus* and their plowing as green fertilizers provide decrease of soil toxicity, increase of organic matter because.

When studying perennial herbs as green manure crops it was found that the soil content of mobile forms of zinc, copper, nickel, cobalt, manganese and iron increases from the phase of stooling to the phase of flowering of the plants under study (Table 8.3).

Thus, zinc concentration in the soil with *Coronilla varia* L. changed according to development phases from 33.8 to 42.8 mg/kg; copper from 12.3 to 15.74 mg/kg; nickel from 13.0 to 15.4 mg/kg; cobalt from 9.6 to 11.2 mg/kg, manganese from 550 to 700 mg/kg; iron from 330 to 440 mg/kg (Table 8.3)

When comparing absorbing properties of legume herbs (*Coronilla varia, Onobrýchis, Medicago, Melilotus*), it is possible to conclude that *Coronilla varia* variegated accumulates the highest amount of microelements, and the application of this culture will contribute to the improvement of topsoil, restoration of its fertility and physical state.

Because of selective ability of alanit under its contact with mowed green mass of *Coronilla varia* cleaning of the biomass from lead, cadmium, copper and zinc occurs. During 5 days the process is accompanied by sclarification and decrease of the metal particles size, which accelerates the oxidation of toxicants. This method is especially effective for acid soils with pH below 5.

After plowing of the total *Coronilla varia* green mass with alanit the soil enrichment with organic and mineral substances occurs. Alanit protects mowed green mass from evaporation absorbing the moisture. Alanit layer keeps nitrogen compounds secreted by *Coronilla varia.* The results of the experiment are reported in the Table 8.4.

Table 8.4 data show that after plowing and partial mineralisation of organic mass in the soil in the optimal variant the amount of heavy metals dramatically decreased down to maximum permissible concentrations, which indicates the effectiveness of alanit application when plowing of *Coronilla varia* as green fertilizer with zeolite containing alanit.

**TABLE 8.3** Accumulation of Heavy Metals (mg/kg) in the Soil by Development Phases of *Coronilla varia* L.

| Development phase | Zn | Cu | Ni | Co | Mn | Fe |
|---|---|---|---|---|---|---|
| Stooling | 33.8 | 12.3 | 13.0 | 9.6 | 550 | 3300 |
| Budding | 44.7 | 20.43 | 16.3 | 11.2 | 680 | 4000 |
| Flowering | 42,8 | 15.47 | 15.4 | 11.2 | 700 | 4400 |
| MPC | 100 | 3 | 2 | 5 | 500 | 100 |

*Note*. MPC – Maximum prevent the concentration.

**TABLE 8.4** Change of Heavy Metals Content (mg/kg) in the Soil in Dependence on the Plowing of *Coronilla varia* Green Mass

| Soil processing | Zn | Cu | Fe | Co | Pb |
|---|---|---|---|---|---|
| Ploughing of *Coronilla varia* mass (without alanit) – control | 95 | 84 | 624 | 0.41 | 5.8 |
| Ploughing of *Coronilla varia* mass with alanit | 83 | 52 | 460 | 0.32 | 5.0 |
| Ploughing of *Coronilla varia* with alanit 3–5 days after mowing (soil sampling 2 months after plowing) | 67 | 31 | 320 | 0.28 | 4.5 |
| Ploughing of *Coronilla varia* mass with alanit 7–10 days after mowing (soil sampling 2 months after plowing) – optimal variant | 65 | 20 | 135 | 0.22 | 3.8 |
| MPC | 100 | 18.8 | 154 | 0.26 | 5.0 |

Change of soil solution pH was detected when plowing without or with alanit and several days after mowing. In all variants with alanit application pH increased in comparison with variant without using alanit (4.84). Maximum value (5.10) and at the same time the best result was obtained when plowing *Coronilla varia* mass with alanit 5 days after mowing (see Table 8.4)

In the experiments with plowing of *Coronilla varia* mass with and without alanit (2–3 t/ha) the dynamics of copper and iron content was followed. Thus, if in the variant "Ploughing *Coronilla varia* + alanit" the copper content was in average 52 mg/kg of dry substance, in the variant "Ploughing *Coronilla varia* + alanit (3 days later, soil sampling 2 months after plowing)" this value decreased to 42 mg/kg, in the variant "Ploughing *Coronilla varia* + alanit (5 days later, soil sampling 2 months after plowing)" – to 31 mg/kg, and when plowing *Coronilla varia* mass with alanit 7 later (soil sampling 2 months after plowing) this value amounted 20 mg/kg of dry substance only.

During the vegetation of sand sainfoin (*Onobrychis* Scop.) the content of HM mobile forms in soil changed (in mg/kg) as follows: zinc from 36.7 to 38.8; copper from 13.7 to 15.9; nickel from 14.6 to 16.3; cobalt from 10 to 8; manganese from 670 to 700; iron from 390 to 410.

Inlucerne *Medicago sativa* L. soil HM mobile forms content also increases from stooling to blooming phase: zinc from 40.1 to 43.2 mg/kg; copper from 12.8 to 15.9 mg/kg; nickel from 13.5 to 16.2 mg/kg; cobalt from 10 to 11.7 mg/kg; manganese from 580 to 710 mg/kg, iron from 360 to 440 mg/kg [5].

It is necessary also to notice maximum concentration of zinc and copper mobile forms in soil was accumulated by the budding phase of legume herbs under study. Thus, *Coronilla varia* takes from soil up to 44.7 mg/kg of zinc and up to 20.43 mg/kg of copper; *Onobrychis viciifolia* absorbs up to 54.0 mg/kg of zinc and up to 31.5 mg/kg of copper; *Medicago sativa* absorbs – up to 99.2 and 32.4 mg/kg of zinc and copper, respectively (see Table 8.2).

The comparison of *Coronilla varia* with other legume cultures showed its advantage in a series of characters. Aboveground mass of *Coronilla varia* in the year of sowing develops slowly. The next year its root system is able to accumulate organic matter and biological nitrogen up to 200 kg/ha.

When sown individually, seeds of annual culture of *Amaranthus* accumulates considerable amount of macro- and microelements (vanadium, manganese, molybdenum, cobalt, copper and other elements). The mowing of aboveground mass was performed in the year of sowing in the phase of milky-wax ripeness in order to part of ripened *Amaranthus* seeds remains in the soil and germs the next year together with neighboring strips of *Coronilla varia*, because at this phase *Amaranthus* shed its ripened seeds and they fall into the soil. When falling into the soil, *Amaranthus* seeds together with aboveground mass accomplish the function of absorbing substances. Under their contact with contaminated soil chemical reactions neutralizing heavy metals and radionucleotides occur.

By the moment of plowing of green mass (grown *Coronilla varia* and *Amaranthus* germed from seeds) an amount of organic substances sufficient to decrease soil toxicity is already accumulated. As it is shown in the Table 8.5, combined sowing of *Amaranthus* and *Coronilla varia* by alternated strips provides the decrease of HM concentration in the soil.

The reported data of Table 8.5 show that lead content in the mixed crop of *Amaranthus* and *Coronilla varia* amounts 30.6 mg/kg. At the plot, where alternated strips sowed these cultures, lead content decreased to 26.4 mg/kg. The advantage of alternated sowing is evident also in other elements (nickel, copper, zinc).

**TABLE 8.5** Soil Heavy Metals Content (mg/kg) in Dependence on the Way of Sowing of Cultures for Green Fertilizer

| Culture | Soil nitrogen content, kg/ha | Ni | Pb | Cu | Zn |
|---|---|---|---|---|---|
| Combined sowing of *Amaranthus* + *Coronilla varia* | 123 | 26.4 | 30.6 | 4.2 | 28.0 |
| *Amaranthus* | 148 | 19.8 | 28.4 | 3.0 | 24.0 |
| *Coronilla varia* | 162 | 15.4 | 32.0 | 3.8 | 32.0 |
| *Amaranthus* + *Coronilla varia*, sown by separate strips | 206 | 13.2 | 22.4 | 2.2 | 23.2 |
| MPC | – | 20.0 | 32.0 | 6.8 | 35.0 |

It was detected that *Coronilla varia* accumulates in the second year of life about 12 t/ha of aboveground mass and stubbly remains, whereas in *Melilotus*, *Medicago* and *Onobrychis* this value does not exceed 10 t/ha.

## 8.4 CONCLUSIONS

1. Heavy metals accumulation in legume herbs is going intensively from stooling to blooming phase.
2. The level of heavy metals accumulation in legume herbs differs not only among species but also among cultivars. Among five cultivars of red clover Vladikavkazsky cultivar accumulates maximum amount of heavy metals.
3. Maximum amount of heavy metals in legumes is accumulated after precursor cultures of corn and mangel-wurzel.
4. Among legume herbs the maximum amount of heavy metals is absorbed by *Coronilla varia*, what provides the improvement of topsoil, restoration of its fertility and physical state.
5. Ploughing of legume herbs for green fertilizer in the mixture with alanit considerably (down to the level of maximum permissible concentration) decrease the amount of heavy metals.
6. The mixture of *Coronilla varia* and *Amaranthus* sown by separate strips provides positive results in soil fertility restoration.

## LIST OF SPECIES

*Astragalus galegiformis* Lam. – Astragalus galegeae
*Coronilla varia* – Coronilla colorful (colorful)
*Galega orientalis* Lam. – Eastern galega
*Medicago hybridum* – Lucerne hybrid
*Medicago sativa* L. – Alfalfa
*Melilotus officinalis* – melilot
*Onobrychis viciifolia* – Onobrychis viciifolia
*Trifolium repense* – White clover

## KEYWORDS


- **alanit**
- ***Astragalus***
- **chloroses**
- **clover**
- ***Coronilla varia***
- **heavy metals**
- **legume herbs**
- ***Medicago hybridum***
- ***Melilotus officinalis***
- **necroses**
- ***Onobrychis viciifolia***
- **plowing red**
- ***Trifolium***


## REFERENCES


1. Alborov, I. D., Zaalishvili, V. B., Tedeyeva, F. G., Popadeykin, V. V., Kasyanenko, A. A., Torbek, V. E. (2013). Ecological risk, principles of the evaluation of natural environment and population health. Vladikavkaz. Vladikavkaz Scientific Center of RAS. 307 p. (In Russian).

2. Topalova, O. V., Pimneva, L. A. (2016). Chemistry of the environment. St-Petersburg – Moscow – Krasnodar. Kubansky State Agrarian University. 160 p. (In Russian).
3. Mazhaysky, Yu.M., Yevtyukhin, V. F. (2014). Basis of agrochemical and agrobiological methods of detoxication of soils contaminated by heavy metals. Agroecology. 1, 50–55.
4. Sokayev, K. E., Khubayeva, G. P. (2014). Ecology of natural environment of Vladikavkaz city and its suburbia. Vladikavkaz. Olymp., 207 p. (In Russian).
5. Biological control of environment/Genetic monitoring. Moscow. Academia Publishing Center. 2010, 136 p. (In Russian).
6. Biological control of environment. Bioindication and biotesting. (2010). [Eds. O. P. Melekhova, E. I. Sorapultsev] Moscow. Academia Publishing Center., 156 p. (In Russian).
7. Zaalishvili, V. B., Osikina, R. V. Patent RF № 2375869. "Invention Method of the evaluation of ecological state of territory". Published 20.12.2009. (In Russian).
8. Velts, N.Yu. Patent RF № 2257597 "Invention Method of the Evaluation Pollution of the Environment by Heavy Metals". Published 27.07.2005 (In Russian).
9. Zaalishvili, V. B., Bekuzarova, S. A., Kozayeva, O. P. Patent RF № 2485477. "Method of the evaluation of technological environment pollution". Published 20.06.2013 (In Russian).
10. Bekuzarova, S. A., Kuznetsov, I. Yu., Gasiyev, V. I. (2014). *Amaranthus* is an universal culture. Vladikavkaz. Scientific-methodic publishing house "Colibri.", 180 p. (In Russian).
11. Bekuzarova, S. A., Shabanova, I. A. (2014). Clover as bioindicator of heavy metals. Proceedings of Gorsky State Agrarian University., Vol. 51–3. pp. 56–63. (In Russian).
12. Methodical instructions on the conductance of complex monitoring of agricultural soils fertility (by Central Institute of Agrochemistry) (2003). 76 p. (In Russian).
13. Sokayev, K. E., Bestayev, V. V. Sokayeva, R. M. (2013). Monitoring of heavy metals and radionucleotides in agrobiocenoses of RSO-Alania [K. L. Khegaturov] Vladikavkaz. North Ossetian State University Publishing house, 50 p.
14. Gukalov, V. N. Heavy metals in the system of agrolandscape. (2010). Krasnodar. Kubansky State Agrarian University, 345 p. (In Russian).

## CHAPTER 9

# INFLUENCE OF ACID SOIL STRESSORS ON PHYTOCENOSES FORMATION

TATYANA L. EGOSHINA,[1,2] LYUDMILA N. SHIKHOVA,[1] and EUGENE M. LISITSYN[1,3]

[1]*Vyatka State Agricultural Academy, 133 Oktyabrsky Prospect, Kirov, 610017, Russia*

[2]*All Russian Institute of Game and Fur Farming, 72, Preobrazhenskaya St., Kirov, 610017, Russia*

[3]*North-East Agricultural Research Institute, 166-A Lenin St., Kirov, 610007, Russia, E-mail: shikhova-l@mail.ru*

## CONTENTS



### ABSTRACT

The Kirov region is located in the east of the European part of Russia in zone of Povolzh'e forests. Considerable extent of the region from the north on the south cause essential distinctions is soil-climatic conditions and natural landscapes in its territory that allows carrying out the analysis of interdependence of development of phytocenoses and chemical properties of soil (in particular – level of its acidity). The raised acidity of soil has as a consequence increase of amount of aluminum and heavy metals available to plants. Objects in all natural zones of the region have been chosen for this purpose. Forest phytocenoses were compared with anthropogenic agro-phytocenoses in similar natural and climatic conditions. As a result of the spent researches the floristic structure of natural phytocenoses and segetal plants of agro-phytocenoses in each of natural zones of the region is defined. Studying of similarity degree of phytocenoses by comparison of Jaccard index has allowed revealing only insignificant degree of similarity between investigated phytocenoses among themselves both for each pair of agro-phytocenoses and for each pair of forest phytocenoses. The biodiversity coefficient fluctuated from 0.63 in agro-phytocenosis from a sub-band of middle taiga to 5.18 in forest phytocenosis from a zone of mixed coniferous-broad-leafed forest. On soils with close agrochemical parameters meadow phytocenoses are characterized by the greatest coefficient of biodiversity (2.58–3.73); forest phytocenoses had lower values of the coefficient (1.07–2.18); and agro-phytocenoses – the lowest one (0.63–0.91). Decrease in coefficient of a biodiversity with increase in acidity of soil is noted for forest and meadow phytocenoses. The floristic structure of weed species was richer in agro-phytocenoses on acid soils; annual plants prevailed among them but presence of gramineous plants is essential also. The biodiversity coefficient of agro-phytocenoses increases at soil acidity decrease. Florae of agro-phytocenoses and natural phytocenoses on acid soils have more boreal character rather than florae of similar phytocenoses on less acid soils. Degree of similarity of such phytocenoses depends significantly on soils pH level. Research of interdependence between soils acidity and quality of vegetative raw materials has allowed to reveal presence of statistically significant interrelation between level of soils acidity and the content of arbutin in leaves of a cowberry (*Vaccinium vitis-idaea* L.), cardiac glycosides in "grass" of a lily-of-the-valley (*Convallaria majalis* L.), and insignificant dependences of concentration of polyphenols in Saint-John's-wort's (*Hypericum perforatum* L.) "grass" on soil acidity.

It is interesting to note there opposite orientation: the content of arbutin increases at strengthening of acidity, and concentration of polyphenols and cardiac glycosides decreases.

## 9.1 INTRODUCTION

The Kirov region is located in the east of the European part of Russia in a zone of Povolzh'e forests. Forests occupy here about 67% of territory of the region. Forests are distributed non-uniformly within a territory. In northern districts of region forests occupy 75–85% of the area [1], in the center of region and in its southern areas – only 7–20% [2]. Main forest-forming species in the region are *Picea excelsa* (L.) Karst., *Picea obovata* Ledeb., *Abies sibirica* Ledeb., and *Pinus silvestris* L. Broad-leaved species of forest trees are also grow here – *Betula pendula* Roth and *B. pubescens* Ehrh., *Populus tremula* L., *Quercus robur* L., *Tilia cordata* Mill.

Among forest types most extended are bilberry spruce forest – 12%, bilberry birch forests – 6.1%, and beadruby-bilberry birch forests – 5.2%. An appreciable role play also haircap-moss spruce forests – 4.3%, haircap-moss birch forests – 4.2%, wood sorrel spruce forests – 3.6%, bilberry pine forests – 3.5%, sphagnum pine forests – 3.0%. In northern part of the region spruce forests and spruce-fir forests prevail with ground cover consists with hygrophilous green and sphagnum mosses. In the central districts of the region spruce-fir forests with green mosses and grass-shrubby layer are extended. In southern districts coniferous species of the first layer the oak and a linden are added forming mixed coniferous-broad leaved forests which have underbrush and a grassy cover various on specific structure [3].

Considerable extent of the region from the north on the south causes essential distinctions of environmental conditions and natural landscapes in its territory. Depending on distribution of various types of forests, 3 sub-bands are divided in a vegetative cover of the region: middle taiga, a southern taiga and mixed coniferous-broad leaved forests. Northern districts of the region are in a sub-band of middle taiga; central districts – in a southern taiga; and southern districts – in a strip of mixed coniferous-broad leaved forests. Mother rocks here are characterized by high diversity. The main part of territory of the region is occupied with soils of podzolic type [4].

The regional flora includes 1.116 species of vascular plants. They unite in 77 families. A leading place on number of genus has families Asteraceae (48 species), Poaceae (44 species), Umbelliferae (30 species), Rosaceae and Brassicaceae (24 species each). The largest on number of species are families Asteraceae (118 species), Poaceae (101 species), Rosaceae (78 species), Cyperaceae (61 species), Fabaceae (57 species), and Caryophyllaceae (53 species). More than half of families are presented by one genus; more than third – by one specie. Among these families are Butomaceae, Oxalidaceae, Celastraceae, Adoxaceae, and Polemoniaceae. The greatest specific diversity had genus Carex (48 species – 4.4%), Salix (19 species – 1.7%), Poligonum (18 species – 1.6%), Ranunculus and Alchemilla (16 species – 1.5% each), Hieracium (15 species – 1.4%), Viola (13 species – 1.2%), Potamogeton and Veronica (12 species – 1.1% each), Rumex (11 species – 1.0%), Stellaria, Dianthus, Potentilla, Trifolium, Vicia, and Galium (10 species – 0.9% each), Juncus, Epilobium, Campanula, and Centaurea (9 species – 0.8% each).

Acid podzolic and sod-podzolic soils of the region are characterized by low pH level, low capacity of absorption, fulvatic structure of humus, and washing water mode. One of the most harmful factors of such acid soils is large amount of exchangeable aluminum. Though aluminum on the nuclear weight formally does not concern heavy metals nevertheless reaction of live organisms to its raised concentrations does not differ in the physiological features from reaction to such heavy metals as lead, cadmium and others. In this connection many researchers unite aluminum with other chemical elements in the modular group of elements named "heavy metals" [5–7]. For this reason consideration of influence of various acidity of soils of the Kirov region caused by presence of mobile ions of trivalent aluminum on floristic and a chemical compound of plants natural and agro-phytocenoses was a main objective of the given work.

## 9.2 MATERIALS AND METHODOLOGY

Plant objects in all three natural zones of the region are chosen for solving of the given task. Whenever possible objects were dated for the soils developed on granulometrically heavy mother rocks – blanket clays and loams. Forest phytocenoses were compared with agro-phytocenoses created by human (farmlands) in similar natural and climatic conditions. Objects of studying represent soil profiles with phytocenoses located on a surface adjoining to them.

Samples for chemical analyzes of soil were selected from soil profiles by genetic horizons. Soil samples were placed in labeled polyethylene packages. Then soil was dried up, mechanically grounded and sifted through a 2 mm sieve. Definition of mobile forms of chemical elements spent in 1.0 M acetic-ammonia solution pH 4.8 by a method of atom-absorption spectrophotometry [8, 9].

Studying of vegetation of trial platforms was spent with use of the adapted techniques of field and laboratory researches developed on the basis of standard geobotanical and resource-studying methods [10, 11] and methods of inspection of rare species of plants [12].

Inspection of vegetative communities was accompanied by full geobotanical description of phytocenosis with the detailed characteristic of specific structure, holistic and merological traits (projective covering, density of crones, an abundance, height, vitality, age, degree of quality of wood plants, phonological phases and some others). Method [13] was used at definition of forest mensuration characteristics.

The complete definition of physics-geographical conditions (level of humidifying, an exposition, a lay of land, characteristic of a dead cover, phytocenosis environment etc.) allows to receive detailed ecological-cenotic characteristic of vegetation at data processing.

Phenological observation over development and condition of plants were spent by a technique [14].

Similarity degree of phytocenoses was estimated with use of Jaccard index [15], the biodiversity coefficient – by Shannon-Wiener Diversity Index [16].

## 9.3 RESULTS AND DISCUSSION

### *9.3.1 ZONE OF MIXED CONIFEROUS-BROAD LEAVED FORESTS*

Forest phytocenosis is presented by fir-linden-spruce mixed herbs forest with an elm and aspen addition. Soil type is gray forest gleyey soil on blanket loess-like clays. Following horizons are noted in a profile: $O_i$ 0–5 cm; $A_h$ 5–19 cm; $A_{he}$ 19–33 cm; AB 33–57 cm; B 57–80 cm; BC 80–90 cm.

#### 9.3.1.1 Agricultural Field Site

Soil type is dark gray forest soil on blanket loess-like loam. Following horizons are noted in a profile: $A_p$ 0–30 cm; $A_{he}$ 30–39 cm; AB 39–60 cm; B 60–96 cm; BC 96–141 cm; C 141–160 cm.

### *9.3.2 SUB-BAND OF A SOUTHERN TAIGA*

Forest phytocenosis is presented by spruce-pine grassy forest with a birch addition. Soil is podzolic loamy, gleyey, on blanket loamy deposits. Next horizons are allocated in a profile: $O_e$ 0–7 cm; $A_h$ 7–20 cm; $A_{he}$ 20–30 cm; AB 30–70 cm.

#### 9.3.2.1 Agricultural Field Site

Soil is sod-light-podzolic, loamy, on blanket loams. A profile structure: $A_p$ 0–22 cm; $A_{he}$ 22–32 cm; AB 32–55 cm; B 55–100 cm; BC 100–140 cm.

### *9.3.3 NORTHERN PART OF A SOUTH TAIGA SUB-BAND*

Forest type is spruce-fir with a birch addition, bilberry-grassy. Soil is strongly podzolic, loamy, on carbonate blanket clay. A soil profile structure: $O_i$ 0–7 cm; $A_h$ 7–30 cm; AB 30–51 cm; B 51–86 cm; BC 86–120 cm.

Agro-phytocenosis on an arable land is presented by wheat which projective covering makes only 30%. Soil is sod-strongly-podzolic, loamy, on blanket loams. Soil profile structure: $A_p$ 0–28 cm; $A_{he}$ 28–48 cm; AB 48–70 cm; B 70–96 cm; BC 96–120 cm; C 120–130 cm.

### *9.3.4 SUB-BAND OF MIDDLE TAIGA*

Forest phytocenosis is presented by a fern-grassy fir-grove. The soil is presented by heavy loamy podzol on blanket loamy sediments. A profile structure: $O_e$ 0–7 cm; $A_h$ 7–50 cm; AB 50–66 cm; B 66–104 cm; BC 104–125 cm; C 125–140 cm.

Agro-phytocenosis is located on sod-strongly-podzolic loamy soil formed on blanket loams. A profile structure: $A_p$ 0–22 cm; $A_{he}$ 22–54 cm; AB 54–70 cm; B 70–108 cm; BC 108–125 cm; C 125–136 cm.

Considerable extent of the Kirov region in meridional direction allows to track gradual change of biogeocenoses from the south on the north.

Small part of territory in the south of the region is located in a *zone of mixed coniferous-broad leaved forests*. Here there are the most fertile soils of the region – gray forest soil. As they are formed on rocks rich in the bases their acidity is insignificant. A profile of forest soil is gleyed therefore reaction of a soil solution is close to the neutral on all profile. High humus content and depth of its penetration in a profile is explained by over-watering.

Forest phytocenoses on gray soils are presented by spruce-linden-fir forests with an elm and aspen addition. Rather thin undergrowth is presented by fir, linden, spruce, an elm. Norway maple having average height about 12 m, density of crones about 0.1 is noted in underbrush. The grassy cover is combined by 22 plant species. Dog's-mercury (*Mercurialis perennis* L.), male shield fern (*Dryopteris filix-mas* (L.) Schott), ashweed (*Aegopodium podagraria* L.), lungwort (*Pulmonaria obscura* Dumort.), wolfsbane monkshood (*Aconitum excelsum* Reichenb.), and bottlebrush (*Equisetum sylvaticum* L.) prevail among them. Participation of other plant species in composition of a vegetative cover is less considerably. The projective covering of each of them does not exceed 1%. These are following plant species: sweet woodruff (*Galium odoratum* (L.) Scop.), saxifrage (*Chrysosplenium alternifolium* L.), touch-me-not (*Impatiens noli-tangere* L.), corydalis (*Corydalis hallerii* (Willd.) Willd.), Alpine circaea (*Circaea alpine* L.), bladder-fern (*Cystopteris sudetica* (A. Brown & Milde) A.P. Khokhr.), meadow geranium (*Geranium pretense* L.), common nettle (*Urtica dioica* L.), easter-bell (*Stellaria holostea* L.), poisonberry (*Actaea rubra* (Ait.) Willd.), bitter peavine (*Lathyrus vernus* (L.) Bernh.), rabbit-meat (*Lamium purpureum* L.), black-headed hawkweed (*Hieracium nigrescens* Willd.), milkwort (*Polygala vulgaris* L.), hairlike sedge (*Carex capillaries* L.), and Filzige Segge (*Carex tomentosa* L.). Nemoral species prevail in phytocenoses structure with an addition of East Siberian and Siberian plant species.

Chemical properties of the studied soil horizons are presented in Table 9.1.

**TABLE 9.1** Chemical Properties of Investigated Soils

| Horizon | $pH_{KCl}$ | Total carbon, % | Humus, % | Mg-equivalent/100 of soil | | | | | |
|---|---|---|---|---|---|---|---|---|---|
| | | | | $Ca^{++}$ | $Mg^{++}$ | $Ca^{++} + Mg^{++}$ | $Al^{+++}$ | $H^{+}$ | Hydrolytic acidity |
| **Sub-band of middle taiga**, arable land, sod-strongly-podzolic loamy soil. | | | | | | | | | |
| $A_p$ | 6.45 | 1.35 | 2.33 | 9.4 | 5.0 | 14.4 | – | – | 0.99 |
| $A_{he}$ | 5.13 | 0.43 | 0.75 | 4.4 | 2.2 | 6.6 | 0.04 | 0 | 1.90 |
| AB | 4.42 | 0.21 | 0.36 | 8.8 | 4.6 | 13.4 | 0.26 | 0.07 | 2.86 |
| B | 4.23 | 0.23 | 0.40 | 13.6 | 7.4 | 21.0 | 0.71 | 0.10 | 4.32 |
| BC | 4.28 | 0.14 | 0.25 | 13.6 | 5.2 | 18.8 | 0.39 | 0.10 | 3.19 |
| C | 4.33 | 0.12 | 0.21 | 12.0 | 7.0 | 19.0 | 0.32 | 0.10 | 2.68 |
| **Sub-band of middle taiga**, forest, heavy loamy podzol soil | | | | | | | | | |
| $O_e$ | 3.74 | – | – | 5.8 | 4.2 | 10.0 | 4.28 | 0.48 | 36.7 |
| $A_h$ | 4.18 | 0.29 | 0.50 | 3.0 | 1.2 | 4.2 | 2.99 | 0.10 | 5.37 |
| AB | 3.82 | 0.13 | 0.23 | 6.8 | 5.2 | 12.0 | 3.32 | 0.13 | 6.25 |
| B | 3.99 | 0.24 | 0.41 | 15.4 | 6.4 | 21.8 | 0.98 | 0.16 | 4.71 |
| BC | 4.25 | 0.22 | 0.32 | 17.2 | 8.0 | 25.2 | 0.23 | 0.16 | 3.13 |
| C | 4.28 | 0.21 | 0.37 | 16.0 | 10.0 | 26.0 | 0.13 | 0.13 | 3.19 |
| **Northern part of a South taiga sub-band**, arable land, strongly podzolic loamy soil. | | | | | | | | | |
| $A_p$ | 5.87 | 1.28 | 2.21 | 11.9 | 1.6 | 13.5 | 0. | 0.07 | 1.10 |
| $A_{he}$ | 4.22 | 0.37 | 0.64 | 5.1 | 1.0 | 6.1 | 1.90 | 1.94 | 4.05 |
| AB | 4.03 | 0.26 | 0.44 | 13.7 | 4.6 | 18.3 | 1.76 | 2.05 | 4.32 |

| | | | | | | | | | |
|---|---|---|---|---|---|---|---|---|---|
| B | 4.10 | 0.25 | 0.43 | 18.0 | 5.9 | 23.9 | 0.91 | 1.12 | 3.79 |
| BC | 4.21 | 0.21 | 0.36 | 18.0 | 7.5 | 25.5 | 0.42 | 0.52 | 2.68 |
| C | 4.34 | 0.22 | 0.38 | 15.2 | 6.1 | 21.3 | 0.21 | 0.28 | 2.31 |
| **Northern part of a South taiga sub-band**, forest, strongly podzolic loamy soil | | | | | | | | | |
| $O_i$ | 4.33 | – | – | – | – | – | – | – | 23.9 |
| $A_h$ | 3.91 | 0.61 | 1.05 | 1.4 | 0.4 | 1.8 | 4.98 | 5.05 | 7.76 |
| AB | 4.09 | 0.37 | 0.64 | 16.0 | 4.6 | 20.6 | 1.49 | 1.77 | 4.42 |
| B | 5.18 | 0.25 | 0.43 | 21.3 | 5.2 | 26.5 | сл. | 0.07 | 1.46 |
| C | 7.07 | 0.18 | 0.31 | – | – | – | 0 | 0 | 0.34 |
| **Sub-band of a southern taiga**, arable land, sod-light-podzolic loamy soil | | | | | | | | | |
| $A_p$ | 4.47 | 1.00 | 1.73 | 7.4 | 3.3 | 10.7 | 0.48 | 0.51 | 3.48 |
| $A_{he}$ | 4.06 | 0.51 | 0.88 | 9.1 | 3.3 | 12.4 | 3.43 | 3.54 | 5.37 |
| AB | 4.01 | 0.33 | 0.34 | 12.5 | 6.9 | 19.4 | 3.34 | 3.61 | 5.25 |
| B | 4.04 | 2.67 | 0.46 | 16.4 | 8.4 | 24.8 | 1.93 | 2.21 | 4.32 |
| C | 4.33 | 0.18 | 0.31 | 16.2 | 6.0 | 22.2 | 0.28 | 0.34 | 1.64 |
| **Sub-band of a southern taiga**, forest, podzolic loamy gleyey soil | | | | | | | | | |
| $O_e$ | 3.95 | – | – | – | – | – | – | – | 28.7 |
| $A_p$ | 3.87 | 1.55 | 2.68 | 4.1 | 1.2 | 5.3 | 5.82 | 5.90 | 11.0 |
| $A_{he}$ | 4.02 | 0.85 | 1.47 | 4.73 | 2.88 | 7.61 | 4.37 | 4.42 | 7.28 |
| AB | 4.34 | 0.26 | 0.45 | 20.2 | 9.7 | 29.9 | 0.53 | 0.53 | 2.99 |

**TABLE 9.1** (Continued)

| Horizon | $pH_{KCl}$ | Total carbon, % | Humus, % | Mg-equivalent/100 of soil | | | | | |
|---|---|---|---|---|---|---|---|---|---|
| | | | | $Ca^{++}$ | $Mg^{++}$ | $Ca^{++} + Mg^{++}$ | $Al^{+++}$ | $H^{+}$ | Hydrolytic acidity |
| **Zone of mixed coniferous-broad leaved forests**, arable land, dark gray forest loam soil | | | | | | | | | |
| $A_p$ | 5.58 | 2.01 | 3.47 | 23.0 | 2.6 | 25.6 | 0 | 0.06 | 1.53 |
| $A_{he}$ | 5.40 | 1.78 | 3.08 | 20.2 | 3.0 | 23.2 | 0 | 0.07 | 3.05 |
| AB | 4.09 | 0.36 | 0.62 | 20.2 | 3.0 | 23.2 | 0.26 | 0.33 | 3.79 |
| B | 4.61 | 0.30 | 0.51 | 25.2 | 3.2 | 28.4 | 0.03 | 0.13 | 2.57 |
| BC | 5.93 | 0.22 | 0.38 | 23.4 | 6.6 | 30.0 | 0 | 0 | 1.08 |
| C | 7.00 | 0.12 | 0.21 | – | – | – | 0 | 0 | 0.13 |
| **Zone of mixed coniferous-broad leaved forests**, forest, gray forest gleyey soil | | | | | | | | | |
| $O_i$ | – | – | – | – | – | – | – | – | 0.44 |
| $A_h$ | 5.73 | – | – | 100.0 | 20.0 | 120.0 | 0.07 | 0.16 | 17.5 |
| $A_{he}$ | 5.78 | 4.04 | 6.98 | 38.8 | 10.0 | 48.8 | 0 | 0.10 | 2.80 |

The arable soil has about 3.5% of humus in arable horizon. Acidity of soil solution reaches the greatest value in horizon AB – pH 4.09. Reaction of solution of humus horizons is lightly acidic – pH 5.40–5.58. Soils have a considerable part of alkaline-earth cations in a soil-absorbing complex. Agro-phytocenosis on gray forest soils has been presented by a corn field where corn domination is caused by anthropogenous intervention. However, the structure of weed species reflects common features of agrophytocenoses of the given zone. Participation of mentioned segetal species in composition of phytocenoses is insignificant. So, the projective covering of a perennial sow thistle (*Sonchus arvensis* L.) and field bindweed (*Convolvulus arvensis* L.) made only 2%; and Canadian thistle (*Cirsium arvense* (L.) Scop.), fumitory (*Fumaria officinalis* L.), garden orache (*Atriplex hortensis* L.), and hedge nettle (*Stachys palustris* L.) is meet only individual. These are species of boreal, steppe, and ancient Mediterranean origins.

As a whole, the floristic structure of phytocenoses of a zone of mixed coniferous-broad leaved forests differs by abundance of nemoral species.

### 9.3.5 SUB-BAND OF A SOUTHERN TAIGA

A basis of biocenoses of the given zone is podzolic soils. They possess high acidity ($pH_{kcl}$ 3.9–4.5) out to high concentration of absorbed aluminum. Degree of expressiveness and capacity of eluvia horizons are considerable. Degree of saturation with bases is not great. The content of humus in arable horizon does not exceed 2%. A profile of forest soil is gleyey therefore the top part of a profile is enriched by organic matter.

Forest phytocenoses developed on the given soils are usually presented by spruce-pine grassy forests with a birch additives; average age – 60–70 years, average height – 25 m, degree of a density of crones – 0.4–0.5.

Individual firs and fur-trees present undergrowth; in underbrush an individual mountain ash (Sorbus) and elder (Sumbucus) are noted. The grass cover is combined by 15 species of plants among which dominate sour trefoil (*Oxalis acetosella* L.), European strawberry (*Fragaria vesca* L.), and hairy wood-rush (*Luzula pilosa* (L.) Willd.). Participation of bird's-eye (*Veronica chamaedrys* L.), killwort (*Chelidonium majus* L.), carpenter's herb (*Ajuga reptans* L.), European wild ginger (*Asarum europaeum* L.), European goldenrod (*Solidago virgaurea* L.), Canadian hawkweed (*Hieracium umbellatum* L.), and spreading millet grass (*Milium effusum* L.) is considerable. Mountain pansy (*Viola montana* L.), panicled bellflower (*Campanula divaricata* Michx.), groundcedar (*Lycopodium complanatum* L.), common St.-John's wort, and sweet woodruff are noted as individual plants. Boreal species prevail among the species composing these phytocenosis; they are supplemented with nemoral sub-oceanic species.

The grassy cover of agro-phytocenosis is made by a red clover (*Trifolium pratense* L.) with insignificant participation of 10 segetal plant species which total projective covering does not exceed 10%. These species are listed in order of decrease in their participation in phytocenosis composition: German camomile (*Tripleurospermum inodorum* (L.) Sch.Bip.), European cinquefoil (*Potentilla goldbachii* Rupr.), narrowleaf hawk's beard (*Crepis tectorum* L.), bee nettle (*Galeopsis speciosa* Mill.), mugwort (*Artemisia vulgaris* L.), field scorpion grass (*Myosotis arvensis* (L.) Hill.), field pansy (*Viola arvensis* Murray), mare's-tail (*Equisetum arvense* L.), low cudweed (*Gnaphalium uliginosum* L.), and perennial sow thistle. Boreal and steppe species prevail in given agro-phytocenosis. As a whole, phytocenoses of sub-bands of a southern taiga have more boreal character in comparison with phytocenoses of Zone of mixed coniferous-broad leaved forests.

### 9.3.6 NORTHERN PART OF A SUB-BAND OF A SOUTHERN TAIGA

Influence of Perm carbonate sediments which are lying down close to a surface and sometimes directly being mother rocks is palpably rather in soil profiles in northeast and east parts of the Kirov region. Degree of expressing of eluvial processes depends on depth of occurrence of the Perm rocks. In spite of the fact that at the bottom part of a profile there are carbonates podzolic process reaches considerable development. The profile has heavy enough podzolic horizon with high acidity ($pH_{KCl}$ 3.91). The absorbing complex is sated by hydrogen and aluminum ions. The content of humus is insignificant.

Forest phytocenoses on the described soils are presented by spruce-fir bilberry-mixed herbs forest with birch addition; middle age of forest is 70 years; average height of trees – 18 m; degree of a density of crones – 0.6. Undergrowth is presented by individual firs and spruces. In underbrush a common juniper (*Juniperus communis* L.), European raspberry (*Rubus idaeus* L.), bush honeysuckle (*Lonicera tatarica* L.), European black currant (*Ribes nigrum* L.) are noted. The grass-subshrub circle is combined by 14 species of plants. Dominants are bilberry (*Vaccinium myrtillus* L.), stone bramble (*Rubus saxatilis* L.), European wild ginger, hairy wood-rush, and oak fern (*Gymnocarpium dryopteris* (L.) Newman). Participation of twinflower (*Linnaea borealis* L.), European strawberry, and herb Paris (*Paris quadrifolia* L.) is considerable. There are individual plants of a cowberry, bottlebrush, sweet woodruff, nodding melick grass (*Melica nutans* L.), wonder violet (*Víola mirabilis* L.), and sour trefoil. The Siberian taiga species prevail in given phytocenoses, but nemoral species are noted as individuals.

The cultivated soil has acid pH on all profile (4.03–4.34) except arable horizon (5.87). An arable land sowed by wheat, the dominating species having a projective covering of 30%, presents agro-phytocenosis. Considerable participation in addition of a grassy cover accepts ruderal plant species, which number in given phytocenosis reaches 17. Co-dominated species are Canadian thistle, perennial sow thistle, field pennycress (*Thlaspi arvense* L.) which projective covering fluctuates from 8 to 10%. It is individual plants of field scorpion grass, German chamomile, wild marigold (*Matricaria suaveolens* (Pursh) Buch.), red clover, bluebottle (*Centaurea cyanus* L.), bee nettle and hemp nettle (*Galeopsis tetrahit* L.), Hawksbeard (*Crepis sibirica* L.), upland cress (*Barbarea vulgaris* R.Br.), garden orache, lesser stitchwort (*Stellaria graminea* L.), tufted vetch (*Vicia cracca* L.), black bindweed (*Polygonum convolvulus* L.), annual meadow grass (*Poa annua* L.). With rare exception the flora of this site is made by boreal species.

### 9.3.7 SUB-BAND OF MIDDLE TAIGA

Podzolic process is expressed most fully and sharply in soils of middle taiga zone. Washing water mode having good drainage along with taiga character of vegetation and climatic conditions leads to development of heavy well-expressed eluvial horizon. Whole profile has strong acidity of soil solution ($pH_{KCl}$ 3.74–4.18). Exchangeable aluminum occupies most proportion of absorbed cations in top horizons at insignificant amount of bases. Most part of organic matter concentrated in forest ground litter. In lower horizons humus content is less than 0.50%

Forest phytocenoses formed in sub-band of middle taiga on loamy podzols are presented by spruce fern-grass forests having average age about 80 years; average tree height is 25 m; degree of density of crones – 0.6. Undergrowth is not thick and is presented by spruce and birch. Underbrush is composed by mountain ash (*Sorbus aucuparia* L.), black alder (*Alnus glutinosa* (L.) Gaertn.), fly honeysuckle (*Lonicera xylosteum* L.), prickly wild rose (*Rosa acicularis* Lindl.), and European raspberry; its average height is about 80 cm. Grass-shrubby layer is composed by 21 plant species. Prickly-toothed fern (*Dryopteris carthusiana* (Vill.) H.P. Fuchs) is dominant species, and ashweed, European wild ginger, sour trefoil, and stone bramble are co-dominant species. Considerable participation in composition of grass-shrubby cover has lesser stitchwort, bifoliate beadruby (*Maianthemum bifolium* (L.) F.W. Schmidt), green-flowered wintergreen (*Pyrola chlorantha* Sw.), European goldenrod, herb Paris, European strawberry, common globeflower (*Trollius europaeus* L.), spreading millet grass, and wood sedge (*Carex sylvatica* Huds.). There are also some plants of wolfs bane monkshood, wood crane (*Geranium sylvaticum* L.), wonder violet, cowberry, bilberry, bottlebrush, and small meadow rue (*Thalictrum simplex* L.). Siberian taiga species prevail in phytocenoses but portion of nemoral species is rather high also.

Arable soil has increased content of humus in plow-layer (up to 2.60%) in compare with eluvial horizon from which it was formed because of systematic input of organic fertilizers and lime. Liming neutralizes soil acidity, and pH of plow-layer increases up to 6.45. Carbonates which move with water flows promotes acidity lowering and considerable decreases content of absorbed aluminum in lower horizons. Thus, forming of a plow-layer and maintenance of its fertility level demand considerable efforts.

Studied agro-phytocenosis is presented by winter rye sowing. There are 14 species of weeds in grass cover, but German chamomile dominates. Co-dominants are bluebottle, wild carrot (*Pastinaca sylvestris* Mill.), and caseweed (*Capsella bursa-pastoris* (L.) Medik.). There are some individual plants of white clover (*Trifolium repens* L.), field pansy, field chickweed (*Cerastium arvense* L.), common plantain (*Plantago major* L.), and devil's grass (*Agropyron repens* (L.) P. Beauv.), hemp nettle, and narrow-leaf hawk's beard. Species of boreal, steppe, and ancient-Mediterranean origin are noted among segetal species of this phytocenosis.

Studied phytocenoses of sub-band of middle taiga are composed by species of more northern origin than phytocenoses of previously described vegetation zones and sub-bands.

Determining of degree of similarity of the phytocenoses with use of Jaccard index for each pair of agro-phytocenoses and forest phytocenoses (Tables 9.2 and 9.3) allows discovering only insignificant degree of similarity.

As it is visible from Tables 9.2 and 9.3 coefficient of biodiversity of studied phytocenoses fluctuated from 0.63 in agro-phytocenosis from a

**TABLE 9.2** Degree of Similarity of Agro-Phytocenoses and Coefficient of Bio-Diversity

| Vegetation zone | 1 | 2 | 3 | 4 | Number of species | Coefficient of bio-diversity |
|---|---|---|---|---|---|---|
| 1 | 1 | 0.06 | 0.08 | 0 | 11 | 0.91 |
| 2 | 0.06 | 1 | 0.14 | 0.16 | 7 | 0.68 |
| 3 | 0.08 | 0 | 1 | 0.11 | 18 | 0.84 |
| 4 | 0.14 | 0.16 | 0.11 | 1 | 14 | 0.63 |

*Note*. Vegetation zone: 1 – Sub-band of a southern taiga; 2 – Zone of mixed coniferous-broad leaved forests; 3 – Northern part of a South taiga sub-band; 4 – Sub-band of middle taiga.

**TABLE 9.3** Degree of Similarity of Forest Phytocenoses and Coefficient of Bio-Diversity

| Vegetation zone | 1 | 2 | 3 | 4 | Number of species | Coefficient of bio-diversity |
|---|---|---|---|---|---|---|
| 1 | 1 | 0.03 | 0.21 | 0.13 | 5 | 1.26 |
| 2 | 0.03 | 1 | 0.12 | 0.06 | 22 | 5.18 |
| 3 | 0.21 | 0.12 | 1 | 0.01 | 14 | 3.26 |
| 4 | 0.13 | 0.06 | 0.01 | 1 | 21 | 2.24 |

*Note*: see Table 9.2.

sub-band of middle taiga to 5.18 in forest phytocenosis from a zone of mixed coniferous-broad leaved forests.

The analysis of 67 geobotanical descriptions of phytocenoses of fields, meadows and forests has allowed to reveal some tendencies in their composition on soils of various acidities.

Comparison of specific structure and the coefficient of a biodiversity calculated on Shannon-Wiener Diversity Index for the investigated vegetative communities has shown that on soils with close agrochemical parameters meadow phytocenoses are characterized with the greatest coefficient of a biodiversity (2.58–3.73); they are followed by forest phytocenoses (1.07–2.18); the lowest coefficient of biodiversity is characteristic for agro-phytocenoses (0.63–0.91).

Decrease in coefficients of a biodiversity with increase in acidity of soil is noted for forest and meadow phytocenoses. This result contradicts the statement [17] about the greatest specific variety of communities of poor soils.

The analysis of contamination of agro-phytocenoses formed by crops of the same species of cultural plants on soils of different acidity has allowed to reveal dependence of floristic structure of segetal plants not only on a type of cover-forming plants but also on level of soils pH. In agro-phytocenoses on acid soils the floristic structure of weed species was richer; among them annual plants prevailed but also presence of cereals is essential. The biodiversity coefficient in agro-phytocenoses increases at acidity decrease.

Appreciable relative density in geographical structure of weed-field flora belongs to species of southern areas (steppe, Mediterranean); at the same time in forest flora they are practically absent. The flora of meadow complexes is intermediate.

Florae of agro-phytocenoses and natural phytocenoses on acid soils have more boreal character than florae of similar phytocenoses on less acid soils. Degree of similarity of the phytocenoses significantly depends on level of soils pH.

It is obviously possible to note distinctions between confinedness of species to soils of various acidity resulted in the literature and their real habitats. So, bee nettle belonged to basiphilous plants are noted on soils with pH 4.47; but mare's-tail traditionally considered as acidophilious forms dense highly productive thickets on soils with pH 5.87. Possibly, this fact indicates absence of researches of edaphic confinedness of a species within all extent of its area and registration of not only abiogenic factors but biogenic as well, and phytocenotic factors first of all.

Course studying of succession process with insignificant pasturable demutation was spent during four growth seasons on a fallow of the third, fourth, fifth, and sixth year of overgrowing on sod-podzolic loamy acid (pH 4.5) soil in a sub-band of a southern taiga of the Kirov region.

The observable association was formed as a result of overgrowing in 1992 of a site of an arable land continuously cultivated not less than 20 years.

During only two growth seasons there was a considerable change in structure of phytocenosis in which process there was a transformation of Canadian thistle – common yarrow (*Achillea millefolium* L.) – cereal association in Canadian thistle – mixed herbs association at considerable decrease in a role of segetal and ruderal species of plants (perennial sow thistle, hemp nettle, sorrel dock (*Rumex acetosella* L.), coltsfoot (*Tussilago farfara* L.), German chamomile, mare's-tail) and increase in value of mixed herb species (hybrid and red clovers, meadow pea (*Lathyrus pratensis* L.), goldilocks (*Ranunculus auricomus* L.), and great oxeye (*Leucanthemum vulgare* Lam.)) at invariable total number of species ($n = 22$). Shannon-Wiener Diversity Index remained stable but similarity degree of phytocenoses by Jaccard Index is rather insignificant (37.5%) that testifies to high degree of distinction in their structure; it allows to assume a course of succession changes in the given type of phytocenosis for the first 2 years of observation from ruderal to meadow and to track the beginning of pasturable demutation of phytocenosis shown in increase of the role of clovers, first of all a hybrid clover, in composition of a vegetative cover.

It is interesting to note considerable decrease in a projective covering of common yarrow, traditionally accepted as the indicator of neutral and near-neutral soils; and loss of sorrel dock and mare's-tail as traditional indicators of acid soils from phytocenosis structure at invariable level of soil acidity. It underlines once again importance of the registration of not only physiological, but also phytocenotic optimum of plants at researches on edaphic confinedness of plants.

Comparison of change in macro- and microelement structure of plants' samples which have been selected in given phytocenosis in 1994 and 1995 has shown that during a process of succession changes of phytocenosis the content in plants of calcium, iron, copper, molybdenum, zinc, and cadmium has increased; the aluminum and lead contents has decreased; the content of magnesium, potassium, and zinc has not changed practically.

In 1996 there were further changes of phytocenosis. The total projective covering of aboveground parts of plants has increased. The total number of plant species has reached 26. The biodiversity index has increased which size has made 2.72. The structure of dominating species and their co-dominants was replaced. Many of ruderal species have dropped out; the role has increased of species of meadow mixed herb and especially of representatives of legume family. The fallow has turned in great oxeye – clover – mixed herb association close by their characteristics to surrounded upland meadows.

In 1997 continuation of succession towards the subsequent approach of a fallow to surrounded meadow phytocenoses was also observed. Total number of plant species has increased up to 28. There was a loss of ruderal species such as perennial sow thistle, cotton burdock (*Arctium tomentosum* Mill.), sorrel dock; the further decrease is noted in participation in community composition of common yarrow, red clover, lesser stitchwort; increase in participation of typically meadow species: meadow pea, great oxeye, mountain clover (*Trifolium montanum* L.), and tufted vetch. There were new species in structure of phytocenosis – fall dandelion (*Leontodon autumnalis* L.), Canadian hawkweed, square stalked St. John's wort (*Hypericum quadrangulum* L.). The fallow has got character of great oxeye–mixed herb – cereal.

The similarity analysis of phytocenoses of different years of observation has not revealed a strong likeness between them; however has shown that pairs of phytocenoses of 1994–1995 and 1995–1996 are closest among themselves (37.5% and 37.1%); most distinct are phytocenoses of 1994–1996 (33.3%).

### 9.3.8 INTERRELATION BETWEEN SOIL ACIDITY AND QUALITY OF VEGETATIVE RAW MATERIALS

Widespread enough in territory of the Kirov region medicinal plant species – cowberry, common St.-John's wort, and lily-of-the-valley – have served as models for studying of interrelation between acidity of soil on which herbs grow and quality of vegetative raw materials.

In the Kirov region the *cowberry* meets in pine forests and spruce-pine cowberry forests stand in which has reached age 30–100 years, heights – 4–25 m, a density of crones 0.4–0.7. The projective covering of grass-subshrub layer does not exceed 50%; cowberry prevails in it. Quite often the bilberry is co-dominant species. The grassy vegetation is presented by the species having an insignificant projective covering or meeting as individual plants. There are yellow rattle (*Rhinanthus minor* L.), ground-cedar (*Lycopodium complanatum* L.), club moss (*Lycopodium annotinum* L.), European pyrole (*Pyrola rotundifolia* L.), green-flowered wintergreen, one-sided wintergreen (*Ramischia secunda* (L.) Garcke), umbellate wintergreen (*Chimaphila umbellata* (L.) W.P.C. Barton), common pussy-toes (*Antennaria dioica* (L.) Gaertn.), European goldenrod, squirrel-ear (*Goodyera repens* (L.) R. Br.), pilosa hawkweed (*Hieracium pilosella* L.), nodding melick grass, spreading millet grass, bush grass (*Calamagrostis epigeios* (L.) Roth.), and some other species. The moss cover almost continuous and is made by haircap moss (*Polytrichum commune* Hedw.), rugose fork-moss (*Dicranum polysetum* Swartz), and mountain fern moss (*Hylocomium splendens* (Hedwig) Schimper). The projective covering of lichens fluctuates from 3 to 50%. Lichens are presented, basically, by species of genus Cladonia: *Cladonia arbuscula* (Wallr.) Flot., *Cladonia alpestris* (L.) Rabenh, *Cladonia ceraspora* Vain.

The cowberry prefers sandy and sandy strongly acid (average value pH 3.8) soils poor in humus (2.4%), phosphorus and potassium (4.4 and 4.7 mg/100 g of soils accordingly), calcium and magnesium (2.0 and 0.75 mg-equivalent/100 g of soils accordingly).

The statistical analysis has allowed to reveal high adaptive ability of a cowberry. Only the value of magnesium content among other elements of soil positively correlates with size of a projective covering ($r = 0.48$) and productivity of leaves ($r = 0.61$).

The projective covering of cowberry plants in different vegetative sub-bands is approximately equal fluctuates from 9.0% to 10.6%. Parameters of plants height (12.8–14.1cm) are so stable. The greatest average productivity of leaves is noted for a sub-band of mixed coniferous-broad leaved forests (69.8 g/m$^2$), the least – in a sub-band of a southern taiga (51.0 g/m$^2$).

The Most fruitful types of forest – a lichen pine forest and a cowberry pine forest (59.9 and 58.3 $g/m^2$ accordingly). The least productivity is in a bilberry spruce forest (38.4 $g/m^2$). In a lichen pine forest the greatest average size of a projective covering – 12.0% is noted. The analysis of participation of a cowberry in phytocenosis composition testifies to their high stability.

Cowberry leaves are the recognized medical product applied at treatment of some diseases. Among biologically active substances of cowberry leaves the important place is taken away arbutin which defining their pharmacological value [18]. The amount of arbutin in August samples of cowberry leaves in studied region fluctuated from 5.2 to 10.0% (Table 9.4) that a little above requirements of State Pharmaceutical Article [18] – not less than 4.5%.

In the plant samples which have been selected in a Zone of mixed coniferous-broad leaved forests the content of arbutin is a little higher than in a sub-band of middle taiga. However, this difference is insignificant statistically.

The average amount of arbutin in cowberry leaves in the Kirov region has made 7.3±0.3% ($V$ = 16.7%). The arbutin content in cowberry leaves significantly depends on soil acidity ($r$ = 0.84 at P = 0.95), increasing with its strengthening.

*Common St.-John's wort* grows on upland mixed herb – cereal and floodplain cereal – mixed herb meadows, fallows, and forest edges. Here common St.-John's wort is dominant or co-dominant species, giving yellow aspect to herbage in flowering. Considerable participation in composition of communities accepts: common yarrow, yellow rattle, ladies'-mantle (*Alchemilla vulgaris* L.), wild madder (*Galium mollugo* L.), European strawberry, turfy hair grass (*Deschampsia caespitosa* (L.) Beauv.), slender foxtail (*Alopecurus myosuroides* Huds.), cock's foot (*Dactylis glomerata* L.), common timothy (*Phleum pratense* L.).

Common St.-John's wort prefers middle- and light-loam, and sandy soils of different degree of podzolic (pH 3.9–5.2) essentially differing on the basic agrochemical parameters. The insignificant interrelation between productivity of a species and soil content of calcium and magnesium ($r$ = 0.38) is noted. In a sub-band of middle taiga common St.-John's wort is close to its phytocenotic optimum on floodplain cereal – mixed herb meadows with sod-alluvial light acidic (pH 5.2) soils. The value of projective

**TABLE 9.4** Content of Arbutin in Cowberry Leaves (% per Dry Matter)

| | | Agrochemical features of soil | | | | | |
|---|---|---|---|---|---|---|---|
| | | P | K | | | Ca | Mg |
| Vegetation zone | Arbutin content | Mg/100 g of soil | | Humus, % | pH | Mg-equivalent/100 g of soil | |
| Zone of mixed coniferous-broad leaved forests | 9.2 | 13.3 | 3.0 | 3.00 | 4.2 | 1.65 | 0.20 |
| | 7.8 | 0.8 | 2.3 | 1.30 | 3.2 | 1.91 | 8.30 |
| | 6.9 | 0.1 | 2.0 | 2.10 | 3.6 | 3.44 | 9.80 |
| | 8.1 | 1.0 | 2.2 | 3.40 | 3.7 | 2.25 | 0.19 |
| | 6.3 | 51.0 | 16.5 | 3.84 | 5.7 | 3.12 | 0.80 |
| Sub-band of a southern taiga | 8.2 | 2.0 | 2.0 | 1.60 | 4.0 | 13.12 | 2.25 |
| | 8.2 | 1.4 | 3.5 | 3.20 | 3.7 | 3.75 | 0.28 |
| | 7.0 | 4.2 | 6.5 | 5.20 | 3.9 | 2.37 | 0.29 |
| | 8.1 | 1.5 | 5.8 | 2.75 | 3.0 | 1.12 | 0.18 |
| | 8.1 | 0.7 | 5.2 | 2.68 | 3.6 | 2.50 | 0.25 |
| | 8.7 | 2.7 | 6.5 | 2.12 | 3.9 | 1.12 | 0.39 |
| | 8.7 | 2.3 | 3.4 | 1.39 | 3.5 | 2.25 | 0.21 |
| | 7.0 | 0.5 | 2.3 | 4.09 | 3.2 | 3.12 | 0.24 |
| | 7.7 | 6.0 | 10.0 | 1.60 | 3.9 | 1.75 | 0.39 |
| | 6.9 | 3.3 | 12.3 | 4.85 | 4.0 | 1.62 | 0.30 |
| Northern part of a South taiga sub-band | 8.7 | 4.8 | 11.5 | 4.12 | 3.4 | 1.87 | 0.60 |
| | 8.0 | 2.1 | 3.0 | 1.35 | 3.5 | 1.25 | 0.27 |
| | 6.4 | 1.4 | 3.3 | 0.65 | 4.2 | 1.37 | 0.40 |
| | 8.6 | 2.5 | 5.1 | 1.12 | 3.9 | 2.00 | 0.60 |
| | 6.9 | 1.3 | 1.3 | 0.78 | 3.6 | 0.62 | 0.20 |
| | 9.7 | 2.0 | 2.0 | 0.43 | 3.7 | 0.87 | 0.20 |
| | 5.2 | 2.5 | 3.3 | 1.03 | 4.0 | 2.12 | 0.47 |
| | 7.5 | 3.0 | 2.2 | 0.28 | 3.7 | 0.75 | 0.37 |
| | 7.5 | 6.2 | 1.3 | 1.25 | 3.4 | 0.75 | 0.20 |
| | 5.2 | 0.3 | 0.5 | 0.28 | 4.1 | 0.62 | 0.13 |
| Sub-band of middle taiga | 9.4 | 4.2 | 5.3 | 2.41 | 4.2 | 3.75 | 0.85 |
| | 8.0 | 3.7 | 1.3 | 0.37 | 3.7 | 1.25 | 0.25 |
| | 9.3 | 10.0 | 1.0 | 0.97 | 4.0 | 1.00 | 0.15 |
| | 9.2 | 15.9 | 1.5 | 0.61 | 4.4 | 0.75 | 0.15 |
| | 10.0 | 5.7 | 2.0 | 0.50 | 4.0 | 1.37 | 0.22 |
| | 5.8 | 10.6 | 2.0 | 0.79 | 4.3 | 0.37 | 0.16 |

*Note*: Means in table were calculated as an average arithmetic of the measured parameters. Their deviations from an average do not exceed 15%.

covering average for a sub-band makes 10%, height of plants – 44.1±1.3 cm, productivity – 13.4±1.7 g/m$^2$.

In a sub-band of a southern taiga common St.-John's wort is most productive on mixed herb upland meadow on sandy acid (pH 3.7–4.1) soil where common burnet saxifrage (*Pimpinella saxifraga* L.) is co-dominant species. On the average for a sub-band phytocenoses are characteristic with a projective covering of common St.-John's wort about 15%, height of plants – 40.3±1.2 cm, productivity – 19.5± 2.6 g/m$^2$.

Zone of mixed coniferous-broad leaved forests common St.-John's wort dominates on water-meadows on loamy acid (pH 4.5) light-podzolic soil where it meets together with European goldenrod, cock's foot, and common timothy. Here common St.-John's wort reaches maximum for area of research of productivity (52.3±2.6 g/m$^2$) and heights 56.7±1.4 cm, the projective covering makes 15%. Value of a projective covering average for zone is 15%, height of plants 51.7±1.2 cm, productivity 28.2±3.3 g/m$^2$.

The correlation analysis of interrelations of agrochemical parameters of soil with degree of participation of plants in composition of phytocenosis allows to characterize the given species as resistant to various abiotic stresses and possessing considerable ecological plasticity.

Common St.-John's wort is a species long since used in medicine. "Grass" of the species, i.e., the top part of a plant in the blossoming condition, consisting of the main axis of an inflorescence and the lateral runaways bearing buds, flowers (being in expanded and deflowered condition), the fastened fruits, and also leaves used in medical practice as raw materials. Pharmacological value of raw materials depends on amount of polyphenols containing in it which sum under requirements of State Pharmaceutical Article [18] should be not less than 1.5% in recalculation on rutin.

In the samples selected for research the content of rutin always corresponded pharmaceutical requirements and fluctuated from 1.9% to 2.8%, increasing a little in direction from the north by the south. The tendency of increase in polyphenols content in "grass" is noted at decrease in soils acidity (Table 9.5).

*Lily-of-the-valley* in the Kirov region is on border of its area. It is devoted to pine terraces forests and to oak floodplain forests of 30–80-year-old age,

**TABLE 9.5** Content of Polyphenols in Above-Ground Mass of Common St.-John's Wort (% per dry matter, counted as rutin)

| Vegetation zone | Agrochemical features of soil | | | | | | Polyphenols content |
|---|---|---|---|---|---|---|---|
| | P | K | Humus, % | pH | Ca | Mg | |
| | Mg/100 g of soil | | | | Mg-equivalent/100 g of soil | | |
| Zone of mixed coniferous-broad leaved forests | 0.2 | 5.5 | 2.60 | 4.5 | 10.83 | 9.80 | 2.9 |
| | 1.2 | 3.2 | 5.90 | 4.7 | 7.12 | 0.74 | 2.8 |
| | 2.4 | 6.5 | 1.59 | 5.0 | 4.37 | 0.24 | 2.5 |
| Sub-band of a southern taiga | 2.5 | 7.0 | 4.80 | 4.0 | 2.00 | 0.53 | 2.0 |
| | 1.2 | 16.0 | 1.87 | 3.8 | 5.62 | 1.51 | 1.9 |
| | 20.0 | 15.0 | 2.23 | 4.6 | 4.75 | 0.65 | 2.1 |
| | 1.2 | 8.1 | 2.45 | 4.3 | 5.75 | 1.03 | 2.2 |
| | 1.2 | 2.8 | 1.10 | 4.2 | 3.87 | 0.39 | 2.4 |
| | 4.3 | 19.5 | 2.12 | 4.1 | 2.50 | 0.55 | 2.1 |
| | 9.5 | 3.0 | 1.47 | 4.1 | 6.87 | 0.67 | 2.1 |
| Northern part of a South taiga sub-band | 6.5 | 6.3 | 1.86 | 4.5 | 4.50 | 0.62 | 2.3 |
| | 6.7 | 15.0 | 15.50 | 4.9 | 68.70 | 14.00 | 2.4 |
| | 58.0 | 14.2 | 1.22 | 5.2 | 30.75 | 7.87 | 2.4 |
| | 2.5 | 10.8 | 1.10 | 3.9 | 4.12 | 0.69 | 2.2 |

*Note*: means in table were calculated as an average arithmetic of the measured parameters. Their deviations from an average do not exceed 15%.

a density of crones 0.4–0.6. Pine cowberry, sour trefoil, lily-of-the-valley, and grassy forests, oak lily-of-the-valley forests, floodplain mixed herb meadows on sod-podzolic sandy and floodplain soddy soils with humus content from 2.1% to 13.7% and pH 3–5 are characteristic for its habitat.

Correlation between height of shoots and potassium content in soil ($r$ = 0.67) is revealed. The projective covering of plants of a lily-of-the-valley in community is connected with humus content ($r$ = 0.53) and calcium content ($r$ = 0.59). Number of shoots and value of their phytomass correlate with the humus content ($r$ = 0.40 and $r$ = 0.43 accordingly). The magnesium content in soil appreciably defines size of aboveground phytomass of shoots, their height, and projective covering ($r$ = 0.85, r = 0.79, and $r$ = –0.62 accordingly). Acidity of soils is not defining for any parameter.

The volume of aboveground phytomass of a lily-of-the-valley in a zone of mixed coniferous-broad leaved forests changes from 8.1 to 102.0 g/m$^2$ (green weight), number of shoots – from 6.1 to 62.4 pieces per m$^2$, a projective covering – from 1.6% to 24.0%. On the average for a zone these parameters make 34.5±6.1 g/m$^2$, 21.0±3.6 pieces per m$^2$, and 9.5±1.7% accordingly. The height of plants is on the average 26.2±0.4 cm.

Lily-of-the-valley is a species with high degree of the ecological plasticity, dominating in variety of vegetative communities. A lily-of-the-valley is moderated or weak acidophilous plant preferring rather rich soils but can growing on poor soil also.

The statistical analysis has revealed high adaptive ability of the species. The analysis of participation of a lily-of-the-valley in composition of phytocenosis testifies their high enough stability. The above-stated data shows weak interrelation of the lily-of-the-valley and common St.-John's wort with environmental conditions and their high stability to abiotic stresses, allowing to exist in wide amplitude of biotic and abiotic factors.

"Grass" of a lily-of-the-valley is a recognized medical product for treatment of variety of diseases. Among biologically active substances of the species the important place is taken by the sum of cardiac glycosides (Convallotoxinum, convallozide, and over 40 other cardioactive glycosides), defining its pharmacological value [18].

Amount of cardiac glycosides in June shoots of a lily-of-the-valley in studied region fluctuated from 2.32 to 3.62% (on air-dry weight) that corresponds to requirements of State Pharmaceutical Articles – not less than 1.8%. The tendency of increase in content of cardiac glycosides in plants of lily-of-the-valley is noted at decrease in soil acidity level.

Thus, even reconnaissance research on interdependence between soil acidity and quality of vegetative raw materials has allowed to reveal presence of statistically significant interrelation between level of soil acidity and the content of arbutin in leaves of a cowberry, cardiac glycosides in "grass" of a lily-of-the-valley and tendency in dependences of concentration of polyphenols in "grass" of common St.-John's wort from soil acidity. It is interesting to note their various directions: the arbutin content increases at strengthening of acidity of soils, but concentration of polyphenols and cardiac glycosides fall.

### 9.3.9 EDIFICATOR PLANTS (SPECIES CAPABLE TO CHANGE CHARACTER OF ENVIRONMENT AND OF SPECIES-COMPANIONS)

According to the approaches accepted in geobotanical researches all studied phytocenoses have been divided into four groups according to acidity of bedding rocks: soil pH less than 4.0; soil pH 4.0–4.9; soil pH 5.0–5.9; soil pH 6.0–6.9.

On strongly acid soils there are noted only forest phytocenoses, presented by sour trefoil-grassy spruce forests. The edificatory plants in them are *Picea abies,* which is replaced in northern districts of the region by *P. obovata* and their hybrid forms. *Abies sibirica* is present as individual. On middle-acid soils all types of phytocenoses are developed. Forest phytocenoses are presented by spruce-fir grassy forests. Edificator species in them are *Picea abies, P.obovata,* their hybrid forms and *Abies sibirica.* Considerable roles in composition of such phytocenoses play *Betula verucosa, Populus tremula.* Next species may be considered as edificators of meadow phytocenoses: *Hypericum maculatum, Leucanthemum vulgare, Gallium mollugo*; in fallow phytocenoses of 1–2 years of overgrowing – *Cirsium arvense, Viola arvensis, Achillea millefolium.* Species-edificators of arable phytocenoses are defined by a species of sowing crop. Most successfully on fields with soils of the given degree of acidity develop crops of *Trifolium pratense* and *Secale cereale.* The given species form continuous poorly weeded agro-phytocenoses. In case of sowing of *Triticum aestivum* on such soils heterogeneity of development of plants and a considerable contamination of sowings by *Trifolium hybridum, Tr. repens, Cirsium arvense* is marked.

On lightly acidic soils all types of phytocenoses are also revealed. Forest phytocenoses are presented by spruce-fir grassy forests with insignificant addition of linden. *P. abies* and *A. sibirica* are species-edificators of such forests. Edificators of meadow phytocenoses are *Trifolium alpestris, Tr. pratense, Viscaria viscosa, Centaurea phrygia, Vicia crassa, Dactylis glomerata*: of fallow phytocenoses – *Tripleurospermum inodorum, Sonchus arvensis, Atriphlex hortensis, Cirsium arvense.* Species-edificators of arable phytocenoses are defined by a species of sowing crop. At non-observance of agricultural technicians among species-edificators appear segetal species: *Barbarea vulgaris, Cirsium arvense, Sonchus arvensis, Convolvulus arvense*.

On soils with pH close to neutral only fallow and arable phytocenoses are noted. *Taraxacum officinale, Sonchus arvensis, Leucanthemum vulgare* serve as edificators on fallow phytocenoses. In agro-phytocenoses along with cultivated plants *Tripleurospermum inodorum, Sonchus arvensis, Atriphlex hortensis, Erysinum cheirantoides* play habitat-forming role.

In the course of regenerative stages of succession phytocenoses there is the time change of edificatory species caused in some degree by acidity of bedding rocks.

*Tripleurospermum inodorum, Sonchus arvensis, Atriphlex hortensis* were edificators on fallow phytocenoses of 1 year of overgrowth on light acidic soil; on neutral soils *Taraxacum officinale* was added to them.

On middle acid soils on fallows of 3 years of overgrowth there are revealed *Cirsium arvense, Achillea millefolium, Sonchus arvensis*; of 4 years of overgrowth – *Cirsium arvense, Trifolium hybridum, Achillea millefolium*; of 5 years of overgrowth – *Leucanthemum vulgare, Trifolium hybridum, Tr. pratense, Lathyrus pratensis*; of 6 years of overgrowth – *Leucanthemum vulgare, Lathyrus pratensis, Vicia crassa, Ranunculus auricomus*. As a whole, for 4 years of observation there was a change of phytocenosis from ruderal to meadow one with attributes of pasturable demutation.

## 9.4 CONCLUSIONS

Forest phytocenoses are more various on specific structures than agro-phytocenoses. The coefficient of a biodiversity of investigated forest phytocenoses fluctuates from 1.07 on podzol loamy soil to 5.18 on gray forest soil; of agro-phytocenoses – from 0.63 on sod-strongly acid soil to 0.91 on sod-light acid soil. For forest phytocenoses decrease in coefficient of a biodiversity is noted with increase in soil acidity.

The segetal flora of agro-phytocenoses on sod-podzolic soils has more southern geographical aspect in comparison with frontier forest flora.

The flora of agro-phytocenoses and forest phytocenoses on acid sod-podzolic soils has more boreal character rather than flora of phytocenoses on less acid gray forest soils. Presence of a significant amount of nemoral species in forest phytocenosis in northern districts of the region is probably explained by features of development of the given territory in Holocene.

It is obviously possible to note distinction between standard and resulted in the literature [18, 20] confinedness of species to soils with various level of acidity and their real habitats. So, mare's-tail belonging to acidophilious species forms thickets on soils with pH 5.87; and bee nettle to basephilious species meets in large number on soils with pH 4.47. Possibly, these facts indicate absence of researches of edaphic confinedness of species within all extent of its area, and registration of not only abiogenic factors but biogenic as well and phytocenotic factors first of all [21, 22].

## KEYWORDS


- **agro-phytocenoses**
- **cardiac glycosides**
- **medicinal grasses**
- **natural-climatic zones**
- **plant communities**
- **specific structure**


## REFERENCES


1. Zubarev, L. A. Higher Plants. The Encyclopedia of the Vyatka Land. Kirov: Vyatka Publishing House. 1997, 7, 333–342 (in Russian).
2. Gorev, G. I. The Tutorial by Definition of Forest Types of the Kirov Region. Kirov: Department of Forestry, 1975, 29 p. (in Russian).
3. Shikhova, L. N., Egoshina, T. L. Heavy Metals in Soils and Plants of North-East of European Part of Russia. Kirov: North-East Agricultural Research Institute, 2004, 263 p. (in Russian).
4. Tyulin, V. V., Guschina, A. M. Features of Soils of the Kirov Region and Their Use at Intensive Agriculture. Kirov: Vyatka State Agricultural Academy, 1991, 92 p. (in Russian).
5. Karyakin, A. V., Gribovskaya, I. F. Method of Optical Spectroscopy and Luminescence in the Analysis of Natural and Sewage Waters. Moscow: Khimiya ("Chemistry' in English), 1987, 304 p. (in Russian).

6. Appenroth, K.-J. Definition of "heavy metals' and their role in biological systems. In, I. Sherameti and, A. Varma (eds.), Soil Heavy metals, Soil Biology. Springer-Verlag Berlin Heidelberg, 2010, 19–30.
7. Hohl, H., Varma, A. Soil: the living matrix. In, I. Sherameti and, A. Varma (eds.), Soil Heavy metals, Soil Biology. Springer-Verlag Berlin Heidelberg, 2010, 1–18.
8. Arinushkina, E. V. Handbook on Chemical Analysis of Soils. Moscow: Publishing House of Moscow University, 1970, 488 p. (in Russian).
9. Methodical Indications for Estimation of Microelements in Soils, Fodders, and Plants by Atomic-Absorption Spectroscopy. Moscow: CINAO (Central Institute of Agrochemical Operation of agriculture) Publ., 1995, 95 p. (in Russian).
10. Krylova, I. L., Shreter, A. I. Methodical Instructions for Definition of Stocks of wild-Growing Herbs. Moscow: VILR (All-Russian Institute of Medicinal Plants), 1971, 31 p. (in Russian).
11. Technique of Definition of Herbs Stocks. Moscow: State Commission on Forestry, 1986, 51 p. (in Russian).
12. The Program and a Technique of Supervision for Ceno-Population of Plant Species of the USSR Red Book. Moscow: State Agro-Industrial Committee, 1986, 34 p. (in Russian).
13. Forest Mensuration Handbook. Moscow: Lesnaya promyshlennost' ("Forest Industry" in English), 1980, 287 p. (in Russian).
14. Bejdeman, I. N. Technique of Studying of Phenology of Plants and Vegetative Communities. Novosibirsk: Nauka ("Science" in English) Publishing, Siberian branch, 1974, 154 p. (in Russian).
15. Chao, A., Chazdon, R. L., Colwell, R. K., Shen, T.-J. A new statistical approach for assessing similarity of species composition with incidence and abundance data. Ecology Letters. 2005, 8, 148–159.
16. Bioindication of Pollution of Land Ecosystems. Moscow: Mir Publishing, 1988, 348 p. (in Russian).
17. Tilman, D. Biodiversity: population versus ecosystem stability. Ecology. 1983, 77, 350–363.
18. The State Pharmacopoeia of the USSR, General methods of the analysis. Medicinal Vegetative Raw Materials. Moscow: Medicine, 1989, 11(2), 196 p.
19. Ramensky, L. G., Tsatsenkin, I. A., Chizhikov, O. N. Ecological estimation of fodder lands on a vegetative cover. Moscow: Selkhozgiz, 1956, 473 p. (in Russian).
20. Ellenberg, H., Weber, H. E., Düll, R., Wirth, V., Werner, W., Paulißen, D. Ziegerwerte der Pflanzen in Mitteleuropa. "Handbook of Plants of Central Europe." Scripta geobotanica. 1991, 18, 1–248 (in German).
21. Voronov, A. G. Geobotany. Moscow: Vyschaya shkola ("High School" in English), 1973, 382 p. (in Russian).
22. Mirkin, B. M. Theoretical Basis of a Modern Phytocenology. Moscow: Nauka ("Science" in English), 1985, 137 p. (in Russian).

## APPENDIX

List of plant species mentioned in the chapter, in Latin alphabetic order.

| Latin name | Common name(s) |
|---|---|
| *Abies sibirica* Ledeb. | siberian fir, |
| *Achillea millefolium* L. | common yarrow |
| *Aconitum excelsum* Reichenb. | hungarian monkshood, wolfsbane monkshood |
| *Actaea rubra* (Ait.) Willd. | red baneberry, chinaberry, doll's eye, poisonberry |
| *Aegopodium podagraria* L. | ground elder, herb gerard, bishop's weed, goutweed, snow-in-the-mountain, English masterwort, wild masterwort, ashweed |
| *Agropyron repens* (L.) P. Beauv. | couch grass, common couch, twitch, quick grass, quitch grass (also just quitch), dog grass, quackgrass, scutch grass, witchgrass, devil's grass |
| *Ajuga reptans* L. | bugle, blue bugle, bugleherb, bugleweed, carpet-weed, carpet bungleweed, common bugle, carpenter's herb |
| *Alchemilla vulgaris* L. | common lady's mantle, ladies'-mantle |
| *Alnus glutinosa* (L.) Gaertn. | common alder, black alder, European alder, alder |
| *Alopecurus myosuroides* Huds. | black grass, brush, slender foxtail |
| *Antennaria dioica* (L.) Gaertn. | common cat's foot, common pussytoes |
| *Arctium tomentosum* Mill. | downy burdock, woolly burdock, cotton burdock |
| *Artemisia vulgaris* L. | felon herb, chrysanthemum weed, wild wormwood, old Uncle Henry, sailor's tobacco, naughty man, old man or St. John's plant common wormwood, mugwort |
| *Asarum europaeum* L. | asarabacca, hazelwort, and wild spikenard, European wild ginger |
| *Atriplex hortensis* L. | garden orache, red orach, mountain spinach, French spinach, orache, arrach |
| *Barbarea vulgaris* R.Br. | bittercress, herb barbara, rocketcress, yellow rocket-cress, winter rocket, wound rocket, upland cress |
| *Betula pendula* Roth | silver birch or warty birch, European white birch |
| *Betula pubescens* Ehrh. | downy birch; also known as moor birch, white birch, European white birch or hairy birch |
| *Calamagrostis epigeios* (L.) Roth. | bush grass, chee reed grass, wood small reed |

| Latin name | Common name(s) |
|---|---|
| *Campanula divaricata* Michx. | Appalachian bellflower, panicled bellflower |
| *Capsella bursa-pastoris* (L.) Medik. | shepherd's-purse, caseweed |
| *Carex capillaries* L. | Hair Sedge, hairlike sedge |
| *Carex sylvatica* Huds. | wood sedge |
| *Carex tomentosa* L. | Downy-fruited Sedge, Filzige Segge |
| *Centaurea cyanus* L. | cornflower, bachelor's button, boutonniere flower, hurtsickle, cyani flower, bluebottle |
| *Cerastium arvense* L. | field mouse-ear, field chickweed |
| *Chelidonium majus* L. | greater celandine or tetterwort, nipplewort, swallow-wort killwort |
| *Chimaphila umbellata* (L.) W.P.C. Barton | common pipsissewa, king's-cure, love-in-winter, pipsissewa, umbellate wintergreen |
| *Chrysosplenium alternifolium* L. | alternate-leaved golden-saxifrage, saxifrage |
| *Circaea alpine* L. | Alpine circaea, Alpine Enchanter's-nightshade, Small Enchanter's Nightshade |
| *Cirsium arvense* (L.) Scop. | Creeping Thistle, Canadian thistle |
| *Cladonia alpestris* (L.) Rabenh | |
| *Cladonia arbuscula* (Wallr.) Flot. | |
| *Cladonia ceraspora* Vain | |
| *Convallaria majalis* L. | May lily, lily-of-the-valley |
| *Convolvulus arvensis* L. | field bindweed |
| *Corydalis hallerii* (Willd.) Willd. | corydalis |
| *Crepis sibirica* L. | Hawksbeard |
| *Crepis tectorum* L. | narrowleaf hawk's beard |
| *Cystopteris sudetica* (A. Brown & Milde) A.P. Khokhr. | bladder-fern |
| *Dactylis glomerata* L. | orchard grass, cock's foot |
| *Deschampsia caespitosa* (L.) Beauv. | lime grass, tufted hair grass, turfy hair grass |
| *Dicranum polysetum* Swartz | rugose fork-moss |
| *Dryopteris carthusiana* (Vill.) H.P. Fuchs | narrow buckler-fern, spinulose woodfern, prickly-toothed fern |

| Latin name | Common name(s) |
| --- | --- |
| *Dryopteris filix-mas* (L.) Schott | male shield fern |
| *Equisetum arvense* L. | field horsetail, common horsetail, mare's-tail |
| *Equisetum sylvaticum* L. | bottlebrush |
| *Fragaria vesca* L. | wild strawberry, woodland strawberry, Alpine strawberry, European strawberry, fraise des bois |
| *Fumaria officinalis* L. | common fumitory, drug fumitory, earth smoke, fumitory |
| *Galeopsis speciosa* Mill. | large-flowered hemp-nettle, Edmonton hempnettle, bee nettle |
| *Galeopsis tetrahit* L. | common hemp-nettle, brittlestem hempnettle, hemp nettle |
| *Galium mollugo* L. | babies'-breath, whip-tongue, white bedstraw, hedge bedstraw, wild madder |
| *Galium odoratum* (L.) Scop. | woodruff, sweet woodruff, wild baby's breath; master of the woods |
| *Geranium pretense* L. | meadow cranesbill, meadow geranium |
| *Geranium sylvaticum* L. | woodland geranium, wood cranesbill, wood crane |
| *Gnaphalium uliginosum* L. | marsh cudweed, low cudweed |
| *Goodyera repens* (L.) R. Br. | creeping lady's-tresses, dwarf rattlesnake plantain, lesser rattlesnake plantain, squirrel-ear |
| *Gymnocarpium dryopteris* (L.) Newman | western oakfern, common oak fern, oak fern, northern oak fern |
| *Hieracium nigrescens* Willd. | black-headed hawkweed, hawkweed |
| *Hieracium pilosella* L. | hawkweed, ling-gowans, mouse bloodwort, mouse-ear, pilosa hawkweed |
| *Hieracium umbellatum* L. | Canada hawkweed, narrowleaf hawkweed, northern hawkweed, Canadian hawkweed |
| *Hylocomium splendens* (Hedwig) Schimper | glittering wood-moss, stairstep moss, splendid feather moss, mountain fern moss |
| *Hypericum perforatum* L. | Perforate St John's-wort, common St John's wort, St.-John's wort |
| *Hypericum quadrangulum* L. | St. Peter's wort, peterwort, square stemmed St. John's Wort, square stalked St. John's wort |
| *Impatiens noli-tangere* L. | touch-me-not, yellow Balsam, jewelweed, western touch-me-not, wild balsam |
| *Juniperus communis* L. | common juniper |

| Latin name | Common name(s) |
|---|---|
| *Lamium purpureum* L. | red deadnettle,https://en.wikipedia.org/wiki/Lamium_purpureum - cite_note-BSBI07-1 purple deadnettle, purple archangel, velikdenche, rabbit-meat |
| *Lathyrus pratensis* L. | meadow vetchling, meadow pea |
| *Lathyrus vernus* (L.) Bernh. | spring vetchling, spring pea, spring vetch, bitter peavine |
| *Leontodon autumnalis* L. | autumn hawkbit, fall dandelion |
| *Leucanthemum vulgare* Lam. | ox-eye daisy, oxeye daisy, great oxeye |
| *Linnaea borealis* L. | twinflower |
| *Lonicera tatarica* L. | Tartarian honeysuckle, bush honeysuckle |
| *Lonicera xylosteum* L. | European fly honeysuckle, dwarf honeysuckle, fly woodbine fly honeysuckle |
| *Luzula pilosa* (L.) Willd. | hairy wood-rush |
| *Lycopodium annotinum* L. | bristly club-moss, stiff clubmoss |
| *Lycopodium complanatum* L. | groundcedar, creeping jenny, northern running-pine |
| *Maianthemum bifolium* (L.) F.W.Schmidt | false lily of the valley, bifoliate bead-ruby |
| *Matricaria suaveolens* (Pursh) Buch. | wild marigold |
| *Melica nutans* L. | mountain melick, nodding melick grass |
| *Mercurialis perennis* L. | dog's-mercury |
| *Milium effusum* L. | American milletgrass, wood millet, spreading millet grass |
| *Myosotis arvensis* (L.) Hill. | field forget-me-not, field scorpion grass |
| *Oxalis acetosella* L. | wood sorrel, common wood sorrel, sour trefoil |
| *Paris quadrifolia* L. | true lover's knot, herb paris |
| *Pastinaca sylvestris* Mill. | wild carrot |
| *Phleum pratense* L. | cat's-tail grass, cattail grass, herd's grass, timothy, herd grass, timothy grass, common timothy, meadow catmint |
| *Picea excelsa* (L.) Karst. | Norway spruce, |
| *Picea obovata* Ledeb. | Siberian spruce |
| *Pimpinella saxifraga* L. | saxifrage bumet, saxifrage pimpinella, common burnet saxifrage |
| *Pinus silvestris* L. | Scots pine |
| *Pirola rotundifolia* L. | European pyrole, European shinleaf, skin leaf |

| Latin name | Common name(s) |
|---|---|
| *Plantago major* L. | broadleaf plantain, greater plantain, common plantain |
| *Poa annua* L. | annual meadow grass, annual bluegrass, poa, meadow grass |
| *Polygala vulgaris* L. | common milkwort, milkwort |
| *Polygonum convolvulus* L. | black-bindweed, wild buckwheat |
| *Polytrichum commune* Hedw. | common hair moss, haircap moss |
| *Populus tremula* L. | aspen, common aspen, Eurasian aspen, European aspen, quaking aspen |
| *Potentilla goldbachii* Rupr. | european cinquefoil, thuringian potentilla |
| *Pulmonaria obscura* Dumort. | unspotted lungwort, suffolk lungwort |
| *Pyrola chlorantha* Sw. | green-flowered wintergreen |
| *Pyrola rotundifolia* L. | round-leaved wintergreen, European pyrole, European shinleaf, skin leaf |
| *Quercus robur* L. | English oak, pedunculate oak, French oak |
| *Ramischia secunda* (L.) Garcke | one-sided wintergreen, serrated wintergreen, shinleaf |
| *Ranunculus auricomus* L. | goldilocks buttercup, goldilocks |
| *Rhinanthus minor* L. | yellow rattle, cockscomb |
| *Ribes nigrum* L. | European black currant |
| *Rosa acicularis* Lindl. | prickly rose, bristly rose, arctic rose, prickly wild rose |
| *Rubus idaeus* L. | red raspberry, European raspberry |
| *Rubus saxatilis* L. | stone bramble |
| *Rumex acetosella* L. | sheep's sorrel, red sorrel, sour weed, field sorrel. sorrel dock |
| *Solidago virgaurea* L. | European goldenrod, woundwort |
| *Sonchus arvensis* L. | corn sow thistle, dindle, field sow thistle, gutweed, swine thistle, tree sow thistle, field sowthistle, field milk thistle, perennial sow thistle |
| *Sorbus aucuparia* L. | rowan, mountain ash |
| *Stachys palustris* L. | marsh woundwort, marsh hedge-nettle, hedge-nettle, hedge nettle |
| *Stellaria graminea* L. | grassleaf starwort, common stitchwort, lesser stitchwort |
| *Stellaria holostea* L. | greater stitchwort, addersmeat, easter-bell |
| *Thalictrum simplex* L. | small meadow rue |

| Latin name | Common name(s) |
|---|---|
| *Thlaspi arvense* L. | field penny-cress |
| *Tilia cordata* Mill. | small-leaved lime, small-leaved linden, little-leaf linden |
| *Trifolium hybridum* L. | hybrid clover, alsike clover |
| *Trifolium montanum* L. | mountain clover |
| *Trifolium pretense* L. | red clover |
| *Trifolium repens* L. | Dutch clover, white clover |
| *Tripleurospermum inodorum* (L.) Sch.Bip. | scentless mayweed, scentless chamomile, wild chamomile, mayweed, false chamomile, Baldr's brow, German chamomile |
| *Trollius europaeus* L. | common globeflower |
| *Tussilago farfara* L. | coltsfoot |
| *Urtica dioica* L. | stinging nettle, common nettle |
| *Vaccinium myrtillus* L. | common bilberry, blue whortleberry, blaeberry, hurtleberry, huckleberry, winberry, fraughan, bilberry |
| *Vaccinium vitis-idaea* L. | cowberry, lingonberry |
| *Veronica chamaedrys* L. | germander speedwell, bird's-eye speedwell, bird's-eye |
| *Vicia cracca* L. | tufted vetch, cow vetch, bird vetch, boreal vetch |
| *Viola arvensis* Murray | field pansy |
| *Víola mirabilis* L. | wonder violet |
| *Viola montana* L. | mountain pansy |

# CHAPTER 10

# FEATURES OF HEAVY METALS ACCUMULATION IN VASCULAR PLANTS

TATYANA L. EGOSHINA,[1,2] LYUDMILA N. SHIKHOVA,[1] and EUGENE M. LISITSYN[1,3]

[1]*Vyatka State Agricultural Academy, 133, Oktyabrsky Ave., Kirov, 610017, Russia*

[2]*All-Russian Institute of Game and Fur Farming, 72, Preobrazhenskaya St., Kirov, 610007, Russia*

[3]*North-East Agricultural Research Institute, 166-a, Lenin St., Kirov, 610007, Russia, E-mail: shikhova-l@mail.ru*

## CONTENTS



### ABSTRACT

For carrying out of researches on features of accumulation of heavy metals in plants over 2700 probes of plant material and over 1000 probes of soil were selected in territories the approached to background, technogenic and roadside territories. The conducted researches have shown that the differentiated distribution of microelements between various organs of plants collected in the same phytocenoses is species-specific. But the general tendency of reduction of concentration of pollutants in fruits is obvious. Ability of different plant species growing under identical soil conditions to accumulate microelements differs strongly also. Interactions between the microelements accumulated in plants, show that these processes can be as antagonistic and synergetic one. Sometimes they are shown in a metabolism of more than two elements. The greatest number of antagonistic reactions was observed for Fe, Mn, Cu, and Zn, which, obviously, are key elements in physiology of plants. Synergic interaction between microelements is not observed usually. Synergism of Cd with such microelements as Pb, Fe, and Ni can be an artifact arising owing to destruction of physiological barriers under the influence of stress caused by superfluous concentration of heavy metals. Carrying out gathering of a plant material for carrying out of the chemical analysis on the content of polluting substances it is important to consider features of their accumulation both by separate species, and various organs of plants.

## 10.1 INTRODUCTION

One of the most important factors influencing content of heavy metals (HM) in plants is geochemical features of territory of plant vegetation. Chemical composition of plants as well as accumulation and circle of an element in biogeocenosis is dependent on concentration of the element in soils, mother rocks and ground waters. At last time input of HM into plants through leaf surface became more important because of high degree of atmosphere pollution with HM.

Processes of absorption, redistributions and accumulations of pollutants by plants are difficult enough and depend on set of factors. Some ways of input of chemical elements and complexes into a plant are known, basic of which are root nutrition, gas exchange, exchange adsorption on a surface of leaves. Development of such evolutionary caused systems as conducting tissues, a protective cuticle layer, physiological-and-morphological adaptations define the difficult and mediated dependence of a chemical compound of tissues of vascular plants on change of chemical quality of environment.

Bioavailability of the microelements arriving from air sources through leaves (foliar absorption) can have considerable influence on magnitude of accumulation of heavy metals at plants. It has also practical value at top-dressing especially for such elements as Fe, Mn, Zn, and Cu. It is considered that foliar absorption consists of two phases – non-metabolic penetration of metal through cuticle, which is considered as the main way of entrance, and metabolic processes, which explain the accumulation of elements opposite to action of concentration gradients. The second group of processes is responsible for transportation of ions through plasmatic membranes and into cell protoplasm.

The microelements absorbed by leaves can be transferred to other vegetative tissues including roots where the superfluous amount of some elements can be reserved. Speed of movement of microelements in tissues changes strongly depending on plant's organ, its age and the element nature. Such elements as Cd, Zn, and Pb, absorbed by aboveground parts of plants, apparently, cannot quickly move to roots whereas Cu is very mobile [1].

The part of microelements grasped by leaves can be washed up with rainwater. Distinctions in efficiency of washing away of different microelements can be compared with their functions or metabolic relations. For example, easily occurring removal of Pb at washing off allows assuming that this element is present basically in the form of a deposit on a leave surface. On the contrary the small share of Cu, Zn, and Cd which can be washed off specifies in considerable penetration of these metals into leaves [2, 3]. Roberts notes essential absorption of Zn, Fe, Cd, and Hg brought through foliage [4]. Washing away of elements from leaves with acid rains can include cation-exchange processes in which $H^+$-ions of rain water replaces with low-size cations kept in linked position on leaves cuticle [5].

Absorption of a dust and the aerosols containing metal particles by leaves depends on the sizes of a catching surface [6]. Researches on dust-detain properties of plants have been spent for the first time by [7] and [8]. The last one on the basis of all-round supervision had been established distinctions in level of coating of leaves depending on position in a crone, character of a structure, age of sheet plates, and their specific accessory. Thus the pine (*Pinus* L.) possessed the most advanced level of dust-detain ability exceeding other tree species (an aspen *Populus* L., a birch *Betula* L., a bird cherry *Padus racemosa* Gilib., an elm *Ulmus* L., a maple *Acer* L., etc.) at 10–15 time. In other article [9] it has been shown that on sites with dense vegetation moving of the particles of a dust weighed in air decreased for 75% in comparison with open sites.

Il'kun [10] describes three basic phases of entrance of toxic gasses in a cell: sorption with a cuticle layer and cells of epidermis—diffusion through stomata slots into leaf and dissolution in the water sating covers of leaf,—movement from an absorption place to conjunctive tissues and accumulation in cells. It, apparently, is the basic way of entrance of toxic gasses to leaves.

The second way of entrance of heavy metals to plants is their absorption from soil. Firsova and co-workers [11] established laws of accumulation of five metals in agricultural plants depending on their content in soil. An accumulation line looks so: Zn > Cu > Pb > Cr > Cd. Thus cadmium is absorbed basically by root system and to a lesser degree by other organs. Calculation of total stocks of metals has shown that in green mass of corn its content is 4 times more than in roots. The species also affect specificity of accumulation: for example grasses absorb more cadmium and lead; potato – copper and zinc. Chrome is absorbed least intensively by all terrestrial plants. Cereal crops (wheat and barley) accumulate 2 times more metals in the aboveground part alienated with harvest than roots. Grasses unlike other plants are characterized by higher stocks of heavy metals in underground mass.

Penetration of toxic substances from soil through root system depends on protective properties of plants. The first barrier for some microelements is selective ability of root absorption [12]. Physiological barrier of absorption serves as the second factor of regulation of microelements accumulation [13]. If these protective mechanisms do not work inflow of toxic substances occurs into least physiologically active organs; more often there are tubers and bulbs of agricultural plants [14, 15].

Character of distribution and accumulation of microelements varies considerably for different elements, plant species and growth seasons. As Ref. [16] has informed that in a phase of intensive growth of spring barley content of Fe and Mn are rather low; but Cu and Zn – very high. While first two elements collect mainly in old leaves and leaf vaginas Cu and Zn are distributed apparently more homogeneously on all barley plant.

## 10.2 MATERIALS AND METHODOLOGY

For studying of features of accumulation of heavy metals in soils and plants plant (over 2700 probes) and soil (over 1000 probes) probes were selected in the approached to background, technogenic and roadside territories. The sites removed from the nearest settlements and asphalted highways not less than on 3–6 km considered as approached to the background. The main bulk of background probes is selected in territory of the Kirov region. Roadside territories were the highways of federal value which are passing in the Kirov region and Republic of Komi. Sampling of probes in roadside territories was spent on transect on 1, 3, 5, 10, 20, 50, 100, 200, 400 m removal from a roadbed. Territories round cities of Kirov and Kirovo-Chepetsk as well as the territories adjoining directly to any industrial enterprises: to city treatment facilities, drilling waste disposal site of factory of processing of nonferrous metals, to Kirovo-Chepetsk chemical industrial complex, Kirovo-Chepetsk thermal power station were surveyed as technogenically polluted territories

The plants growing on types of soils most widespread in the region were investigated: podzolic, sod-podzolic, gray wood and alluvial sod under natural vegetation (wood, a meadow).

Description of soil horizons see Appendices in article Lyudmila N. Shikhova, Olga A. Zubkova, Eugene M. Lisitsyn "Dynamics of organic matter content in sod-podzolic soils differ in degree of cultivation" in this book.

Plant samples were selected in dry weather. Plants of the same species from the different trial areas were in the same phenophase. Plant mass were collected being guided by rules of gathering of raw materials for each species [17]. Not less than 30 individuals of each plants species of grass-subshrub layer collected in a territory of the trial area in regular intervals. At sampling of plants their pollution by soil was excluded. For cleaning of roots and rhizomes from soil they were washed out with water or carefully cleared with a brush. Plants were dried on air in well-aired room or in special drying cases. Then probes of plants were packed into a pure dense paper or polyethylene packages and stored in laboratory conditions

The content of heavy metals was defined in the prepared probes after corresponding processing by a method of atomic-absorption spectrophotometry. The content of macro- and microelements is defined in air-dry raw materials in mg/kg [18].

## 10.3 RESULTS AND DISCUSSION

According to floristic division into districts [19] flora of the Kirov region concerns the Holarctic kingdom, Boreal sub-kingdom, circumpolar areas, and the North European province.

Under the rough data the flora of an area totals 1116 species of vascular plants [20]. They unite in 77 families. A leading place on number of genus take families Asteraceae (48 genus), Poaceae (44 genus), Umbelliferae (30 genus), Rosaceae, and Brassicaceae (on 24 genus). The largest on number of species are families Asteraceae (118 species), Poaceae (101 species), Rosaceae (78 species), Cyperaceae (61 species), Fabaceae (57 species), and Caryophyllaceae (53 species) (Table 10.1).

**TABLE 10.1** Abundant of Main Families of Flora of Kirov Region, Russia [20]

| | Number of | | Percent of | |
|---|---|---|---|---|
| **Family** | **Species** | **Genus** | **Species** | **Genus** |
| Asteraceae | 118 | 48 | 10.8 | 10.7 |
| Poaceae | 101 | 44 | 9.2 | 9.8 |
| Rosaceae | 78 | 24 | 7.1 | 5.3 |
| Cyperaceae | 61 | 7 | 5.6 | 1.6 |
| Fabaceae | 57 | 20 | 5.2 | 4.5 |
| Caryophyllaceae | 53 | 20 | 4.8 | 4.5 |
| Brassicaceae | 46 | 24 | 4.2 | 5.3 |
| Scrophylariaceae | 41 | 11 | 3.8 | 2.4 |
| Ranunculaceae | 38 | 15 | 3.5 | 3.3 |
| Lamiaceae | 34 | 20 | 3.1 | 4.5 |
| Total | 607 | 233 | 57.3 | 51.9 |

More than half of families (49) are presented by one genus; more than third (25) by one species. Among these families are Butomaceae, Oxalidaceae, Celastraceae, Adoxaceae, and Polemoniaceae.

Many plant species meeting in a territory of the Kirov region have economic value. From 1116 species of the vascular plants growing in territory of the Kirov region 228 species (20.3%) possess curative properties [20]. In the Kirov region there are 28 species of berry and fruit plants. Seven species from them have considerable stocks and stocks of industrial value: a cranberry, a cowberry, a bilberry, a raspberry, a black currant, a bird cherry, a mountain ash; 11 species have small stocks (are suitable for the local use): two species of wild strawberry, a blueberry, cloudberries, stone berry, two species of currant, a honeysuckle, a guelder-rose, and filbert. Some species are rare or protected (black crowberry etc.). About 67 species of wild-growing plants can be used as food. More than 100 species of plants are used as melliferous and beebread-containing. Over 300 species of wild-growing plants are fodder for domestic and wild animals.

The resulted materials show that more than half of species of vascular plants of flora of the Kirov region is used as medicinal, food, fodder, melliferous and beebread-containing plants [21].

Conducted researches have shown that differentiated distribution of microelements between various organs of the plants collected in the same phytocenosis is species-specific; it is well visible from Table 10.2. But the general tendencies of reduction of pollutant's concentration in fruits are obvious (Figure 10.1).

At definition of biological availability of microelements it is necessary to consider species-specific properties of plants. They vary strongly enough depending on soil conditions and plants state. Ability of different

**TABLE 10.2** Element Composition of Rhizomes and Leaves of Spatter Dock (Nuphar lutea (L.) Sm.) Growing in the Same Water Basin, mg/kg of Air-Dry Mass (Lake Ivanovskoye, Kirov Region)

| Organ of plant | Mg | Cu | Ni | Cd | Pb | Zn | Fe | Cr | Mn |
|---|---|---|---|---|---|---|---|---|---|
| Rhizome | 98.4 | 3.58 | 2.0 | 0.18 | 1.5 | 30.8 | 53.6 | 2.79 | 62.9 |
| Leaf | 28.4 | 1.64 | 4.1 | 0.2 | 1.8 | 19.8 | 121.4 | 2.8 | 47.4 |

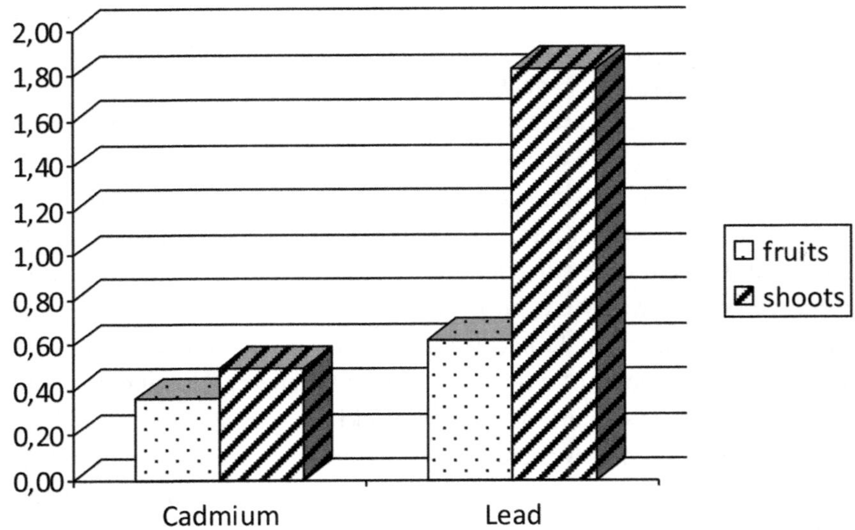


**FIGURE 10.1** Concentration of cadmium and lead in shoots and fruits of a bilberry growing in a bilberry spruce forest, mg/kg of air-dry mass (Zuevka district, Kirov region).

species of vascular plants growing in identical soil conditions to accumulate microelements differs strongly also (Table 10.3).

The understanding of value of some microelements for normal growth and development of plants has developed only in XX century. Now only for ten microelements it is known that they are vital to all plants; and for several it is proved that they are necessary for a small number of species. For other elements it is known that they have stimulating effect on plant growth, but their functions are not established yet (Table 10.4).

Prominent feature of physiology of many elements is their necessity for growth of plants in small doses and toxic action on cells at high concentration. Microelements vital for plants are such which cannot be replaced by other elements in their specific biochemical role and which have direct influence on an organism, i.e., without them it cannot neither grow, nor finish some metabolic cycles. There are elements that demand the additional data for proof their necessity. Lead and cadmium may present such elements.

Bowen [22] classified functions and forms of elements in organisms and has divided microelements present at plants into following groups:

**TABLE 10.3** Element Composition of Aboveground Mass of Some Plant Species Growing in a Same Phytocenosis, mg/kg of Air-Dry Mass (Deposit; Falenki District of the Kirov Region, Soils pH 6.9)

| Element | Dindle *(Sonchus arvensis)* | Blowball *(Taraxacum)* | Common plantain *(Plantago maior)* | Low cudweed *(Gnaphalium uliginosum)* | German camomile *(Tripleurospermum inodorum)* |
|---|---|---|---|---|---|
| Ca | 215.54 | 45.87 | 205.84 | 327.43 | 149.69 |
| Mg | 56.21 | 8.25 | 36.00 | 29.00 | 76.81 |
| Al | 3.52 | 3.21 | 4.24 | 5.81 | 4.18 |
| Cu | 5.37 | 6.47 | 4.95 | 6.24 | 5.01 |
| Ni | 1.23 | 1.24 | 1.21 | 1.03 | 1.30 |
| Cd | 0.10 | 1.21 | 0.23 | 0.21 | 0.19 |
| Pb | 2.30 | 1.84 | 2.00 | 2.13 | 2.43 |
| Zn | 33.80 | 27.95 | 21.80 | 17.80 | 31.18 |
| Mo | 0.41 | 0.35 | 0.45 | 0.31 | 0.53 |
| Fe | 189.51 | 186.91 | 86.50 | 65.80 | 131.21 |

1. Entering into bearing skeleton – Si, Fe; sometimes Ba and Sr.
2. Linked in various small molecules including antibiotics and porphyrin —As, B, Br, Cu, Co, F, Fe, Hg, I, Se, Si, and V.
3. Linked with large molecules mainly proteins including enzymes having catalytic properties – Co, Cr, Cu, Fe, Mn, Mo, Se, Ni, and Zn.
4. Fixed in large molecules having functions of accumulation, carrying over or unknown functions – Cd, Co, Cu, Fe, Mn, Ni, Se, and Zn.
5. Linked with organelles (e.g., mitochondrion, chloroplasts, some enzymatic systems) – Cu, Fe, Mn, Mo, and Zn.

Thus under the data available in scientific literature microelements participate in key metabolic events such as breath, photosynthesis, fixing and assimilation of some main nutrients (e.g., N, S). Microelements-metals of transitive group of periodic system activate enzymes or enter in metal-enzymes into systems of electrons carrying over (Si, Fe, Mn, and Zn); and also catalyze changes of valence in substances of a substratum (Cu, Co, Fe, and Mo). There are indications that some microelements (Al, Cu, Co, Mo, Mn, and Zn) carry out, probably, specific functions in protective mechanisms at cold-resistant and drought-resistant varieties of plants [23, 24].

**TABLE 10.4** Intracell Location and Main Function Of Microelements Vital for Plants (Source: [1] with Additions)

| Element | In what cell component enters | In what processes participates |
|---|---|---|
| Al | – | Govern colloidal properties of cell; possible activates some dehydrogenases and oxidases |
| Co | Co-enzyme cobamide | Symbiotic fixing of nitrogen; stimulation of oxidation-reduction reactions at chlorophyll and protein synthesis |
| Cu | Various oxidases, plasto-cyanins, and ceruloplas-min | Oxidation, photosynthesis, metabolism of pro-teins and carbohydrates, probably, participates in symbiotic fixing of nitrogen and oxidation-reduction reactions |
| Fe | Haemoproteins and other Fe-proteins, dehydroge-nases, ferrdoxins | Photosynthesis, nitrogen fixing, oxidation-reduction reactions |
| Mn | Many enzymatic systems | Oxygen photoproduction in chloroplasts and indirectly in $NO_3$ restoration |
| Mo | Nitrate-reductase, nitroge-nase, oxidases, molybde-num-ferrdoxin | Fixing $N_{2,}$ restoration $NO_3$, oxidation-reduction reactions |
| Ni | Urease (in shoots of *Canavalia*) | Probably, in reactions with participation of hydrogenase and in moving of nitrogen |
| Zn | Anhydrases, dehydroge-nases, proteinases and peptidases | Metabolism of carbohydrates and proteins |
| Cr | – | Possibly, participation in a glucose metabolism |

The requirement of plants as a whole and their separate species in certain microcomponents of nutrition is perfectly shown by Refs. [25] and [26]. If receipt of any necessary microelement is not enough plant growth deviates norm or stops and further development of a plant, in particular its metabolic cycles, are broken. Though symptoms of insufficiency cannot be bringing together they can be rather characteristic for some exact elements. Bergmann and Chumakov [27] result extensive data on symptoms of insufficiency (and also about some symptoms of toxicity) at agricultural plants. Descriptions of symptoms of insufficiency (see Table 10.4) show that the most frequent symptom is chlorosis. Symptoms found out visually are very important for diagnostics of insufficiency. However, alterations of metabolic processes and losses in biomass production occurring thereof can come before insufficiency symptoms become appreciable. For working out of the best methods of diagnostics biochemical indicators (Table 10.5) have been offered by a number of authors [28–30]. They are based on enzymatic trials, which are the sensitive test for latent insufficiency of the given nutritious microcomponent. Activity of some enzymes correlates basically with level of content of Cu, Fe, and Mo in plant tissues. However, practical use of enzymatic trials is rather limited because of large variability of enzymatic activity and technical difficulties of its definition.

Most widely used tests are analyzes of soils and plants. More exact diagnosis of critical levels of some microelements in plant tissues could be received at change of relations of antagonistic elements as it is shown on an example of relation Fe/Zn in maize [31].

**TABLE 10.5** Symptoms of Insufficiency of Nutritional Microcomponents in Some Agricultural Crops (Source: [1] With Additions)

| Element | Symptoms | Crops sensitive to insufficiency of an element |
|---|---|---|
| Cu | Wilt, melanism, white braided tops, reduction of panicle formation, disturbance of lignification | Cereals (oats), sunflower, spinach, Lucerne (alfalfa) |
| Fe | Interfibril chlorosis of young leaves | Fruit trees (citruses), grapes, some calcifugal species |
| Mn | Chlorosis spots and necrosis of young leaves, weakened turgor | Cereals (oats), legumes, fruit trees (apple-trees, cherries, citruses) |
| Mo | Chlorosis of leaf edges, whiptail of leaves and disturbance of curling of cauliflower, "fiery" edges and deformations of leaves caused by $NO_3^-$ surplus, destruction of germinal tissues | Cabbage and relative species, legumes |
| Zn | Interfibril chlorosis, a growth stop, rosetteness in trees, violet-red points on leaves | Cereal (corn), legumes, grasses, hop, flax, grapes, fruit trees (citruses) |
| Cr | Withering of aboveground part, damage of root system, chlorosis of young leaves, chlorotic spots and browning of leaves | Cereals |

Nevertheless analyzes of plant tissues can be used successfully for definition of deficiency of elements if they are compared to their contents in normal tissues of genotypes or plant species in the same organs and at the same stage of development. Extensive literature is published in different countries concerning diagnostics of deficiency of microelements and its correction by means of those or other micro-fertilizers. Last review of the current information on microcomponents has been published by Ref. [32]. It testifies to necessity of application of micro-fertilizers for a number of agricultural crops.

Both deficiency and toxicity of microelements for plants more often are results of complex interaction of some factors, which vary depending on specific properties of environment. However, many observations and experiments spent on various types of soils in different countries have made it clear that genesis and properties of soils are the primary factors governing occurrence of deficiency of microelements. Observed insufficiency of elements is usually connected with the extremely acid or alkaline soils having an adverse water regime and a lot of phosphates, nitrogen, and calcium as well as oxides of Fe and Mn.

Metabolic disturbances in plants are caused not only by a lack of microcomponents of nutrition but also by their surplus. As a whole, plants are resistant to increased rather than against the lowered concentration of elements. The great number of works about harmful action of surplus of microcomponents is published currently however the nature of these effects is still badly studied. According to review articles [22, 33, 34] the main reactions linked with toxic action of surplus of elements are the following:

1. Change in permeability of cellular membranes – Ag, Au, Br, Cd, Cu, F, Hg, I, and Pb.
2. Reactions of thiol groups with cations – Ag, Hg, and Pb.
3. Competition with vital metabolites – As, Se, Te, W, and F.
4. High affinity to phosphatic groups and the active centers in ADP and ATP – Al, Be, Sc, Y, Zr, lantanoids and, possibly, all heavy metals.
5. Replacement of the vital ions (mainly macro-cations) – Cs, Li, Rb, Se, and Sr.

6. Capture of positions in molecules occupied with the vital functional groups, such, as phosphate and nitrate – As, F, B, Br, Se, Te, and W.

The estimation of toxic concentration and action of microelements on plants is very difficult because it depends on such set of factors that they cannot be compared in a uniform linear scale. Proportions in which ions and their complexes are present in a solution concern to the most important factors. For example, As and Se toxicity goes down considerably in the presence of surplus of phosphate or sulfate; metal-organic complexes can be much more toxic than simple cations of the same element as well as much less toxic. It is necessary to notice also that some complexes, for example oxygen-contained compounds of elements, can be more poisonous rather than their simple cations.

Despite divergences in published data on levels of toxicity it is possible to ascertain that the most poisonous both for the higher plants and for a number of microorganisms are Hg, Cu, Ni, Pb, Co, Cd, and possibly, Ag, Be and Sn.

Though plants adapt quickly for chemical stresses all of them can be rather sensitive to surplus of a certain microelement. Resistibility of plants to action of heavy metals has special value. Development of tolerance to metals occurs quickly enough and as it is known has a genetic basis. Evolutionary changes caused by heavy metals are found out in a considerable number of species growing on soils enriched by metals. Such changes distinguish these plants from populations of the same species growing on nonpollinated soils [35]. Higher plants species which are finding out tolerance to microelements belong usually to following families: Caryophyllaceae, Cruciferae, Cyperaceae, Gramineae, Leguminosae, and Chenopodiaceae.

Speaking about laws of accumulation of heavy metals by plants it is necessary to mention works [36, 37] devoted to theoretical working out of phyto-indicatory researches in connection with heterogeneity of geochemical conditions of a spreading surface and a soil cover. They had been selected two groups of plant species: adapted for change of concentration of chemical elements and not adapted for it. Within the first group the subgroup is distinguished of the plant species strongly concentrating chemical elements even at the normal concentrations of microelements in environment (habitual concentrators); and a subgroup of plant species level of content of elements in which corresponds to their concentration in environment (unusual concentrators). In practice of other researches a division of plant species into similar groups also is extended. So, among the plants growing in Czechia on two sites, characterized by the lead concentrations in soil of 85.22 and 40.13 mg/kg accordingly three groups of species differing with level of pollution have been distinguished [38]. Such species as *Polygonum aviculare, Taraxacum officinale. Ranunculus repens, Planiago major,* and *Calamagrostis epigeios* consist first group in which lead accumulation did not depend on its concentration in soil (are listed as reduction of concentration of lead) and varied from 19.20 till 8.21 mg/g. In second group *Alchemilla monticola, Poientilla anserina,* and *Angelica sylvestris* have been jointed accumulating lead to proportionally its content in soil within concentration of 8.46–19.5 mg/g. Species *Achillea millefolium, Jacea pannonica, Arctiwn iappa, Tussilago farfara,* and *Heracleum sphondylium* were included into 3rd group concentration of lead in which reached the maximum values on sites with its minimum concentration in soil (up to 14.95–29.70 mg/g).

The top critical level of content of an element is equal to its least concentration in plant tissues at which there are toxic effects. Macnicol and Beckett [39] have spent processing of a great number of the published data for the purpose of an estimation of critical levels on a number of elements from which A1, Cd, Si, Mn, Ni, and Zn are most well studied (Table 10.6). They noticed that these values for each element are rather changeable that reflects, on the one hand, influence of interaction with other elements, and with another hand – increase in resistibility of plants to high concentrations of elements in tissues.

Mechanisms of resistibility of plants to action of microelements were a subject of many detailed researches which have shown that it can be observed as highly specific and group tolerance to metals [34, 40–43]. In the specified works the possible mechanisms participating in creation of

**TABLE 10.6** Approximate Toxic Concentration of Microelements in Mature Tissues of Leaves (mg/kg of Dry Matter) (After Ref. [39])

| Element | Cd | Co | Cr | Cu | Mn | Mo | Ni | Pb | Zn |
|---|---|---|---|---|---|---|---|---|---|
| Concentration | 5–30 | 15–50 | 5–30 | 20–100 | 300–500 | 10–50 | 10–100 | 30–300 | 100–400 |

tolerance to metals are summarized. Authors distinguished external factors such as low solubility and low mobility of cations in media surrounding plant roots; and also antagonistic action of ions of metals. True tolerance, however, is connected with internal factors. It does not represent a certain uniform mechanism and includes some metabolic processes: selective absorption of ions; the lowered permeability of membranes or other distinctions in their structure and functions; immobilization of ions in roots, leaves and seeds; removal of ions from metabolic processes by sedimentation (formation of stocks) in fixed and/or insoluble forms in various organs and organelles; change of character of a metabolism – strengthening of action of enzymatic systems which are exposed to inhibition; increase in content of antagonistic metabolites or restoration of metabolic chains at the expense of the admission of inhibited positions; adaptation to replacement of a physiological element in enzyme by toxic one; removal of ions from plants at washing away through leaves, sap exudation, defoliation, and exudation through roots.

Some authors [40, 43] produced proof that tolerant plants can be stimulated in their development by the raised amount of metals that testify to their physiological requirement in surplus of certain metals in comparison with the basic genotypes or species of plants. However, many moments are not clear yet in physiology of plant's tolerance to metals. Resistance of plants to the raised concentrations of microelements and their ability to accumulate extremely high concentration of microelements (Table 10.7) can represent the high hazard of people health as suppose penetration of pollution into food chains.

Equation of a chemical compound of live organisms is the basic condition of their normal growth and development. Interaction of chemical elements has the same value for plant physiology as the deficiency and toxicity phenomena. Interaction between chemical elements can be antagonistic or synergistic; and its unbalanced reactions can serve as the reason of chemical stresses at plants.

The antagonism arises when joint physiological action of one or more elements is less than sum of action of the elements taken separately; and synergism – when joint action is higher. Such interactions can be linked with ability of one element to inhibit or to stimulate absorption of other elements by plants. All these reactions are rather changeable. They can

**TABLE 10.7** The Greatest Concentration of Some Metals Which Have Been Found Out in Aboveground Parts at Various Plant Species in North-East of the European Part of Russia

| **Plant species** | **Mg** | **Cu** | **Ni** | **Cd** | **Pb** | **Zn** | **Fe** | **Cr** | **Mn** |
|---|---|---|---|---|---|---|---|---|---|
| Cirsium *arvense* (L.) Scop. | 668.50 | 9.06 | – | – | – | – | – | – | – |
| *Vicia cracca* L. | – | – | – | 0.90 | – | – | – | 5.56 | – |
| *Melilotus officinalis* (L.) Lam. | – | 8.69 | – | – | – | – | – | – | 133.30 |
| *Salix caprea* L. | 310.8 | 14.25 | 3.40 | 4.21 | 3.44 | 60.80 | – | – | – |
| *Salix viminalis* L. (*S. gmelinii* Pall.) | – | 12.81 | – | 0.90 | 3.20 | 75.80 | – | – | 146.30 |
| *Salix acutifolia* Willd. | 201.50 | 146.49 | – | 1.25 | – | 64.80 | 102.00 | 3.42 | – |
| *Salix myrsinites* L. | 204.5 | – | – | 0.80 | – | 82.40 | – | – | – |
| *Salix pentandra* L. | 301.50 | 44.15 | 3.80 | 0.58 | 3.50 | 83.30 | 115.40 | – | 117.51 |
| *Salix triandra* L. v. *glaucophylla* | 235.40 | – | – | 0.96 | 3.80 | 73.50 | 102.80 | – | – |
| *Salix aurita* L. | 345.80 | – | 3.40 | 0.90 | – | 69.40 | – | – | 175.20 |
| *Salix hexandra* Ehrh. | 302.40 | 117.2 | 5.84 | 1.20 | 3.50 | 59.40 | – | 7.71 | – |
| *Salix phylicifolia* L. | – | – | 5.21 | – | 3.50 | 56.80 | 115.80 | – | – |
| *Scirpus lacustris* L. | – | – | 5.40 | 0.90 | 3.50 | – | 119.80 | – | 316.19 |
| *Carex acuta* L. | 454.80 | – | – | 0.50 | – | – | – | 9.06 | – |
| Carex *nirga* (L.) Reichard | 521.40 | 14.38 | – | 0.60 | – | – | – | 8.13 | – |
| *Tanacetum vulgare* L. | – | 241.66 | – | 0.80 | 3.44 | – | – | 4.83 | – |
| *Sorbus aucuparia* L., fruits | – | – | – | 0.42 | – | 81.10 | – | – | – |
| *Sorbus aucuparia* L., shoots | – | – | – | 0.48 | – | – | 115.2 | – | 107.94 |
| *Rosa cinnamomea* L., fruits | – | – | 8.00 | 0.98 | – | – | 196.40 | – | – |

occur within cells, on a surface of membranes, and also in the environment surrounding plant roots. Interaction processes are governed by many factors and their mechanisms are still poor studied though some data is available nevertheless [34, 44, 45]. Perhaps, antagonistic-synergetic interactions of elements depend on level of their content in soil and can change their character at high concentration of pollutants.

Interactions between macro- and microelements, studied by a number of researchers [32, 46–49] show that Ca, P, and Mg are the main antagonistic elements concerning absorption and metabolism of many microelements. However, synergetic effects were observed sometimes for antagonistic pairs of elements that are connected, possibly, with specific reactions at separate genotypes or species of plants.

Antagonistic effects are realized by two ways more often: the macrocomponent can inhibit microelement absorption or, on the contrary, a microelement inhibits macrocomponent absorption.

Interactions between the microelements observed in plants show how much difficult are these processes as they can be that antagonistic as well as synergetic. Sometimes they are shown in a metabolism of more than two elements simultaneously. The greatest number of antagonistic reactions was observed for Fe, Mn, Cu and Zn which obviously are key elements in plant physiology (Tables 10.8 and 10.9). Functions of these microelements are linked with processes of absorption and with enzymatic reactions. Among other microelements in antagonistic relations to these elements often there are Cr, Mo, and Se.

**TABLE 10.8** Coefficients of Pair Correlation Between Concentrations of Some Heavy Metals in Shoots of Cowberry in Non-Pollinated (Background) Locations

| Element | Cu | Ni | Cd | Pb | Zn | Fe | Cr |
|---|---|---|---|---|---|---|---|
| Ni | 0.41 | | | | | | |
| Cd | 0.02 | –0.03 | | | | | |
| Pb | –0.23 | 0.05 | 0.21 | | | | |
| Zn | 0.16 | 0.15 | –0.14 | –0.34 | | | |
| Fe | –0.30 | –0.04 | –0.09 | –0.07 | 0.20 | | |
| Cr | –0.35 | –0.07 | –0.02 | 0.31 | –0.35 | 0.14 | |
| Mn | 0.14 | 0.07 | 0.10 | 0.37 | –0.44 | 0.14 | 0.29 |

**TABLE 10.9** Coefficients of Pair Correlation Between Concentrations of Some Heavy Metals in Shoots of Cowberry in Pollinated Locations

| Element | Cu | Ni | Cd | Pb | Zn | Fe | Cr |
|---|---|---|---|---|---|---|---|
| Ni | –0.62 | | | | | | |
| Cd | 0.97 | –0.44 | | | | | |
| Pb | 0.94 | –0.34 | 0.99 | | | | |
| Zn | –0.24 | 0.90 | –0.02 | 0.09 | | | |
| Fe | –0.96 | 0.43 | –1.00 | –0.98 | 0.06 | | |
| Cr | –0.62 | 0.78 | 0.88 | –0.34 | 0.73 | 0.39 | |
| Mn | –0.77 | 0.04 | –0.91 | 0.21 | –0.40 | 0.89 | 0.02 |

Synergetic interaction between microelements is not observed usually. Synergism of Cd with such microelements as Pb, Fe, and Ni, maybe artifact arising owing to destruction of physiological barriers under the influence of stress caused by superfluous concentration of heavy metals. Besides, some reactions occurring in the environment surrounding roots and influencing consumption of microelements by roots, apparently, are not connected directly with metabolic interactions; however these two types of reactions are hard for distinguishing.

## 10.4 CONCLUSIONS

So, carrying out gathering of a plant material for carrying out of chemical analysis on the content of polluting substances it is important to consider features of their accumulation both by separate species and various organs of plants. Comparison of levels of accumulation of metals between two species of a dandelion – *Taraxacum dahlstedtii* and *T. pectinatiforme* has shown that the first species under conditions of low pollution of soil accumulates toxicants in smaller amount than the second species. At high levels of pollution the opposite effect is observed. The established threshold level of the content of heavy metals in soil at which the root barrier limiting accumulation of heavy metals by plant tissues is makes for zinc 70, for cadmium – 0.3, for copper – 40 mg/kg.

## KEYWORDS


- **beebread-containing plants**
- **microelements**
- **phytocenoses**
- **pollution**
- **spatter dock**
- **species-specificity**


## REFERENCES


1. Kabata-Pendias, A. Trace Elements in Soils and Plants, Fourth Edition. Boca Raton, Florida. CRC Press. 2010, 548 p.
2. Little, P., Martin, M. H. A survey of zinc, lead and cadmium in soil and natural vegetation around a smelting complex. Environmental Pollution. 1972, 3, 241–254.
3. Kabata-Pendias, A. Leaching of micro- and macro-elements in columns with soil derived from granite. Pamietnik Pulawski (Polish Journal of Agronomy). 1969, 38, p. 111.
4. Roberts, T. M. A review of some biological effects of lead emissions from primary and secondary smelters. International Conference on Heavy Metals, Toronto, October 27, 1975, p. 503.
5. Wood, T., Bormann, F. H. Increases in foliar leaching caused by acidification of an artificial mist. Ambio. 1975, 4, 169–175.
6. Chernen'kova, T. V. Reaction of forest plants on industrial pollination. Moscow: Nauka ("Science" in Russian), 2002, 91 p. (in Russian).
7. Yaroslavtsev, G. D. Dust-protective features of same woody plants. Herald of Academy of Sciences of Turkmen Soviet Socialistic Republic. 1954, 5, 40–51 (in Russian).
8. Efimov, M. F. Influence of dust on plant growth. Botanic journal. 1959, 44(6), p. 822–824 (in Russian).
9. Khan, A. M., Pandey, V., Yunus, M., Ahmad, K. J. Plants as dust scavenders – a case study. Indian Forester. 1989, 115(9), p. 670–672.
10. Il'kun, G. M. Atmosphere pollutants and plants. Kiev: Naukova Dumka ("Scientific Thought" in Ukrainian), 1978, 248 p. (in Russian).
11. Firsova, V. P., Pavlova, T. S., Totishchev, V. V., Prokopovitch, E. V. Comparative study of heavy metal content in forest, meadow, and arable soils of forest-steppe ZaUral'e. Ecology. 1997, 2, 96–101 (in Russian).
12. Tarabrin, V. P. Physiology of woody plant resistance under condition of environmental pollution with heavy metals. Microelements in the environment. Kiev: Naukova Dumka ("Scientifi Thought" in Ukrainian), p. 17 (in Russian).

13. Mayer, R., Heinriches, H. Gehalte von Baumwurzeln an chemischen elementen einschliblich Schwermetallen aus Luftverunreiningungen. Zeitschrift für Pflanzenernährung und Bodenkunde. 1981, Bd. 144. № 6. S. 637–646 [Concentrations of chemical elements in tree roots including heavy metals from air pollution]. Journal of plant nutrition and soil science. 1981, 144(6), 637–646 [in German].
14. Garmash, G. A. Content of lead and cadmium in different parts of potato and vegetables grown in soil contaminated with these metals. Chemical elements in a system soil – plant. Novosibirsk: Siberian Branch of Russian Academy of Sciences, 1981, 105–110 (in Russian).
15. Krasnitsky, V. M. Estimation of agricultural production and soils of Omsk region. Agrochemical herald. 2002, 2, 7–10 (in Russian).
16. Scheffer, K., Stach, W., Vardakis, F. Ober die Verteilung der Schwermetallen Eisen. Mangan, Kupfer und Zink in Sommergesternpflanzen. Landwirtschaftliche Forschung. 1978, 31(1), p. 156–163 [The distribution of iron, manganese, copper and zinc in summer barley plants] [in German].
17. Rules for gathering of raw material of medicinal plants. Moscow: Kolos ("Ear" in English), 1989, 406 p. (in Russian).
18. Methodical guide for estimation of heavy metals in soils of agricultural lands and production of plant industry. Moscow: CINAO (Central research institute for agrochemical service of agriculture), 1992, 61 p. (in Russian).
19. Takhtajan, A. L. Floristic areas of Earth. Leningrad: Nauka ("Science" in English), 1978, 248 p. (in Russian).
20. Egoshina, T. L. Basic medicinal plants of Kirov region (ecological-and-cenotic characteristics). PhD Thesis. Saint-Petersburg: Botany Institute, 1999, 22 p. (in Russian).
21. Egoshina, T. L. Monitoring of resources of medicinal plants. Ecological botany: science, education, applied aspects. Materials of International conference. Syktyvkar: Syktyvkar State University, 2002, p. 93–94 (in Russian).
22. Bowen, H. J. M. Enviromental Chemistry of the Elements. Academic Press, New York, 1979, 333 p.
23. Marutian, S. A. Activity of micro and macroelements in vine shoots during non-growing season. 3rd Coll. Le Controle de l'Alimentation des Plantes Cultivees, Budapest, September 4, 1972, p. 763.
24. Shkol'nik, M. Ya. Microelements in plant life. Leningrad: Nauka ("Science" in English), 1974, 324 p. (in Russian).
25. Hewitt, E. J. Sand and Water Culture Methods Used in the Study of Plant Nutrition. Commonwealth Agriculture Bureaux, Bucks, UK, 1966, 547 p.
26. Chapman, H. D. (Ed.) Diagnostic Criteria for Plants and Soils. University of California, Riverside, Calif., 1972, 793 p.
27. Bergmann, W., Cumakov, A., Diagnosis of Nutrient Requirement by Plants. G. Fischer Verlag, Jena, and Priroda, Bratislava, 1977, 295 p.
28. Ruszkowska, M., Lyszcz, S., Sykut, S. The activity of catechol oxidase in sunflower leaves as indicator of copper supply in plants. Polish Journal of Soil Science. 1975, 8, 67–76.
29. Rajaratinam, J. A., Lowry, J. B., Hock, L. I. New method for assessing boron status of the oil palm. Plant Soil. 1974, Vol. 40, p. 417.
30. Gartrell, J. W., Robson, A. D., Loneragan, J. F. A new tissue test for accurate diagnosis of copper deficiency in cereals. Journal of agriculture, Western Australia. 1979, 20, 86–88.

31. Nambiar, K. K. M., Motiramani, D. P., Tissue Fe/Zn ratio as a diagnostic tool for prediction of Zn deficiency in crop plants. Plant and Soil. 1981, 60, 357–367.
32. Mengel, K., Kirkby, E. A., Principles of Plant Nutrition. International Potash Institute, Worblaufen-Bern, 1978, 593 p.
33. Peterson, P. J. Unusual accumulations of elements by plants and animals. Science Progress, Oxford. 1971, 59, 505–526.
34. Foy, C. D., Chaney, R. L., White, M. C. The physiology of metal toxicity in plants. Annual Review of Plant Physiology. 1978, 29, 511–566.
35. Bezel' V. S., Zhuikova, T. V., Pozolotina, V. N. Structure of ceno-populations of dindle and specificity of accumulation of heavy metals. Ecology. 1998, 5, 376–382 (in Russian).
36. Koval'sky, V. V., Petrunina, N. S. Geochemical ecology and evolutionary variation of plants. Reports of Russian Academy of Sciences. 1964, 159(5), 1175–1178 (in Russian).
37. Koval'sky, V. V. Geochemical ecology. Moscow: Nauka ("Science" in English), 1974, 298 p. (in Russian).
38. Bednarova, J. Hromadeni olova vybranymi populacemi roslin. Acta Universitatis Palackianae Olomucensis, Facultas rerum naturalium. Biologie. 1998, 93(28), 21–25 [Accumulation of Sn by some populations of plants].
39. Macnicol, R. D., Beckett, P. H. T. Critical tissue concentrations of potentially toxic elements. Plant and Soil. 1985, 85, 107–129.
40. Antonovics, J., Bradshaw, A. D., Turner, R. G. Heavy metal tolerance in plants. Advances in Ecological Research. 1971, 7, 1–85.
41. Bradshaw, A. D. The evolution of metal tolerance and its significance for vegetation establishment on metal contaminated sites. International Conference on Heavy Metals, Toronto, October 27, 1975, p. 599.
42. Simon, E. Cadmium tolerance in populations of Agrostis tenuis and Festuca ovina. Nature (London), 1977, 265, 328–330.
43. Cox, R. M., Hutchinson, T. C. Multiple metal tolerance in the grass Deschampsia cespitosa, L. Beuv. from the Sudbury Smelting area. New Phytologist. 1980, Vol. 84, p. 631–647.
44. Olsen, S. R. Micronutrient interactions. In: Micronutrients in Agriculture. Mortvedt, J. J., Giordano, P. M., Lindsay, W. L. (Eds.), Soil Science Society of America, Madison, Wis., 1972, 243–261.
45. Wallace, A. Regulation of the Micronutrient Status of Plants by Chelating Adents and Other Factors. Los Angeles, 1971, 309 p.
46. Gadd, G. M., Griffiths, A. J. Microorganisms and heavy metal toxicity. Microbial Ecology. 1978, 4, 303–317.
47. Kabata-Pendias, A., Pendias, H. Trace Elements in the Biological Environment. Warsaw, Publishing Geology (Wyd. Geol. in Polish), 1979, 300 p.
48. Mortvedt, J. J., Mays, D. A., Osborn, G. Uptake by wheat of cadmium and other heavy metal contaminants in phosphate fertilizers. Journal of Environmental Quality. 1981, 10(2), 193–197.
49. Roques, A., Kerjean, M., Auclair, D. The effect of atmospheric pollution by fluorine and sulfur dioxide on the reproductive organs of Pinus sylvestris in forest near Roumare (in France). Environmental Pollution. Series A, Ecological and Biological. 1980, 21, 191–201.

## CHAPTER 11

# DYNAMICS OF CADMIUM AND LEAD IN ACID SOD-PODZOLIC SOILS AND REACTION OF CEREAL PLANTS ON THEM

EUGENE M. LISITSYN[1, 2] and LYUDMILA N. SHIKHOVA[1]

[1]*Vyatka State Agricultural Academy, 133 Oktyabrsky Prospect, Kirov, 610017, Russia*

[2]*North-East Agricultural Research Institute of Rosselkhozacademy, 166-a Lenin St., Kirov, 610007, Russia, E-mail: shikhova-l@mail.ru*

## CONTENTS



### ABSTRACT

Estimation of total content of heavy metal in different horizons of acid sod-podzolic soils of Kirov region (Russia) covered with natural vegetation (forest, meadow) and used as arable land is spent by a method of atomic-absorption spectroscopy. The content of total cadmium in different horizons of sod-podzolic and podzolic loamy soils fluctuated from 0.66 to 1.11 mg/kg. The content of exchangeable cadmium in soils fluctuated from 0.01 to 0.30 mg/kg in different horizons. The greatest content of exchangeable Cd is noted in accumulating-eluvial parts of a profile and in mother rock, the least – in illuvial horizons. In humus horizons of meadow and fallow land soils its content is statistically higher than in arable land soils (0.17±0.02 and 0.08±0.02 mg/kg, accordingly). Ratio of exchangeable cadmium in its total content varied largely. In the top part of a profile it fluctuated from 10 to 40%, in illuvial parts drop down to 1–11% and increased again at approach to mother rock. During growth season content of exchangeable Cd in the top horizons of forest podzolic soil can vary by 4–5 times sometimes reaching the critical values close to MAC (0.2 mg/kg). In fallow land sod-podzolic soil it maximum is revealed in the middle of a season (1.0 mg/kg) that is almost 10 times higher than in autumn. On an arable land under clover crops total cadmium content is minimum in the beginning and the middle of a season and sharply rises in the autumn to the values close to MAC (0.8 mg/kg). It is not revealed any significant correlations between content of cadmium with other agrochemical properties of arable land soils. In covering loams of different districts of Kirov region the amount of total Pb fluctuated from 5.0 to 43.0 mg/kg. The least content of total lead is characteristic to granulometricaly light mother rocks. Significantly higher content of exchangeable forms of lead is in loamy soils (0.45–3.16 mg/kg) than in sandy one (0.45–2.22 mg/kg). The content of exchangeable lead in alluvial sandy deposits does not exceed 0.95 mg/kg. Arable horizons of loamy agricultural soils contain 26–30 mg/kg Pb. Sod-podzolic soils of region accumulate insignificant amounts of exchangeable Pb. Forest podzolic and sod-podzolic soils contain about 1 mg/kg of exchangeable lead in litter horizon. Sod-podzolic arable and meadow soils contained 1.86±0.13 mg/kg of exchangeable lead in humus horizons. High significant coefficients of correlation between contents of exchangeable lead and exchangeable acidity (from 0.81 to 0.99) are received. Influence of acidity on input of HM in plants, interactions of aluminum stress and action of HM has been investigated with use of methods of sand and soil culture. Treatment with cadmium and lead leads to accumulation of heavy metals by plants.

This accumulation reaches 30–35 mg/kg of dry mass. The most amount of cadmium is collected in roots and stems of cereals, and slightly less – in leaves. The basic part of lead was absorbed by roots. In variants with HM treatments the delay of germination for 2–3 days was marked; leaf blades were narrower than in control variants. By 5th week of growth the bottom leaves especially in a variant with lead treatment began to dry up and die off. Decrease in plant height is noted for all oats (by 41–61%) and barley varieties (by 40–48%). The greatest depression under the influence of HM is noted at such parameter as plant dry mass. At cadmium stress depression of plant dry mass makes 25.0–64.2%. Joint influence of cadmium and aluminum has appeared more considerably than toxic action of sole cadmium. Deviation from control values in a pigment complex of oats lies within limits 12.2–50.4% and of barley – within limits 18.2–66.0%. Changes had the general character and near identical intensity for a chlorophyll of both fraction (*a* and *b*). Changes in carotenoids content have similar character. As a whole level of Al-resistance and resistance to toxic action of high concentration of HM coincide on the majority of considered parameters.

## 11.1 INTRODUCTION

Numerous facts of environmental contamination by heavy metals (HM) as a result of dispersion of industrial emissions through atmosphere or in the form of a waste (slag, slime) and the polluted industrial waters are described in the literature [1, 2]. The Kirov region is included into first ten subjects of the Russian Federation on a ratio of farmland with excess of maximum allowable concentration (MAC) on heavy metals [3]. The volume of anthropogenous input of the most toxic metals – lead and cadmium – into environment reaches considerable values (451.64 and 4.05 t, accordingly) [4]. The number of the publications devoted to problems of technogenic pollution of soils of Kirov region with HM for the last years has considerably increased. But it is not always possible to be guided confidently in this data as the territory of region is poorly studied for background content of HM in soils and mother rock.

Long application of mineral fertilizers leads to accumulation of metals in soil [5–8]. Cadmium input is 2 times more than its removal by agricultural plants. It leads to excess of MAC of metals in soils and plants. Under the available data, concentration of cadmium, nickel, lead and chrome in agricultural production of some regions is already above their MAC [9]. Background content of Cd in soil is appreciably defined by soil type. In podzolic and sod-podzolic soils it makes 0.70–2.31 mg/kg, in gray wood soil – 0.65 mg/kg [10]. Concentration of metal in humus horizon of sod-podzolic loamy soils of the European Russia is about 0.14 mg/kg, of chernozems – 0.24 mg/kg [9]. The content of the element in humus horizons of the basic types of soils of Western Siberia is 0.074 mg/kg [11]. Loamy chernozems of southwest Altai contain 0.135 mg/kg of cadmium, alluvial sandy and sandy soils – 0.121 mg/kg, sod-podzolic sandy soils – 0.116 mg/kg [12]. Alluvial soils of Middle Ob' contain 0.1–1.8 mg/kg of cadmium in the top horizons [12].

The highest mobility and availability to plants cadmium possesses in acid soils with low pH level, low capacity of absorption, fulvatic structure of humus, and washing water mode [13, 14]. Lowering of mobility of cadmium in arable soils is promoted by periodic liming [15–17].

Many researchers studying toxicity of heavy metals for agricultural crops notice that cadmium is 2–20 times more toxic then other metals. At comparison of metals on toxicity in their equal doses they have following order: Cd > Zn > Cu > Pb or Cd > Ni > Cu > Zn > Cr > Pb [17, 18]. Investigations of cadmium treatment in a wide range of doses have allowed to establish that in each botanical family there are plants with various level of resistance to this element [17].

In soils of Russian Plain the background content of lead fluctuates from 2.6 to 43.0 mg/kg. On the average, top horizons of sod-podzolic and gray wood soils contain 14.6 and 17.2 mg/kg Pb according to [19] and up to 26–37 mg/kg in arable loamy soils [20]. Sod-podzolic soils of Udmurtiya contain total Pb: 13.9 mg/kg – in sandy and sandy loam soils; 25.6 mg/kg – in loamy soils. In a litter and humus horizon of sod-podzolic soil of Moscow Region there are 36 and 41 mg/kg of total lead according to Refs. [21, 22]. There is data that in acid soils essential part of Pb (II), up to 70%, is in an exchange condition, apparently, as a part of organic-mineral complexes [23]. In acid and strong acid media ions of Pb actively migrate within a soil profile and can be taken out for its limits. Movement of Pb in depth of a profile occurs in a chelate form.

The greatest concentration of lead is found out in the top layer of non-processed soils enriched with organic substance [21, 24]. The second maximum of lead content meets in illuvial horizons of sod-podzolic soils [10].

At last time lead involves a great attention as one of the main components of chemical pollution of environment and as an element, toxic for plants. Though there is no data testifying that lead is vital for growth of any kinds of plants, it is a lot of publications shown stimulating action of low concentration of some lead salts (mainly $Pb(NO_3)_2$) on plant growth. Moreover, effects are described of braking of plants metabolism because of low levels of lead in medium.

Values of MAC of lead for plants lie in wide enough intervals – from 0.5–1.2 mg/kg [25] to 10.0–20.0 mg/kg [26]. The natural content of lead in above ground parts of grasses makes from 1.5 to 40.0 mg/kg of dry matter [24].

Use of possibilities of plants to resist against stress action in a zone of roots (lowered pH of soil solution, presence of toxic ions of metals) is a theoretical basis of the ecological breeding directed on increase of resistance of agricultural crops to action of edaphic stressors [27].

Therefore, the purpose of the given work was (1) estimation of background contents of exchangeable ions of HM (cadmium and lead) in acid sod-podzolic soils of the Kirov region; (2) studying of relations between HM content and some properties of soils; and (3) interrelations of resistance of oats and barley plants to aluminum stress and action of heavy metals.

## 11.2 MATERIAL AND TECHNIQUE

Objects of research were as well as various varieties of oats and barley, used in agricultural production on these soils. For the characteristic of the profile content of heavy metals soil samples were taken by soil horizons to 70–150 cm depth from soil cuts in different areas of surveyed territory. Soil samples were processed according to the standard methods [28]. Estimation of total content of HM is spent by a method of atomic-absorption spectroscopy. Exchangeable fractions of HM were determined in acetate-ammonium buffer solution, pH 4.8 [28, 29].

During growth season (May–September) sample of soils from three top horizons of arable and fallow land podzolic loamy soil were collected monthly.

For studying of influence of acidity on input of HM in plants, interactions of aluminum stress and action of HM on parameters of resistance of oats and barley plants a series of experiments has been made with use of methods of sand and soil culture. Previously level of potential aluminum resistance of some oats and barley varieties was established [30] by method described earlier [31]. The oats and barley varieties different by level of aluminum resistance (in decreasing mode) – oats Krechet > Fakir > Ulov and barley Dina > Abava > Elf were objects of research.

The scheme of experiment in sand culture:

1. Control – without HM, a background – a full dose of Knop's solution;
2. Background + cadmium sulfate (100 μM for oats, 50 μM for barley);
3. Background + lead chloride (500 μM for an oats and barley).

Eight seeds of each variety were sown in pots with 0.5 kg of sand; after 7 days five most developed plants were left in each pot. Duration of experiment was 5 weeks (up to tillering stage). Analyzed traits: height of plants, volume of root system, cation exchange capacity of roots, dry weight of leaves, stems and roots, specific leaf area, content of pigments in leaves [32], the cadmium and lead content in leaves, stems and roots. Each variant of experiment had six replications for each variety.

The scheme of experiment in soil culture:

1. Control soil (without entering HM);
2. Control soil + 1 mg/kg of cadmium;
3. Control soil + 5 mg/kg of cadmium;
4. Control soil +10 mg/kg of cadmium;
5. Control soil + 60 mg/kg of lead;
6. Control soil + 300 mg/kg of lead;
7. Control soil + 600 mg/kg of lead.

Each pot contains 3.5 kg of soil (agrozem with the high content of organic substance (3.5%) and pH 7.25). Initially soil was carefully sifted, acidified with solutions of acids (sulfuric and hydrochloric depending on a pollution variant), treated with corresponding quantity of salt of metal (cadmium sulfate or lead chloride) and mixed.

Ten seeds of each variety of oats or barley were sown in each pot; seven most developed plants were left 7 days later. Duration of experiment is 5 weeks (up to tillering stage). Analyzed traits: height of plants, dry weight of leaves, stems and roots, a ratio of leaves, stems and roots in a plant, the content of pigments, the cadmium and lead content in leaves, stem and roots. Each variant of experiment had six replications for each variety.

For studying of joint action of high HM content and aluminum on plants a series of experiments in sand culture has been made. The scheme of experiment included:

1. Control – without HM, a background – a full dose of Knop's solution;
2. Background + aluminum sulfate (1 mM);
3. Background + aluminum sulfate (1 mM) + cadmium sulfate (100 μM for oats, 50 μM for barley).

Each variant of experiment had 6 replications for each variety.

At tillering, ear tube formation, and earing stages samples of leaves were collected for estimation content of pigments in leaves in triple replications was spent. The content and a ratio of a chlorophyll *a* and *b,* carotenoids in acetone extract [32] were defined.

Statistical processing of the data is spent by methods of dispersion and correlation analyzes with application of software packages STATGRAPHICS *Plus* for Windows 5.1.

## 11.3 RESULTS AND DISCUSSION

### *11.3.1 SEASONAL AND PROFILE DYNAMICS OF CADMIUM*

The territory of the Kirov region is characterized with various enough soil cover. The basic part of territory of area is occupied with soils of podzolic type. The area of podzolic and sod-podzolic soils makes 77.9% of total region area [33]. Sod-podzolic and podzolic soils of heavy mechanical structure are generated basically on covering deposits. In natural state they possess low enough level of fertility, have high acidity, the low contents of organic substance of fulvatic type, a low saturation with bases of a soil-absorbing complex.

The content of total cadmium in soils of the Kirov region is in the limits known for other regions, though some above than across Russia as a whole (Table 11.1).

**TABLE 11.1** Content of Cd in Profiles of Some Podzolic and Sod-Podzolic Loamy Soils of the Kirov Region

| Location | Soil horizon | Depth, cm | pH | Content of Cd (mg/kg) | |
|---|---|---|---|---|---|
| | | | | Total | Exchangeable |
| Murashi district, arable land | $A_p$ | 0–24 | 5.70 | 0.88 | 0.13 |
| | $A_{he}$ | 24–33 | 3.96 | 0.78 | 0.10 |
| | AB | 33–53 | 3.80 | 0.97 | 0.04 |
| | B | 53–86 | 3.82 | 0.74 | 0.04 |
| | BC | 86–118 | 3.85 | 0.93 | 0.01 |
| | C | 118–130 | 3.94 | 0.76 | 0.16 |
| Murashi district, meadow | $O_e$ | 0–4 | 4.59 | 0.26 | 0.11 |
| | $A_h$ | 4–15 | 4.05 | 0.70 | 0.29 |
| | $A_{he}$ | 15–20 | 3.87 | 0.86 | 0.10 |
| | AB | 20–36 | 3.80 | 0.86 | 0.05 |
| | B | 36–70 | 3.84 | 0.77 | 0.02 |
| Murashi district, forest | $O_i$ | 3–6 | 3.89 | 0.73 | 0.15 |
| | $A_{he}$ | 6–35 | 3.85 | 0.81 | 0.23 |
| | AB | 35–70 | 3.76 | 1.08 | 0.04 |
| | B | 70–80 | 3.88 | 0.72 | 0.08 |
| | BC | 80–115 | 3.69 | 0.66 | 0.12 |
| Afanasevo district, arable land | $A_p$ | 0–28 | 5.87 | 0.94 | 0.04 |
| | $A_{he}$ | 28–48 | 4.22 | 0.91 | 0.21 |
| | AB | 48–70 | 4.03 | 0.79 | 0.11 |
| | B | 70–96 | 4.10 | 0.96 | 0.11 |
| | BC | 96–120 | 4.21 | 1.20 | 0.13 |
| | C | 120–150 | 4.34 | 1.01 | 0.13 |
| Afanasevo district, forest | $O_i$ | 0–7 | 4.33 | 1.03 | 0.13 |
| | $A_{he}$ | 7–30 | 3.91 | 1.03 | 0.19 |
| | AB | 30–51 | 4.09 | 1.02 | 0.01 |
| | B | 51–86 | 5.18 | 1.08 | 0.05 |
| | BC | 86–120 | 7.07 | 1.02 | 0.07 |
| Falenki district, forest | $O_i$ | 0–7 | 3.95 | 0.84 | 0.07 |
| | $A_{he}$ | 7–20 | 3.87 | 0.79 | 0.15 |
| | AB | 20–30 | 4.02 | 0.69 | 0.01 |
| | B | 30–70 | 4.34 | 0.74 | 0.12 |

**TABLE 11.1** (Continued).

| Location | Soil horizon | Depth, cm | pH | Content of Cd (mg/kg) Total | Exchangeable |
|---|---|---|---|---|---|
| Zuevka district, fallow land | $A_h$ | 0–27 | 4.54 | 0.82 | 0.07 |
| | $A_{he}$ | 27–45 | 4.04 | 0.74 | 0.19 |
| | AB | 45–70 | 4.05 | 0.84 | 0.05 |
| | B | 70–103 | 4.09 | 1.09 | 0.10 |
| | BC | 103–120 | 4.38 | 0.88 | 0.13 |
| | C | 120–130 | 4.51 | 0.99 | 0.12 |

For different horizons of sod-podzolic and podzolic loamy soils its content fluctuated from 0.66 to 1.11 mg/kg. Biogenic accumulation of total Cd in the studied profiles is expressed slightly that some authors explain with considerable migration of cadmium downwards on a profile in soils of humid landscapes [24]. Eluvial-illuvial redistribution of total Cd in profiles was often observed.

The content of exchangeable cadmium in soils fluctuated from 0.01 to 0.30 mg/kg in different horizons. The greatest content of exchangeable Cd is noted in accumulating-eluvial parts of a profile and in mother rock, the least – in illuvial horizons. In some humus and eluvial horizons the content of exchangeable cadmium exceeds known level of MAC equal to 0.2 mg/kg. Redistribution of Cd on a profile is considerable influenced with exchangeable organic substance; cadmium content is higher in those horizons where content of labile organic matter is higher too.

Ratio of exchangeable cadmium in its total content varied largely. In the top part of a profile it fluctuated from 10 to 40%, in illuvial parts drop down to 1–11% and increased again at approach to mother rock. Statistically significant correlation between the content of total and exchangeable cadmium was absent. Content of exchangeable cadmium in organogenic horizons of the studied podzolic and sod-podzolic soils is insignificant as a whole. However, during growth season content of exchangeable Cd in the top horizons of forest podzolic soil can vary by 4–5 times sometimes reaching the critical values close to MAC (Figure 11.1).

Sod-podzolic soils of Kirov region are characterized by the minimum content of exchangeable cadmium (Table 11.1). Thus in humus horizons

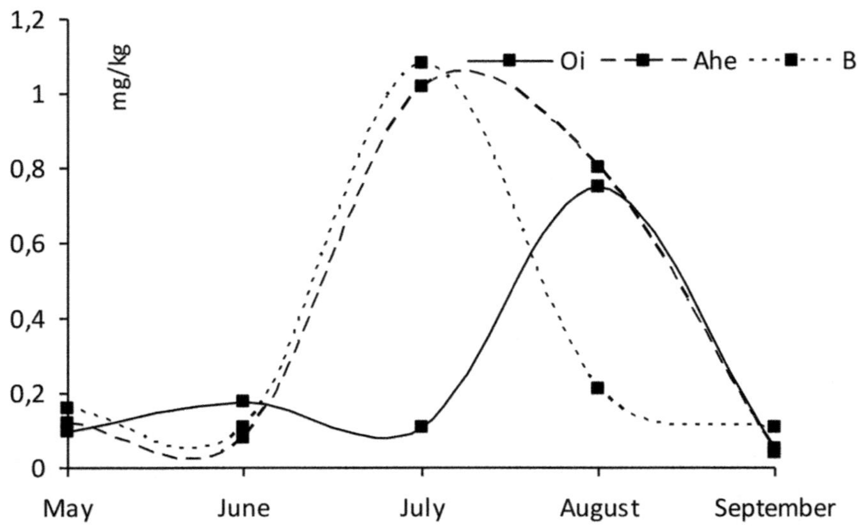


**FIGURE 11.1** Seasonal dynamics of content of exchangeable Cd in forest podzolic soil of the middle taiga.

of meadow and fallow land soils its content is statistically higher than in arable land soils (0.17±0.02 and 0.08±0.02 mg/kg accordingly). The similar content of exchangeable cadmium in arable layer of sod-podzolic loamy soil was noted by [34] – 0.11 mg/kg. Content of exchangeable Cd in arable horizons of sod-podzolic soils of Yarano-Vyatka interfluve is 0.09–0.28 mg/kg, of sod-gleyey soils – 0.12–0.19 mg/kg [35]. Unfortunately authors do not specify in what extractant exchangeable cadmium was defined.

Distribution of exchangeable cadmium on soil profiles of arable sod-podzolic soils in different districts of Kirov region has the same features as on forest sod-podzolic and podzolic soils (Figure 11.2).

Arable horizons as a rule contain a little amount of exchangeable Cd. The maximum content is in eluvial horizons and the minimum one is almost always in illuvial parts of profiles.

It is not revealed any significant correlations between content of cadmium with other agrochemical properties of arable land soils probably because of constant anthropogenous influences. In meadow soils content of exchangeable Cd positively correlates with pH ($r = 0.60$) that obviously is connected with conditions of mineralization of organic matter. Besides it

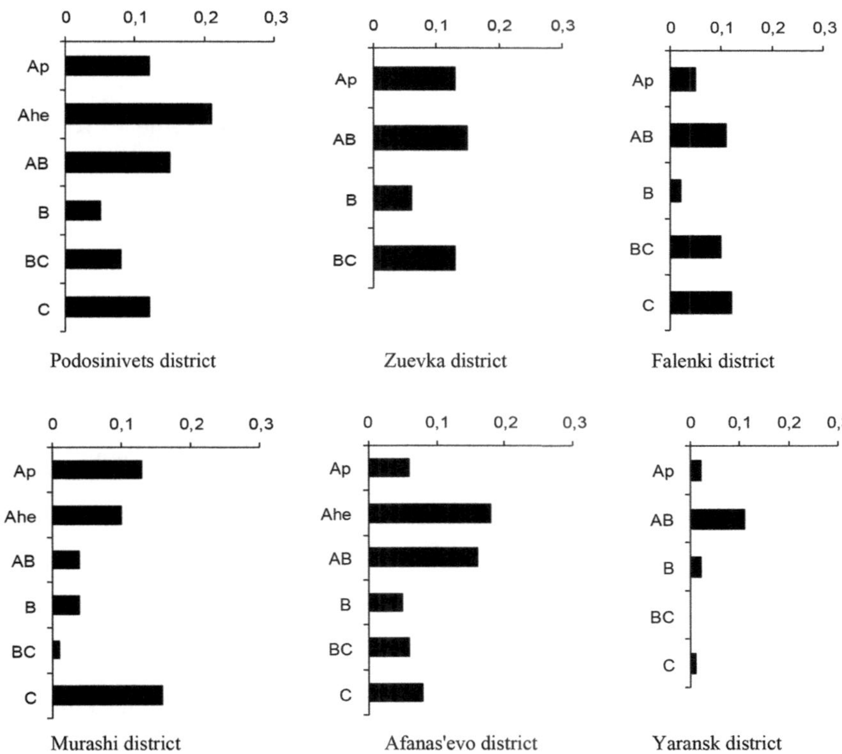


**FIGURE 11.2** Content of exchangeable Cd in profiles of sod-podzolic arable soils in different districts of Kirov region (mg/kg).

is known that exchangeable cadmium is presented in loamy soils basically as exchangeable forms therefore there is a competition between cadmium and calcium, magnesium for exchange sites in soil-absorbing complex. Coefficient of pair correlation between content of Cd on one hand and calcium and magnesium on the other hand is as much as $r = -0.57$.

High coefficient of variation of exchangeable Cd content in soils may be partly explained with fluctuation of the content of the element during a growth season. For example in fallow land sod-podzolic soil it maximum is revealed in the middle of a season (1.0 mg/kg) that is almost 10 times higher than its content in autumn (Figure 11.3).

On an arable land under clover crops cadmium content is minimum in the beginning and the middle of a season and sharply rises in the autumn

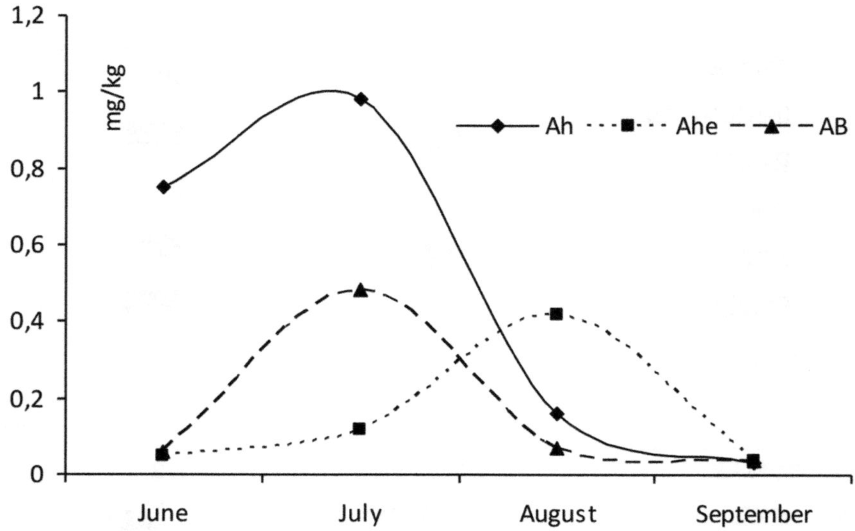


**FIGURE 11.3** Seasonal dynamics of content of exchangeable Cd in fallow land sod-podzolic soil of the Middle taiga.

to the values close to MAC (0.8 mg/kg). Therefore, soil samples collected at different terms can strongly differ on content of exchangeable cadmium.

In fallow land soil where soil processes come nearer to natural high significant coefficients of correlation between content of exchangeable cadmium and content of total and labile carbon (0.64 and 0.73, accordingly) are noted. For arable soil there is not revealed any significant relations between these parameters; apparently it is caused by anthropogenous influence.

### *11.3.2 SEASONAL AND PROFILE DYNAMICS OF LEAD*

In mother rocks and soils of the Kirov region the content of lead (Table 11.2) did not fall outside the limits known for other regions.

In covering loams of different districts of Kirov region the amount of total Pb fluctuated from 5.0 to 43.0 mg/kg. Eluvial-illuvial horizons of Perm deposits of Kotelnich and Orlov districts contain a little total Pb – 7.8 and 9.0 mg/kg accordingly. The least content of total lead as one would expect

**TABLE 11.2** Content of Pb in profiles of some podzolic and sod-podzolic loamy soils of the Kirov region.

| Location | Soil horizon | Depth, cm | pH | Content of Pb (mg/kg) | |
|---|---|---|---|---|---|
| | | | | Total | Exchangeable |
| Murashi district, arable land | $A_p$ | 0–24 | 5.70 | 30 | 1.38 |
| | $A_{he}$ | 24–33 | 3.96 | 17 | 1.25 |
| | AB | 33–53 | 3.80 | 15 | 1.40 |
| | B | 53–86 | 3.82 | 26 | 3.20 |
| | BC | 86–118 | 3.85 | 23 | 3.11 |
| | C | 118–130 | 3.94 | 22 | 4.25 |
| Murashi district, meadow | $O_e$ | 0–4 | 4.59 | 24 | 2.15 |
| | $A_h$ | 4–15 | 4.05 | 20 | 2.00 |
| | $A_{he}$ | 15–20 | 3.87 | 16 | 1.44 |
| | AB | 20–36 | 3.80 | 29 | 1.68 |
| | B | 36–70 | 3.84 | 15 | 3.26 |
| Murashi district, forest | $O_i$ | 3–6 | 3.89 | 27 | 6.75 |
| | $A_{he}$ | 6–35 | 3.85 | 24 | 5.50 |
| | AB | 35–70 | 3.76 | 20 | 1.42 |
| | B | 70–80 | 3.88 | 20 | 1.42 |
| | BC | 80–115 | 3.69 | 20 | 3.20 |
| Afanasevo district, arable land | $A_p$ | 0–28 | 5.87 | 26 | 1.28 |
| | $A_{he}$ | 28–48 | 4.22 | 34 | 1.45 |
| | AB | 48–70 | 4.03 | 31 | 0.40 |
| | B | 70–96 | 4.10 | 32 | 0.40 |
| | BC | 96–120 | 4.21 | 25 | 1.11 |
| | C | 120–150 | 4.34 | 38 | 1.00 |
| Afanasevo district, forest | $O_i$ | 0–7 | 4.33 | 32 | 2.42 |
| | $A_{he}$ | 7–30 | 3.91 | 37 | 2.42 |
| | AB | 30–51 | 4.09 | 27 | 2.62 |
| | B | 51–86 | 5.18 | 33 | 2.62 |
| | BC | 86–120 | 7.07 | 35 | 2.41 |
| Falenki district, forest | $O_i$ | 0–7 | 3.95 | 22 | 2.51 |
| | $A_{he}$ | 7–20 | 3.87 | 20 | 1.45 |
| | AB | 20–30 | 4.02 | 14 | 0.84 |
| | B | 30–70 | 4.34 | 5 | 1.12 |

**TABLE 11.2** (Continued).

| Location | Soil horizon | Depth, cm | pH | Content of Pb (mg/kg) | |
|---|---|---|---|---|---|
| | | | | Total | Exchangeable |
| Zuevka district, fallow land | $A_h$ | 0–27 | 4.54 | 28 | 2.50 |
| | $A_{he}$ | 27–45 | 4.04 | 19 | 0.75 |
| | AB | 45–70 | 4.05 | 25 | 0.50 |
| | B | 70–103 | 4.09 | 27 | 1.20 |
| | BC | 103–120 | 4.38 | 30 | 1.56 |
| | C | 120–130 | 4.51 | 20 | 2.41 |

is characteristic to granulometrically light mother rocks. Fluvial-glacial and ancient alluvial deposits of Kotelnich, Orlov and Orichi districts of the Kirov region contain 1.9 to 3.75 mg/kg of lead. Significantly higher content of exchangeable forms of lead is also in loamy soils (0.45–3.16 mg/kg) than in sandy one (0.45–2.22 mg/kg). The content of exchangeable lead in modern alluvial sandy deposits does not exceed 0.95 mg/kg. It is not revealed statistically significant correlations between the contents of total and exchangeable Pb in mother rocks. The amount of exchangeable Pb both in loamy and in light soils is not connected significantly with their content of organic substance, pH value, and level of exchange acidity. As a whole the content of total and exchangeable Pb in basic mother rocks of region can be characterized as an average, which is not beyond literary known concentration for sedimentary rocks.

Arable horizons of loamy agricultural soils contain 26–30 mg/kg Pb. Sod-podzolic soils of region accumulate insignificant amounts of exchangeable Pb. Forest podzolic and sod-podzolic soils contain about 1 mg/kg of exchangeable lead in litter horizon (Table 11.2). Thus litter horizons of loamy soils contain statistically more lead than litter horizons of sandy soils. Element distribution on profiles of forest soils has chaotic enough character. Samples of loamy soils contain as a whole more exchangeable fractions of an element than of light soils.

The maximum content of an element is not always marked in the top horizons. It is obvious that absence of laws of distribution of an element on a profile is caused by seasonal dynamics of its content. The amount of lead during a season can differ by 2–3 times (Figure 11.3). Loamy forest soils contain significantly more exchangeable lead than sandy soils.

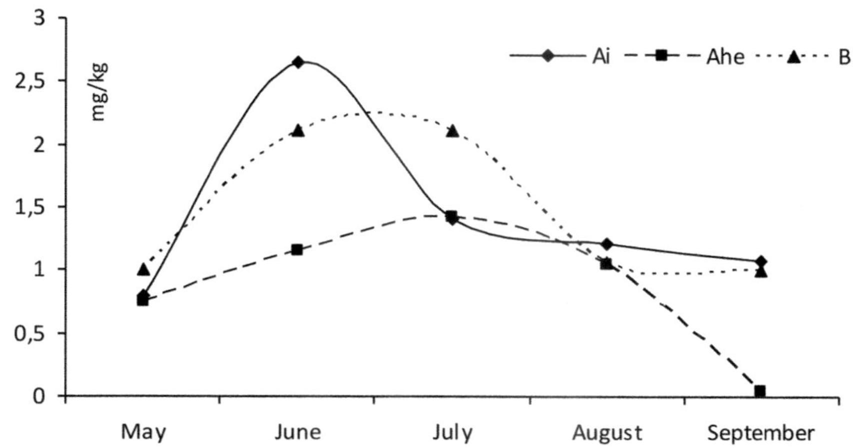


**FIGURE 11.3** Seasonal dynamics of exchangeable Pb content in podzolic soils of the middle taiga.

Sod-podzolic arable and meadow soils contained 1.86±0.13 mg/kg of exchangeable lead in humus horizons on the average (Table 11.2). Element distribution on profiles of arable soils has not strongly pronounced biogenic-accumulative character. Sometimes the top part of a profile is poor in exchangeable lead in comparison with underlying horizons. The increase in content of exchangeable fractions of an element in mother rocks is more expressed. The minimum of content of an element is almost always noticed in eluvial horizons.

In arable soils it is not revealed statistically significant correlations between content of exchangeable Pb and pH, total carbon, exchangeable acidity, and humidity also.

Seasonal dynamics of exchangeable lead in arable and fallow land sod-podzolic soil are various (Figures 11.4 and 11.5). Dynamics of the content of exchangeable Pb significantly correlates with ratio of labile organic matter in humus horizons of both locations but with different signs. The correlation coefficient is equal to $r = -0.90$ on arable land and $r = 0.95$ on a fallow land. Such different values of correlation coefficients possibly testify various dynamics of organic matter in annually processed and fallow soils and with different structure of organic matter.

In both sites high significant coefficients of correlation between contents of exchangeable lead and exchangeable acidity (from 0.81 to 0.99)

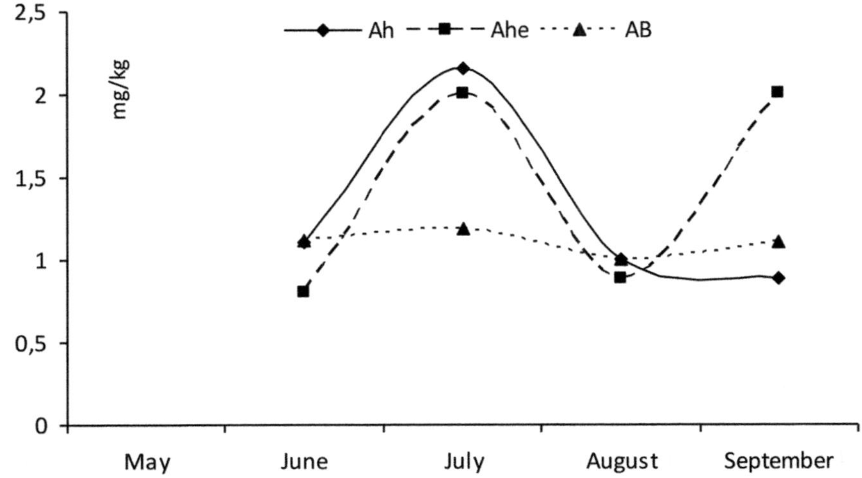


**FIGURE 11.4** Seasonal dynamics of exchangeable Pb in fallow land podzolic soil of the middle taiga.

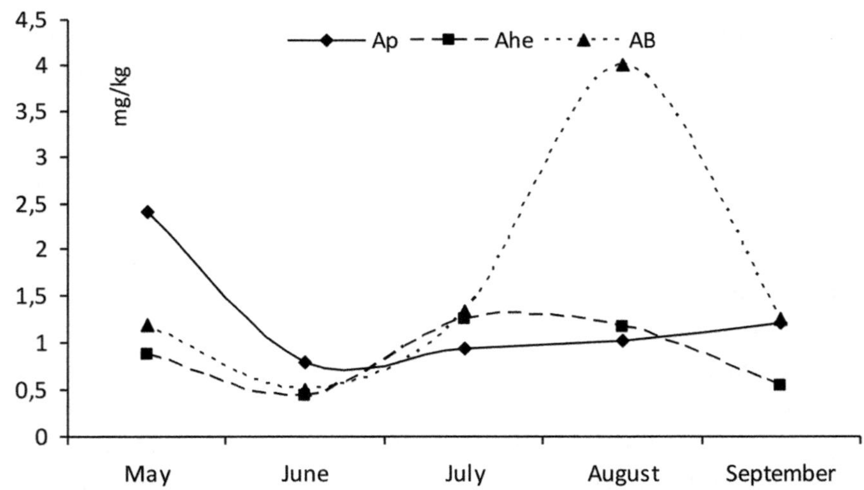


**FIGURE 11.5** Seasonal dynamics of exchangeable Pb in arable podzolic soil of the middle taiga.

are received. It is caused by a competition between ions $Al^{3+}$, $H^+$ and $Pb^{2+}$ for exchange sites in soil-absorbing complexes when the acidity increase conducts to replacement of $Pb^{2+}$ into solution.

### 11.3.3 INFLUENCE OF IONS OF HEAVY METALS OF OATS AND BARLEY PLANTS

Data of Table 11.3 indicate that input of cadmium and lead (in a dose of 4.5 and 41.0 mg/kg of sand accordingly) in the conditions of sandy culture leads to accumulation of heavy metals by plants. This accumulation exceed control level by some ten times and reaches 30–35 mg/kg of dry mass. Distribution of heavy metals on parts of oats and barley plants is different. The most amount of cadmium is collected in roots and stems of cereals, and slightly less – in leaves. The basic part of lead was absorbed by roots. Varieties of oats and barley could not be differentiated on degree of heavy metal accumulation in their vegetative organs.

During growth season plants of oats and barley in variants of heavy metal treatment did not differ significantly from control variants on development of morphometric traits. Unique difference is noted at oats plants. It consisted in various terms of approach of a full phase of panicle formation. In variants with lead treatment approach of this phase has been noted for 3 days before control variants, and at cadmium treatment – for 3 days later.

**TABLE 11.3** Accumulation of Heavy Metals in Plants of Oats and Barley

| | | Cd, mg/kg of dry mass | | | Pb, mg/kg of dry mass | | |
|---|---|---|---|---|---|---|---|
| **Variety** | **Treatment** | **Roots** | **Stems** | **Leaves** | **Roots** | **Stems** | **Leaves** |
| ***Oats*** | | | | | | | |
| Ulov | Control | 1.08 | 0.60 | 0.42 | 27.61 | 4.13 | 3.44 |
| | Heavy metal | 13.45 | 27.08 | 24.22 | 65.46 | 11.11 | 9.49 |
| Fakir | Control | 0.27 | 0.52 | 0.49 | 21.24 | 4.10 | 4.29 |
| | Heavy metal | 11.70 | 26.11 | 21.63 | 61.41 | 29.99 | 8.63 |
| Krechet | Control | 0.43 | 0.40 | 0.47 | 40.12 | 2.84 | 3.42 |
| | Heavy metal | 17.73 | 38.92 | 21.44 | 56.65 | 13.53 | 9.49 |
| ***Barley*** | | | | | | | |
| Elf | Control | 0.75 | 0.44 | 0.17 | 15.05 | 3.61 | 6.87 |
| | Heavy metal | 28.34 | 49.15 | 17.17 | 63.72 | 21.25 | 8.71 |
| Abava | Control | 0.57 | 0.66 | 0.28 | 30.95 | 7.14 | 6.55 |
| | Heavy metal | 24.16 | 13.08 | 12.90 | 48.12 | 20.21 | 12.47 |
| Dina | Control | 0.47 | 0.23 | 0.41 | 35.81 | 8.22 | 3.45 |
| | Heavy metal | 33.30 | 36.41 | 8.14 | 59.34 | 24.3 | 11.35 |

Experiments in sand culture have shown that mutual action of aluminum and cadmium on plants of oats is shown in decrease in plant height by 28.3–32.3%. Significant decrease in plant height among barley was observed at variety Dina (in cadmium + aluminum treatment – by 6.8%, in aluminum treatment – by 16.2%) and Elf (in cadmium + aluminum treatment – by 7.6%).

Significant decrease in dry mass of plants was observed for oats Fakir and Krechet (by 25.2 and 34.8% accordingly) at mutual treatment with cadmium and aluminum (Figure 11.6). Thus the mass shortage is basically characteristic for such parts of plant as stems and roots; but the mass ratio of leaves is increased in comparison with a control treatment. At barley decrease in dry mass of plants is noted also at mutual action of cadmium and aluminum for varieties Elf (by 11.8% in comparison with the control treatment) and Dina (by 19.3%) (Figure 11.6). Unlike oats, in barley depression is characteristic for roots and leaves – change in stems has not concerned; in a background of decrease in total mass of plant the mass ratio of stems has even grown. At oats Ulov and barley Abava the mass of plants did not change significantly.

In a pigment complex of leaves significant disturbances have occurred only in oats Fakir in Cd+Al treatment: content of chlorophyll *b* increase by 28.3%, but content of chlorophyll *a* has decreased a little; therefore change of the content of total chlorophyll ($a + b$) has appeared insignificant.

Among barley disturbances in pigment content of leaves are revealed for varieties Abava and Dina but changes are characteristic only for variants of aluminum treatment and have diverse character: variety Abava has increased content of chlorophyll *a* and at the expense of it the sums of

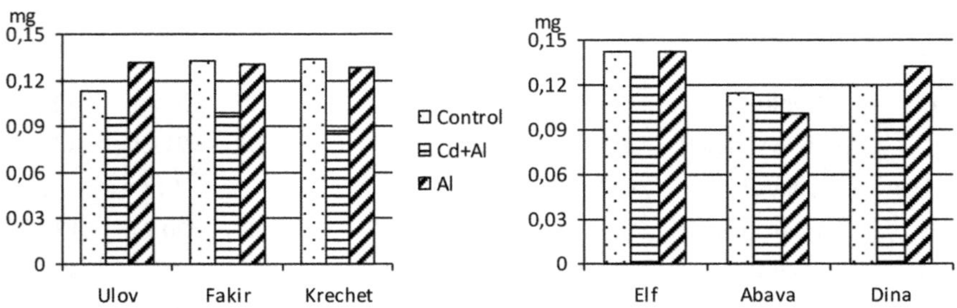


**FIGURE 11.6** Influence of Cd and Al on dry mass of oats and barley plants.

chlorophyll; variety Dina on the contrary has lowered it. At mutual influence of cadmium and aluminum there were not any statistically significant changes in a pigment complex of barley.

Both in oats and in barley carotenoids of leaves have appeared more stable – their changes are insignificant.

Possibly used concentration of cadmium (4.5 mg/kg of sand) and aluminum (11 mg/kg of sand) are insufficient for depression of development of cereal plants under conditions of sand culture. Considering that MAC of total cadmium in sandy and sandy loam soils makes 0.5 mg/kg, it is possible to notice that nine fold excess of MAC does not lead to considerable depression of oats and barley plants' growth and development even in sand culture when there is no practically absorbing ability (buffer action of soil in relation to pollutant) and all metal is in the exchangeable form accessible to plants. Possibly given cereal crops possess effective mechanisms of cadmium detoxication.

To overcome protective mechanisms of plants concerning heavy metals concentration of cadmium and lead has been increased up to 10 and 640 mg/kg of sand accordingly that exceed their MAC levels by 20 times.

With increase in heavy metal and Al concentration depression of almost all estimated parameters began to have significant character. In variants with HM treatments the delay of germination for 2–3 days was marked; leaf blades were narrower than in control variants. By 5th week of growth the bottom leaves especially in a variant with lead treatment began to dry up and die off. Decrease in plant height in comparison with the control treatment is noted for all oats varieties (Ulov has lowered plant height on the average by 61%, Fakir – by 49%, and Krechet – by 41%) and barley varieties (Elf – by 48%, Abava – by 40%. and Dina – by 42%) (Figure 11.7).

Joint action of HM and aluminum was observed only in barley (basically in cadmium + aluminum treatment). In variants with lead treatment for barley and in all treatments for oats it is noted significant difference between action of single metal and joint action of metal and aluminum. Thus varieties of oats and barley are classified by degree of growth depression under condition of sand culture in the same order as by their resistance to high acidity and aluminum content in natural soils.

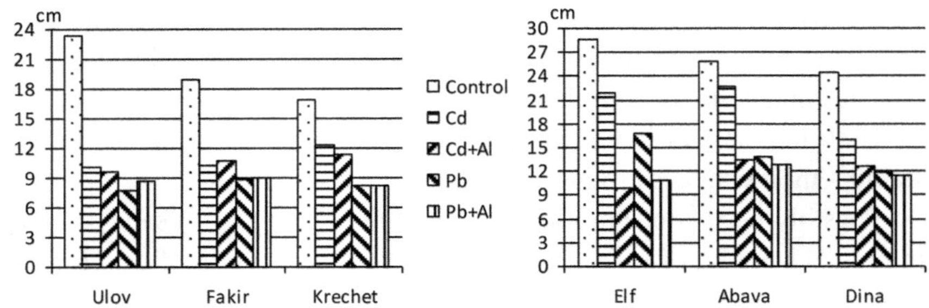


**FIGURE 11.7** Influence of Al and high doses of Cd and Pb on height of plants of barley and oats.

The greatest depression under the influence of HM is noted at such parameter as plant dry mass (Figure 11.8). Oats plants demonstrate decrease in plant dry mass in lead and lead + aluminum treatments by 76.2–85.8%; variety Ulov has most suffered and Krechet – the least. At cadmium stress depression of plant dry mass is less and makes from 64.2% in variety Ulov (cadmium + aluminum treatment) to 25% in variety Krechet in the same variant. Thus resistance of oats varieties estimated by dry mass of plants coincides with their field aluminum resistance. The difference between treatments with sole HM and HM + Al is insignificant except for a case of variety Krechet considered above.

A same different pattern is observed in barley (Figure 11.8). Resistance of varieties to HM action by given trait does not coincide with their Al-resistance: barley Elf has less suffered. The greatest decrease in plant dry mass is noted in cadmium + aluminum, lead, and lead + aluminum treatments – by 63.3–87.1%. In sole cadmium treatment depression of plant

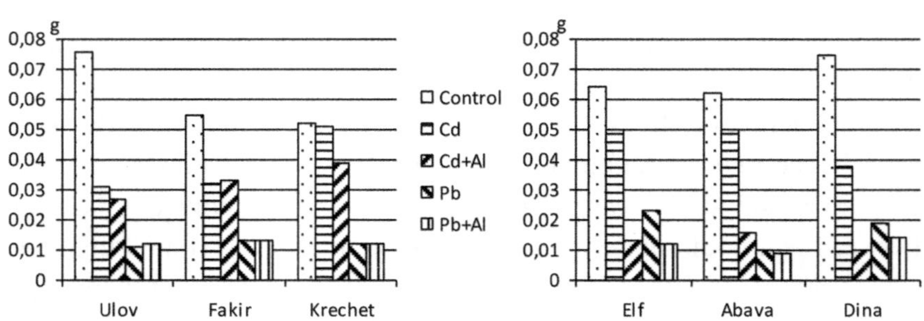


**FIGURE 11.8** Influence of Al and high doses of Cd and Pb on dry mass of plants of oats and barley.

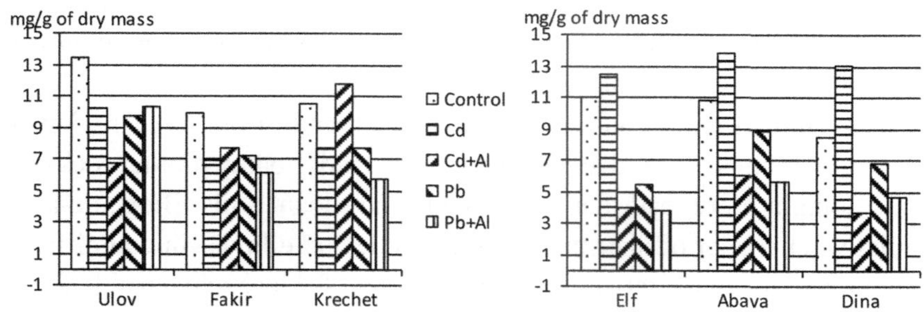


**FIGURE 11.9** Influence of Al and high doses of Cd and Pb on content of total chlorophyll in oats and barley leaves.

mass makes no more than 50%. Joint influence of cadmium and aluminum has appeared more considerably than toxic action of sole cadmium.

Under the influence of high HM concentration changes have occurred in a pigment complex of leaves as well. Basically decrease in content of chlorophyll takes place but stimulation effect however was marked too.

Deviation from control values in a pigment complex of oats lays within limits 12.2% to 50.4% and is observed in different variants of treatment. Joint action of HM and aluminum was showed both in decrease, and in increase of chlorophyll content; these changes had the general character for chlorophyll of both fractions (*a* and *b*) (Figure 11.9); however content of chlorophyll *a* varied in a greater degree to what significant decrease in the ratio *a/b* in comparison with control values (by 7.1–18.0%) testifies.

In a pigment complex of barley a deviation from control values lays within limits 18.2% to 66.0%. Under cadmium treatment stimulation of synthetic of pigments was observed but in all other treatments – its depression. Changes had the general character and near identical intensity for a chlorophyll of both fraction (*a* and *b*). Changes in carotenoids content have similar character. Pigment complex of barley Abava and Dina was more resistant to toxic action of HM that as a whole coincides with their characteristic as edaphic resistant varieties.

## 11.4 CONCLUSION

Thus soils of Kirov region are characterized in comparison with other part of the European Russia by raised content of total Cd and insignificant amount of exchangeable cadmium. High content of exchangeable and total

Cd in some soil samples are caused possibly with anthropogenic influence. Exchangeable cadmium accumulates in top accumulating-eluvial parts of soil profile as a part of labile organic matter.

The studied region does not differ considerably from other regions of Russia on content of total and exchangeable lead. Loamy mother rocks and soils contain more lead than sandy and sandy loam soils. Biogenic-accumulative character of distribution on a profile and elluvial-iluvial differentiation is more characteristic for total lead than for its exchangeable forms. Profile distribution of exchangeable lead is chaotic that is caused by considerable seasonal dynamics of its content. It was not possible to reveal accurate dependences between the content of lead and the content of organic matter. High correlation coefficients between content of organic matter and the content and mobility of lead in arable and fallow land sod-podzolic soils serve as indirect evidence of considerable influence of organic matter on the content of lead at studying of their seasonal dynamics.

Such heavy metals as cadmium and lead have obvious toxic effect on cereal crops only at very high concentration exceeding MAC in some tenfold. The effect of joint treatment with high doses of HM and aluminum as a rule is shown on barley and is more characteristic for cadmium + the aluminum treatment; barley Abava which are intermediate on Al-resistance has appeared the most resistant to joint toxic action of HM and aluminum.

## KEYWORDS

- **acid soil**
- **aluminum**
- **barley**
- **cadmium**
- **carotenoids**
- **chlorophyll**
- **lead**
- **oats**
- **seasonal and profile dynamics**
- **soil horizons**

## REFERENCES

1. Abii, T. A., Okorie, D. O. Assessment of the level of heavy metals [Cu, Pb, Cd and Cr] contamination in four popular vegetables sold in urban and rural markets of Abia State Nigeria. Continental Journal of Water Air and Soil pollution (Nigeria). 2011, 2(1), 42–47.
2. Machado, S., Rabelo, T. S., Portella, R. B., Carvalho, M. F., Magna, G. A. M. A Study of the Routes of Contamination by Lead and Cadmium in Santo Amaro, Brazil. Environmental Technology. 2013, 34, 559–571.
3. State (national) Report on Condition and Use of Land in Russian Federation for 1995, Moscow: Russlit Publ., 1996, 120 p. (in Russian).
4. Burkov, N. A. Applied Ecology. Kirov: Publishing House Vyatka, 2005, 271 p. (in Russian).
5. Oyedele, D. J., Asonugho, C., Awotoye, O. O. Heavy Metals in Soil and Accumulation by Edible Vegetables After Phosphate Fertilizer Application. Electronic Journal of Environmental Agricultural Food Chemistry 2006, 5(4), 1446–1453.
6. Guzman, E. T. R., Regil, E. O., Gutierrez, L. R. R., Albericl, M. V. E., Hernandez, A. R., Regil, E. D. Contamination of corn growing areas due to intense fertilization in the high plane of Mexico. Water, Air and Soil Pollution. 2006, 175, 77–98.
7. Mendes, A. M. S., Duda, G. P., Nascimento, G. W. A., Silva, M. O. Bioavailability of cadmium and lead in a soil amended with phosphorus fertilizers. Scientia Agricola. 2006, 63, 328–332.
8. Dissanayake, C. B., Chandrajith, R. Phosphate mineral fertilizers, trace metals and human health. J. Natn. Sci. Foundation Sri Lanka. 2009, 37(3), 153–165.
9. Dobrovolsky, V. V. Ed. Zink and Cadmium in the Environment. Moscow: Nauka, 1992, 200 p. (in Russian).
10. Zolotareva, B. N., Scripnichenko, I. I., Geletjuk, N. I., Sigaeva, E. V., Piunova, V. V. Content and distribution of heavy metals (lead, cadmium, and mercury) in soils of European USSR. Genesis, Fertility and Amelioration of Soils. Pushchino: Scientific Center of Academy of Sciences, 1980, 77–90 (in Russian).
11. Ilin, V. B. Estimation of Mass-Flow of Heavy Metals in System Soil-Agricultural Crop. Agrochemistry. 2006, 3, 52–59 (in Russian).
12. Gorjunova, T. A. Heavy Metals (Cd, Pb, Cu, Zn) in Soils and Plants of South-West Part of Altai Territory. Siberian Ecological Journal. 2001, 2, 181–190 (in Russian).
13. Titova, V. I. Optimization of plant nutrition and ecological-agrochemical estimation of fertilizers use on soils with high contents of exchangeable phosphorus. DSc Thesis. Saint-Petersburg, 1998, 340 p. (in Russian).
14. Gavrilova, I. P., Bogdanova, M. V., Samonova, O. A. Experience of Square Estimation of Degree of Russian Soil Pollution by Heavy Metals. Herald of Moscow University. Soil Science. 1995, 1, 48–53 (in Russian).
15. Mineev, V. G., Gomonova, N. F. Estimation of Agrochemistry Ecological Functions on Behavior of Cd in Agrocenosis on Sod-Podzolic Soil. Herald of Moscow University. Soil Science. 1999, 1, 46–50 (in Russian).
16. Tsiganjuk, S. I. Influence of Long-Term Application of Phosphorus and Lime Fertilizers on Accumulation of Heavy Metals in Soil and Plant Products. PhD Thesis.

Moscow: All Russian Institute of Fertilizers and Agrochemistry, 1994, 126 p. (in Russian).
17. Yagodin, B. A., Govorina, V. V., Vinogradova, S. B., Zamarajev, A. G., Chapovskaya, G. V. Accumulation of Cd and Pb by Some Agricultural Crops on Sod-Podzolic Soils of Different Degree of Cultivation. Izvestiya of Timiryazev Agricultural Academy. 1995, Issue 2, 85–100 (in Russian).
18. Zyrin, N. G., Sadovnikova, L. K. Ed. Chemistry of Heavy Metals, Arsenic, and Molybdenum in Soils. Moscow: Publishing House of Moscow University, 1985, 208 p. (in Russian).
19. Zyrin, N. G., Chebotaryova, N. A. About forms of copper, zinc, and lead complexes in soils and their availability to plants. Content and Form of Microelements Complexes in Soils. Moscow: Publishing House of Moscow University, 1979, p. 350–386 (in Russian)
20. Egoshina, T. L., Shikhova, L. N. Lead in Soils and Plants of North-East of European Russia. Herald of Orenburg State University. 2008, 10, 135–141 (in Russian).
21. Travnikova, L. S., Kakhnovich, Z. N., Bolshakov, V. A., Kogut, B. M., Sorokin, S. E., Ismagilova, N. H., Titova, N. A. Importance of analysis of organic-mineral fractions for estimation of pollution of sod-podzolic soils with heavy metals. Pochvovedenie (Soil Science). 2000, 1, 92–101 (in Russian).
22. Volgin, D. A. Background level and the content of heavy metals in a soil cover of the Moscow area. Herald of Moscow State Region University. Geography. 2011, 1, 26–33 (in Russian).
23. Ponizovsky, A. A., Mironenko, E. V. Mechanisms of lead (II) absorption by soils. Pochvovedenie (Soil Science). 2001, 4, 418–429 (in Russian).
24. Kabata-Pendias, A., Trace Elements in Soils and Plants, Fourth Edition. CRC Press. Boca Raton FL. 2010, 548 p.
25. Lukina, N. V., Nikonov, V. V. State of spruce biogeocenoses of North under conditions of technogenic pollution. Apatity: Academy of Science of USSR, 1993, 132 p. (in Russian).
26. Sauerbeck, D. Welche Schwerrmetallgehalte in Pflanzen durfen nicht uberschritten werden, um Wachstumsbeeintrachtigungen zu vermeiden? Landwirtschaftliche Forschung: Kongressband. 1982, 16, 59–72 [in German].
27. Lisitsyn, E. M., Shchennikova, I. N., Shupletsova, O. N. Cultivation of barley on acid sod-podzolic soils of north-east of Europe. In: Barley: Production, Cultivation and Uses. Ed. Steven, B. Elfson. New York: Nova Publ. 2011, 49–92.
28. Arinushkina, E. V. Handbook on Chemical Analysis of Soils. Moscow: Publishing House of Moscow University, 1970, 488 p. (in Russian).
29. Methodical Indications for Estimation of Microelements in Soils, Fodders, and Plants by Atomic-Absorption Spectroscopy. Moscow: CINAO Publ., 1995, 95 p. (in Russian).
30. Lisitsyn, E. M., Batalova, G., Shchennikova, I. N. Creation of oats and barley varieties for acid soils. Theory and Practice. Palmarium Academic Publishing, Saarbrucken, Germany, 2012, 228 p. (in Russian).
31. Lisitsyn, E. M. Intravarietal Level of Aluminum Resistance in Cereal Crops. Journal of Plant Nutrition. 2000, 23(6), 793–804.

32. Lichtenthaler, H. K., Bushmann, C. Chlorophylls and Carotenoids: Measurement and Characterization by UV-VIS Spectroscopy. Current Protocols in Food Analytical Chemistry. 2001, F4.3.1–F4.3.8.
33. Tjulin, V. V., Gushchina, A. M. Features of Soils of Kirov Region and Their Use at Intense Agriculture. Kirov: Publishing House Vyatka, 1991, 94 p. (in Russian).
34. Yulushev, I. G. System of Fertilizers Use in Crop Rotation. Kirov: Publishing House Vyatka, 1999, 154 p. (in Russian).
35. Prokashev, A. M. Soils with Complex Organic Profiles of South of Kirov Region. Kirov: Publishing House Vyatka, 1999, 174 p. (in Russian).

## CHAPTER 12

# COMPARATIVE CYTOGENETIC ANALYSIS OF INDIGENOUS AND INTRODUCED SPECIES OF WOODY PLANTS IN CONDITIONS OF ANTHROPOGENIC POLLUTION

TATYANA V. BARANOVA and VLADISLAV N. KALAEV

*Botanical Garden of Voronezh State University, 1, Botanical Garden St., Voronezh, 394068, Russia, Tel.: +(4732) 51-84-38, E-mail: tanyavostric@rambler.ru, Dr_Huixs@mail.ru*

## CONTENTS



## ABSTRACT

The cytogenetic characteristics of *Betula pendula* and *Rhododendron ledebourii* seed progeny in areas with different levels of anthropogenic pollution were studied. A significant number of cytogenetic disturbances and an increase of the proportion of cells at prophase have been revealed, which leads to a shift of the peak of mitotic activity. Genome instability indicates the stress state in the seed progeny and maternal plants *Betula pendula* and *Rhododendron ledebourii*, being a proof of adverse environmental conditions in the study areas. Similar cytogenetic reactions for anthropogenic factors of indigenous and introduced species have been noted.

## 12.1 INTRODUCTION

Nowadays the cytogenetic characteristics of woody plants are actively studied: *Betula pendula* [1–3], *Quercus robur* [4, 5], *Pinus silvestris* [6–9], *Pinus sibirica* [10], species of the Pinaceae family [11–13] and other conifer species [14, 15]. The results are used in the monitoring of environmental pollution and afforestation [3, 16–20]. However, the comparison of cytogenetic responses of indigenous and introduced plant species in similar conditions of anthropogenic pollution wasn't performed. Finding a solution to this problem will help to specify variants of adaptive responses of indigenous and introduced species. For this purpose we used a local species – weeping birch (*Betula pendula* Roth.) and an introduced plant under conditions of Central Black Earth Region – Ledebour rhododendron (*Rhododendron ledebourii* Pojark.).

*Betula pendula* is a typical representative and an indigenous species for the Central Black Earth area, it is common in our forests, light-requiring, relatively drought-resistant and widely used in city landscaping. As *Betula pendula* is a perennial plant, it may experience chronic impact of environmental mutagens, accumulate some doses of mutagens and be conveniently used in cytogenetic studies, being one of the most sensitive species to bio-indication due to the high intensity of photosynthesis. Recently weeping birch has been often used as a test object in the monitoring and cytogenetic studies [21–23] due to the fact that it has several advantages compared to other plants. Ledebour rhododendron (*Rhododendron ledebourii* Pojark.) is an introduced plant in the Central Black Earth area. Its natural habitat includes Siberia (Altai and Sayan Mountains), Far East,

Mongolia [24]. It is successfully introduced as a highly ornamental plant under conditions of Voronezh [25].

The purpose of our research was to compare the cytogenetic characteristics of seed progeny of *Betula pendula* and *Rhododendron ledebourii*, growing in areas with different levels of anthropogenic pressure.

## 12.2 MATERIALS AND METHODOLOGY

The research was carried out in anthropogenically polluted areas of Voronezh: Levoberezhny district (experimental area 1, located 15 km from the reference area) and Kominternovsky district (experimental area 2, located 7 km from the reference area). The experimental areas are 8 km from each other. As a reference (ecologically safe) area we used the B.M. Kozo-Polyansky Botanical Garden of Voronezh State University (VSU) [26].

Soils of Botanical Garden of VSU are characterized by a high content of humus, heavy texture, neutral media reaction and maximum buffer value. According to Fedorova and Shunelko [27], recreational area, which includes "Dynamo" Central Park of Culture and Rest and the Area of Sanatorium named after M. Gorky, has the following content of heavy metals (HM) in soils: Pb – 6.07 mg/kg of dry soil; Cr – 1.38; Cu – 0.55; Zn – 5.48; Ni – 0.87, respectively [27]. The area of Botanical Garden of VSU also can be related to the recreational area with similar characteristics. It includes a geographical park located near "Dynamo" park and is merging into it. Soils of both experimental areas are contaminated with heavy metals.

In the industrial area (near the Voronezh Synthetic Rubber Plant) similar values were revealed: Pb – 24.09 mg/kg of dry soil; Cr – 2.0; Cu – 1.95; Zn – 19.64; Ni – 2.24. Maximum concentrations of Ni are observed on Leninskiy Prospekt (the Levoberezhny district of Voronezh, where the Synthetic Rubber Plant is located). Nickel has a similar spread with another metal, which also has a small atomic mass – iron [27]. Thus, the concentrations of HM in soils of contaminated area are 3–4 times higher than those in soils of clear area.

As the material for the research, we used seeds collected from 10 plants of weeping birch (*Betula pendula* Roth.) and rhododendron Ledebour

(*Rhododendron ledebourii* Pojark.). The seeds were germinated on wet filter paper in the Petri dishes at 20°C. In 6 days, when the length of roots reached 0.5–1 cm, young seedlings were fixed with the 3:1 mixture of ethanol and glacial acetic acid (10 mL of the solution per 50 seedlings) at 9 a.m., when the meristem of *B. pendula* shows the maximum mitotic activity [2]. The fixed material was kept refrigerated at the temperature of 4°C. Wittmann's method [28, 29] was used for permanent micropreparation (one seedling per one slide). 40 microscopic preparations of seedlings from each experimental district were studied under the LABOVAL-4 microscope (Carl Zeiss, Jena) at magnification of 40×1.5×10 and 100×1.5×10. For each specimen, we recorded the total number of cells (at least 500) and an estimated mitotic activity, which was calculated as mitotic index (MI – the proportion of dividing cells to the total number of analyzed cells). We calculated the proportion of cells in prophase stage in percentages. The number of mitotic pathologies (MP) and persistent nucleoli (PN), we detected at the stages of metaphase, anaphase, telophase and calculated their percentage. These characteristics were called as the proportion of MP, the proportion of cells with PN.

According to the classification by Alov [30, 31], all mitotic pathologies were divided in three groups: (i) the pathologies related to chromosomal damage (delay of mitosis in the prophase; disturbances of spiralization or despiralization of chromosomes; early disjunction of chromatids; chromosome fragmentation and pulverization; bridges; lagging chromosomes; formation of micronuclei; irregularities of chromosomal segregation; agglutinations of chromosomes); (ii) the pathologies related to injury of mitotic spindle (delay of mitosis in the metaphase; C-mitosis; dispersion of chromosomes in the metaphase; multipolar mitosis; three-group metaphase; asymmetric; monocentric mitosis); (iii) the disturbances of cytotomy (cytokinesis) (delay or absent of cytotomy; precocious cytotomy).

The results were statistically analyzed using STADIA 6.0 Professional for Windows software package (InCo Products, 1997). The data grouping and processing were done using the method described by Kulaichev [32]. For comparison of birch cytogenetic characteristics the following criteria were used: the Van-der-Vaerden rank X-test for frequency of persistent nucleoli, pathological mitosis and the Student's *t*-test for mitotic indexes [32]. The proportion of cells with different types of pathological mitosis was compared using their angular transformation and Yeits correction according to Lakin [33] (http://www.novayagazeta.ru/inquests/69364.html).

## 12.3 RESULTS AND DISCUSSION

The results of the mitotic cytology study of *Betula pendula* trees from the ecologically safe area are shown in Table 12.1.

The mitotic activity of weeping birch peaked at 9 a.m. Table 12.1 shows an increase of the mitotic index (MI) at 9 a.m., which is reliably different from ones at other time of fixation. There were found no statistically reliable differences between MI at 9 a.m. and 11 a.m. fixations in the experiment and in the control. MI of *Betula pendula* test sample at 10 a.m. increases due to the proportion of cells at the prophase stage, which is significantly higher than at 9 a.m. and 11 a.m. (Table 12.2). This causes an increase of the mitotic index and 10-hour shift of the activity peak. The spectrum of mitotic pathologies in the root meristem of plantlets from the *Betula pendula* experimental sampling was represented by chromosome lagging at anaphase and metakinesis (72%), agglutination of chromatin (21%), fragmentation of the nucleus (7%). It was noticed that *Betula pendula* has the reliably higher proportion of cells with a persistent nucleolus at metaphase and anaphase in the experiment. PN at metaphase and anaphase were found in the root cells of the *Betula pendula* plantlets in the control. Bridges in the spectrum of MP reached 67%, chromosome lagging at anaphase and metakinesis – 33%.

**TABLE 12.1** Cytogenetic Indices of Seed Progeny of Weeping Birch, Growing in the Ecologically Safe Area

| Time of fixation, a.m. | MI, % | Proportion of cells at prophase, % | Proportion of MP, % | Proportion of cells with PN, % |
|---|---|---|---|---|
| 9 | 9.5 ± 0.2 | 14.0 ± 1.6 | 1.1 ± 0.6 | 5.3 ± 1.3 |
| 10 | 7.5 ± 0.1* | 12.3 ± 2.0 | – | – |
| 11 | 6.7 ± 0.1* | 12.5 ± 1.6 | – | – |

*Note*: MP – mitotic pathologies, PN – persistent nucleoli; * differences with the results of fixation at 9 a.m. are reliable ($P < 0.05$).

**TABLE 12.2** Cytogenetic Indices of Seed Progeny of Weeping Birch, Growing in the Ecologically Polluted Area

| Time of fixation, a.m. | MI, % | Proportion of cells at prophase, % | Proportion of MP, % | Proportion of cells with PN, % |
|---|---|---|---|---|
| 9 | 9.3 ± 0.1 | 18.2 ± 2.6 | 7.6 ± 0.5* | 10.1 ± 1.5* |
| 10 | 11.4 ± 0.3* | 27.6 ± 3.5* | – | – |
| 11 | 6.5 ± 0.1 | 13.1 ± 1.1 | – | – |

*Note*. MP – mitotic pathologies, PN – persistent nucleoli; * differences with the control are reliable ($P < 0.05$).

MI in root meristem cells of *Rhododendron ledebourii* seed progeny plantlets was reliably higher than in the control, which was combined with a higher proportion of prophases. Moreover the proportion of cells at metaphase, anaphase-telophase stages was reducing, which indicates a decrease of the number of mitotic divisions. Reliable differences with the control were observed for the proportion of cells at all stages of the mitotic cycle (Table 12.3).

Table 12.4 shows that the proportion of MP of plantlets of the *Rhododendron ledebourii* experimental sampling was slightly higher than in the control. The chromosome lagging at anaphase and metakinesis was mainly observed in the spectrum of disturbances (93.8% in the control and 85.8% in the experiment), but agglutination of chromatin reached 5.5% in the control and 12.4% in the experiment. Individual cases of nuclear fragmentation were noted in the experiment. The presence of such mitotic

**TABLE 12.3** Cytogenetic Characteristics of Rhododendron ledebourii in the Experiment and in the Control

| Species | MI, % | MI excluding prophases, % | Proportion of cells at prophase, % | Proportion of cells at metaphase, % | Proportion of cells at anaphase-telophase, % |
|---|---|---|---|---|---|
| *Rh. ledebourii* (control) | 6.1±0.6 | 3.8±0.4 | 37.9±1.9 | 18.7±2.1 | 43.4±1.7 |
| *Rh. ledebourii* (experiment) | 7.5±0.4* | 3.5±0.2 | 52.2±1.4** | 11.1±0.8** | 33.7±1.5** |

*Note*. * differences with the control are reliable ($P < 0.05$); ** differences with the control are reliable ($P < 0.01$).

**TABLE 12.4** Proportion of Cytogenetic Abnormalities of Rhododendron Species in the Experiment and in the Control

| Species | Proportion of MP, % | Proportion of cells with PN, % |
|---|---|---|
| *Rh. ledebourii* Pojark. (control) | 5.2±1.1 | 10.9±1.3 |
| *Rh. ledebourii* Pojark. (experiment) | 5.5±1.0 | 9±0.7 |

*Note*. MP – mitotic pathologies, PN – persistent nucleoli.

pathologies as nuclear fragmentation in the root meristem cells of *Betula pendula* and *Rhododendron ledebourii* seed plantlets indicates the ongoing process of apoptosis.

The increase of MI in the root meristem cells of *Betula pendula* and *Rhododendron ledebourii* seed plantlets in the contaminated area is caused by the increase of the prophase proportion, which was noted previously during studies of *Betula pendula* in stress (under the influence of technogenic pollution, chemicals, sewage, etc.) [22, 23]. The observed fact of the increase of the proportion of cells at prophase stage and the actual inhibition of mitotic activity of the seed progeny in experimental areas can be explained by the influence of stress factors (e.g., chemical mutagens) on the objects of research. Their effect is to delay cells at prophase. The simultaneous exit from this state can cause a synchronization of mitosis and the shift of mitotic activity peak [34]. This effect was observed during cytogenetic studies of other objects under the conditions of technogenic pressure: *Quercus robur* L. [5, 16], *Pinus silvestris* L. [6].

According to Alov [30, 31], the increase of proportion of cells at the prophase stage can be considered as mitotic pathology associated with chromosomal aberrations, which implies serious disturbances of genetic material and an inability of cells to move to the next stage of mitosis. It may be the absence of protein synthesis of cleavage spindle or chromosome damages. Disturbances, such as agglutination, don't let chromosomes move to the next stage and eventually cause the delay of cells in prophase, therefore the number of damaged cells increases and MI rises. Agglutination (clumping) of chromatin also can be seen at the stage of prophase, but the full spectrum of disturbances of the genetic apparatus is detected only by the analysis of the later phases of the mitotic cycle. Chromatin clumping, which indicates a high toxicity of the substance activity, represents an unrepairable effect leading to cell death [35]. During the study of the effect of heavy metal ions on the *Allium cepa* roots, it was noted that this chromosomal aberration is typical only for lethal concentrations of the tested compounds [36].

Cytogenetic characteristics (MI, MP) in the cells of root meristem of the *B. pendula* and *Rh. ledebourii* experimental sampling differ from the norm, which indicates the instability of its genetic system in the polluted area as compared to the environmentally safe one. The predominance of chromosome lagging at anaphase and metakinesis in the spectrum of disturbances indicates a low intensity of reparative processes of *Betula pendula* and *Rhododendron ledebourii* [37, 38]. The presence of such genetic material damages as agglutination of chromatin and fragmentation of the nucleus in the spectrum of cytogenetic abnormalities makes it possible to assert the presence of apoptosis in the cell population of root meristem of seed plantlets collected in the area with high technogenic pressure on plants [39].

It is known that apoptosis – genetically programd cell death, is a natural biological process, which releases a multicellular organism from weakened, aged or damaged cells and is triggered by specific external or intracellular signals [39–41]. They may be presented by non-specific factors such as temperature, toxic agents, oxidants, free radicals, gamma and UV radiation, bacterial toxins, etc. [40, 41]. Individual cells or small groups of cells from a tissue cell population are typically exposed to apoptosis, without developing inflammatory responses [40]. Apoptosis is a quick process (about 15–120 min), which makes the cytological identification difficult. The increase of the dose of the relevant agent develops cell necrosis – an unregulated death of a large group of cells accompanied by the development of the inflammatory response [39]. Thus, mechanisms of cell hypersensitivity were revealed. They have shown that the less time is between cell division (i.e., the faster the process of proliferation), the less time is left for reparation and the more special significance has the protection of the organism through the apoptosis of damaged cells [42]. Therefore, we can assume that the low proportion of MP in root meristem cells of *Rhododendron ledebourii* is caused by apoptosis in the area with increased technogenic pressure, which is the studied area. For example, a pronounced character of apoptosis of individual seed plantlets and a lack of necrosis were revealed in the study of Scots pine seed progeny from the area of Novolipetsk *metallurgical complex*. Apparently, this fact indicates that the damaging influence of exogenous toxic agents is significant, but it is still under control of repair systems and it has not reached the threshold values for the development of mitotic catastrophe and necrosis [39]. The presence of nuclear fragmentation, agglutination, condensation (pyknosis) of chromatin, which were observed in the spectrum of disturbances of *Rhododendron ledebourii*, proves the flow of apoptotic process.

An increase of the proportion of cells with the PN in root meristem cells is associated with anthropogenic impact on the objects of study. Phenomenon of the appearance of the persistent nucleoli at such mitosis phases as metaphase, anaphase and telophase is not typical for the normal course of division, because, as a rule, the nucleolus disappears at prophase and recovers only at late telophase [43]. PN look like round or drop-shaped structures connected with the chromosomes at metaphase, anaphase and early telophase of mitosis, when the nuclear membrane is absent and chromosomes are significantly reduced. Under conditions of the increased level of pollution, the activity of unique genes is suppressed and the activity of genes of ribosomal cistrons increases due to the weakening of the link between DNA – protein in nucleolar organizer area [44]. This is one of the adaptive mechanisms that manifest themselves at the molecular level. The action of these genes is similar to the mechanism of the appearance of heat shock proteins: in stressful conditions, a puffing near moderately repetitive DNA sequences occurs, in our case – the puffing of ribosomal genes. The presence of MP during the cell division can be considered as an individual case of the nucleolar activity. We assume that such cytogenetic response is an adaptive mechanism for stress to maintain a sufficient level of protein synthesis under the anthropogenic pressure. Perhaps it leads to the reduction of frequency of MP.

The predominance of chromosome bridges in the spectrum of mitotic pathologies suggests an active work of repair system [37, 38]. Bridges in the experimental and control samplings of *Rhododendron ledebourii* were not observed. The proportion of cells with MP in the experiment was reduced, which has statistically relevant differences with the control. This may indicate a decrease of the metabolic activity of the root meristem cells of *Rhododendron ledebourii* seed plantlets due to the anthropogenic stress.

Obviously, the trees from experimental sampling, which grow in the environmentally polluted area, are exposed to the damaging effect of industrial emissions and automobile exhaust gases.

## 12.4 CONCLUSIONS

The results of cytological study show genome instability of *Betula pendula* and *Rhododendron ledebourii* seed progeny, collected in the ecologically polluted areas. Genome instability indicates the state of stress of *Betula pendula* and *Rhododendron ledebourii* seed progeny and maternal plants, which proves the unfavorable environmental conditions in the studied area. Cytogenetic responses of indigenous and introduced species are similar in nature, which is shown as the increase of MI due to the increase of the proportion of cells at prophase, the presence of cells with MP and similar disturbances of mitosis. It proves the versatility of their responses to environmental factors.

## KEYWORDS

- **cytogenetic disturbances**
- **mitotic activity**
- **mitotic pathologies**
- **persistent nucleoli**
- **prophase**

## REFERENCES

1. Varduni, T. V. Rearrangements of chromosomes in the cells of higher plants as an indicator for monitoring environmental mutagens. Abstract of Ph. D. thesis, Voronezh, Russian Federation, 1997, 23 p. (In Russian).
2. Vostrikova, T. V. Cytoecology of weeping birch (Betula pendula Roth.). Abstract of PhD thesis, Voronezh, Russian Federation, 2002, 24 p. (In Russian).
3. Kalaev, V. N., Karpova, S. S. The Influence of Air pollution on Cytogenetic Characteristics of Birch Seed Progeny. Forest Genetics. 2003, 10(1), 11–18.

4. Butorina, A. K. Cytogenetics of forest woody plants (in connection with questions of their evolution and breeding). PhD Dissertation, Voronezh, Russian Federation, 1990, 368 p. (In Russian).
5. Kalaev, V. N. Cytogenetic responses of deciduous woody plants to stressful conditions and prospects of their use to assess environmental genotoxicity. PhD Dissertation, Voronezh, Russian Federation, 2009, 414 p. (In Russian).
6. Doroshev, S. A. The influence of anthropogenic stressors on cytogenetic variability of common pine (Pinus sylvestris). Abstract of PhD thesis, Voronezh, Russian Federation, 2004, 23 p. (In Russian).
7. Mironov, A. N. Cytogenetic effects from exposure to ionizing radiation and pulsed magnetic fields on woody plants. Abstract of PhD thesis, Moscow, Russian Federation, 2002, 24 pp. (In Russian).
8. Muratova, E. N. Features meiosis of Scots pine (Pinus sylvestris, L.) near the northern border of its range. Russian Journal of Developmental Biology. 1995, 26(2), 158–169. (In Russian).
9. Muratova, E. N., Sedel'nikova, T. S. Karyological study of wetland and upland populations of Scots pine (Pinus sylvestris, L.). Russian Journal of Ecology. 1993(6), 41–50. (In Russian).
10. Sedel'nikova, T. S., Muratova, E. N. Specific Karyological Features of Siberian Stone Pine (Pinus sibirica Du Tour) in Western Siberian Bogs. Russian Journal of Ecology. 2002, 33(5), 303–308.
11. Butorina, A. K., Evstratov, N. The first detected case of amitosis of pine. Forest Genetics. 1996, 3, 137–139.
12. Sedel'nikova, T. S. Wetland and upland differentiation of populations of species of the family Pinaceae Lindl. (reproductive and karyotypic features): Abstract of PhD thesis, Tomsk, Russian Federation, 2008, 35 p. (In Russian).
13. Korshikov, I. I., Tkacheva Yu.A., Privalikhin, S. N. Cytogenetic Abnormalities in Norway Spruce (Picea abies (L.) Karst.) Seedlings from Natural Populations and an Introduction Plantation. Cytology and Genetics. 2012, 46(5), 280–284.
14. Muratova, E. N. In chromosome gymnosperms. Biology Bulletin Reviews. 2000, 120(5), 452–465. (In Russian).
15. Sedel'nikova, T. S., Muratova, E. N., Pimenov, A. V. Variability of Chromosome Numbers in Gymnosperms. Biology Bulletin Reviews. 2011, 1(2), 100–109.
16. Butorina, A. K., Kalaev, V. N., Vostrikova, T. V., Myagkova, O. E. Cytogenetic characteristics of seed progeny of Quercus robur, L., Pinus sylvestris, L., and Betula pendula Roth under conditions of anthropogenic contamination in the city of Voronezh. Cytology. 2000, 42(2), 196–201. (In Russian).
17. Geras'kin, S. A., Vasil'ev, D. V., Dikarev, V. G., Udalova, A. A., Evseeva, T. I., Dikareva, N. S., Zimin, V. L. Bioindication-based Estimation of Technogenic Impact on Pinus sylvestris, L. Populations in the Vicinity of a Radioactive Waste Storage Facility. Russian Journal of Ecology. 2005, 36(4), 249–258.
18. Oudalova, A. A., Geras'kin, S. A. The Time Dynamics and Ecological Genetic Variation of Cytogenetic Effects in the Scots Pine Populations Experiencing Anthropogenic Impact. Biology Bulletin Reviews. 2012, 2(3), 254–267.

19. Kalaev, V. N., Butorina, A. K. Cytogenetic Effect of Radiation in Seed of Oak (Quercus robur, L.) Trees Growing on Sites Contaminated by Chernobyl Fallout. Silvae Genetica. 2006, 55(3), 93–101.
20. Karpova, S. S., Kalaev, V. N., Artyukhov, V. G., Trofimova, V. A., Ostashkova, L. G., Savko, A. D. The Use of Nucleolar Morphological Characteristics of Birch Seedlings for the Assessment of Environmental Pollution. Biology Bulletin. 2006, 33(1), 73–80.
21. Vostrikova, T. V. The Cytology of Mitosis in Weeping Birch (Betula pendula Roth.). Cytology. 1999, 41(12), 1058.
22. Vostrikova, T. V. Instability of Cytogenetic Parameters and Genome Instability in Betula pendula Roth. Russian Journal of Ecology. 2007, 38(2), 80–84.
23. Vostrikova, T. V., Butorina, A. K. Cytogenetic Responses of Birch to Stress Factors. Biology Bulletin. 2006, 33(2), 185–190.
24. Alexandrova, M. S. Rhododendrons. 2003, Moscow: Phyton, 192 p. (In Russian).
25. Moseeva, E. V., Baranova, T. V., Voronin, A. A., Kuznetsov, B. I. Features of seed propagation of the genus rhododendron (Rhododendron, L.). Problems of Regional Ecology. 2012(4), 100–103. (In Russian).
26. Mamchik, N. P., Kurolap, S. A., Klepikov, O. V., Chubirko, M. I., Yakimchuk, R. V., Kolnet, I. V., Barvitenko, N. T., Fedotov, V. I., Korystin, S. I., Kravets, B. B. Ecology and Monitoring of Voronezh-city Health. Voronezh. Voronezh State University, 1997, 180 p. (In Russian).
27. Fedorova, A. I., Shunelko, E. V. Pollution of surface soil horizons of Voronezh heavy metal. Bulletin of Voronezh State University. Series of Geography and Geoecology 2003, 1, 74–82. (In Russian).
28. Wittmann, W. Aceto-iron-haemotoxylin for staining chromosomes in squashes of plant material. Stain Technology. 1962, 3(1), 27–30.
29. Kalaev, V. N., Karpova, S. S., Artyukhov, V. G. Cytogenetic Characteristics of Weeping Birch (Betula pendula Roth.). Seed Progeny in Different Ecological Conditions. Bioremediation, biodiversity & bioavailability. 2010, 4(1), 77–83.
30. Alov, I. A. Cytophisiology and Pathology of Mitosis. 1972, Moscow: Medicine, 264 p. (In Russian).
31. Alov, I. A. Pathology mitosis. Bulletin of Medical Sciences. 1965, № 11. P. 58–66.
32. Kulaichev, A. P. Methods and Means of Data Processing in the Windows Stadia 6.0. Moscow. Informatics and Computers. 1996, 257 p. (In Russian).
33. Lakin, G. F. Biometrics. Moscow. Graduate School, 1990, 352 p. (In Russian).
34. Mitrofanov Yu.A. Radiosensitivity of cells in different phases of the mitotic cycle. Advances in modern genetics. Moscow, 1969, 125–160. (In Russian).
35. Fiskesjo, G. Allium test. Methods in Molecular Biology. Vitrotoxicity Testing Protocols. S. O'Hare, C. K. Alberwill (Ed.). Totowa. 1995, 119–127.
36. Dovgalyuk, A. I., Kalinyak, T. B., Blume Ya.B. Cytogenetic effects of toxic metal salts in the cells of the apical meristem of Allium cepa, L. roots. Cytology and Genetics. 2001, 35(2), 3–10. (In Russian).
37. Akopyan, E. M. Influence of different types of ionizing radiations on chromosome aberrations origin in Vicia. Russian Journal of Genetics. 1967, 3(5), 45–51. (In Russian).

38. Simakov, E. A. About post-irradiation reparation of cytogenetic damages in plantlets of different potatoes' forms. Radiobiology. 1983, 23(5), 703–706. (In Russian).
39. Mashkina, O. S., Kalaev, V. N., Muraya, L. S., Lelikova, E. S. Cytogenetic responses of Scots pine seed progeny to the combined anthropogenic pollution in the vicinity of Novolipetsk *metallurgical complex*. Ecological Genetics. 2009, 7(3), 17–29. (In Russian).
40. Umanskiy, S. R. Apoptosis: molecular and cellular mechanisms. Molecular Biology. 1996, 30(3), 487–502. (In Russian).
41. Manskih, V. N. Ways of cell death and their biological significance. Cytology. 2007, 49(11), 909–915. (In Russian).
42. Velegzhaninov, I. Modern understanding of the mechanisms of low doses of ionizing radiation on cells and organisms. Bulletin IB. 2009, 9, 15–21. (In Russian).
43. Chentsov, Yu. S. General cytology. 1984, Moscow: MSU, 350 p. (In Russian).
44. Butorina, A. K., Isakov, Yu. N. Puffing of chromosomes in the metaphase–telophase of the mitotic cycle in Quercus robur. Reports of the Academy of Science USSR. 1989, 308(4), 987–988. (In Russian).

## CHAPTER 13

# FEATURES OF PATHOGENESIS OF CELLS UNDER THE INFLUENCE OF HEAVY METALS

NATALIA K. BELISHEVA

*Kola Science Centre, Russian Academy of Science, 14A, Fersman St., Apatity, Murmansk Region, 184211, Russia, Tel.: +79113039033, +78155579452, E-mail: natalybelisheva@mail.ru*

## CONTENTS

## ABSTRACT

The comparative characteristics of effects of metals (Pb, As, Hg, Cd, Ni, Cr, V) interaction with cell structures are reported. It is shown that the disturbance of homeostasis at the different levels of cellular structures organization, displayed in the structural disorganization, chaotization of cellular functions and genome destabilization, is a characteristic peculiarity of cell pathogenesis. These phenomena condition toxic and carcinogenic activity of metals, which could take the role of initiators, as well as promoters of carcinogenesis.

## 13.1 INTRODUCTION

Environmental contamination by heavy metals leads not only to the distortion of natural microelements combination in ecosystems, but affects on the health of living people, as well as further generations [1–3]. This follows from the capability of a series of metals to induce mutagenesis in somatic and reproductive cells [4]. Unfortunately, in many cases the mechanism of pathogenic effect of metals on human body remains unclear [4–6]. Low knowledge of this issue is related with that the pathogenic effect of metals is often evaluated basing only on the final clinic results of the chronical contact of a human with metals within the framework of his professional activity. However, clinic display of the pathology may be the result of a cascade of secondary disorders, arisen under primary contact of target cells with metal, as well as of the cumulative effect of combined effect of metals and other toxic agents [7]. Moreover, a large number of studies show that similar system and organs lesions appear under the effect of different metals [4–6], whereas the same metal may have different toxic potential in dependence on the form of its contact with cell [4, 6].

That it is why it is necessary to search the cause of multiple pathogenic effects, which appear in the body under the contact with metals, at the cellular level. The analysis of peculiarities of cellular pathogenesis under effect of different metals will allow detect specific and nonspecific mechanisms of cellular structures damage, which will contribute to the development of methods of protection from toxic effect evoked by the interaction of target cells with metals.

## 13.2 GENERAL DISORDERS AT THE LEVEL OF ORGANISM EVOKED BY CONTACT WITH METALS

The analysis of the effects of the most widely spread metals (Pb, As, Hg, Cd, Ni, Cr, V) in toxic doses shows that all of them (in different measure) lead to the destructive and dysfunctional disorders [1–6]. These effects touch all vitally important systems of the body [4]: central and peripheral nervous systems [8–15]; cardiovascular and reproductive systems [4–6, 11, 15–17] and they are displayed at the level of different organs: placenta, lungs, bone tissue, kidneys, liver, spleen [4, 8–17]. Among the above mentioned set of metals, the ions of bivalent mercury and its compounds have the highest toxicity and breadth of spectrum of effects [8–10], vanadium has the most specific effect [18] and nickel has the most expressed carcinogenic activity [4, 6, 13, 14].

Let's consider the peculiarities of cellular pathogenesis under the interaction of these metals with target cells and let's try to detect the disorders, which could determine general and specific characteristics of the effect of metals.

## 13.3 GENERAL CONCEPTION ABOUT STRUCTURE AND FUNCTIONAL DISORDERS ARISING AT THE CELLULAR LEVEL UNDER EXPOSURE OF METALS

In biological systems, heavy metals affect cellular organelles and components such as cell membrane, mitochondrial, lysosome, endoplasmic reticulum, nuclei, and some enzymes involved in metabolism, detoxification, and damage repair [4, 19].

Metal ions interact with cellular components such as DNA and nuclear proteins, causing DNA damage and conformational changes that may lead to cell cycle modulation, carcinogenesis or apoptosis [19–22]. Several studies have demonstrated that reactive oxygen species (ROS) production and oxidative stress play a key role in the toxicity and carcinogenicity of metals such as arsenic [23–25] cadmium [26–29], chromium [6, 30], lead [31], and mercury [31–33]. Because of their high degree of toxicity, these five elements rank among the priority metals that are of great public health significance [33, 34]. They are all systemic toxicants that induce the organ damage, even at lower levels of exposure. According to the United States Environmental Protection Agency (U.S. EPA), and the International Agency for Research on Cancer (IARC), these metals are also classified as either "known" or "probable" human carcinogens based on epidemiological and experimental studies showing an association between exposure and cancer incidence in humans and animals [4].

The penetration of metal in the cell evokes the activation of intracellular systems, which function consists in the nonspecific defense, in particular against the effect of xenobiotics. The following ways of defense are possible in dependence on the form of metal contact with cell (particle or ion of the metal):

1. phagocytosis, accompanied by the activation of lysosomal enzymes dissolve metallic particles;
2. encapsulation of insoluble particles;
3. metabolic inactivation of soluted form of metal.

The first way allows not only to free from the accumulation of alien agents, but also initiate the process of detoxication and metabolism of xenobiotics (switch on the third way).

The effectiveness of cell defense against the destructive invasion is determined by the coordination of the functioning of all mechanisms, sustaining homeostasis. In the case of structural disorders in bio-molecules performing the protection (lysosomal enzymes, regulators of genes activity, enzymes of energy supply, proteins of cytoskeleton systems etc.) it is possible to expect destructive manifestations at the sub-molecular level (membranes of the cell surface, nucleus, endoplasm reticulum, cellular organelles; disorders of mitotic spindle structure and chromatin organization) [3–6, 35–41].

In dependence on the defense effectiveness the interaction of metal with cell either will not lead to any considerable disorders [3, 4] or it may display as toxic effects and as extreme form of pathology, namely malignant transformation of cells [6, 39, 41].

### 13.3.1 STRUCTURE FUNCTIONAL DISORDERS UNDER TOXIC EFFECTS OF METALS

According to Martin [2], it is easy to imagine mechanisms making metal ions toxic, but it is hard to indicate it to any particular metal.

Broad spectrum of structural and functional disturbances in different systems of the body under exposure of metals [3–6] leads to the supposition that different mechanisms of cellular pathogenesis under contact with these metals should exist. It is possible to explain these differences basing on the series of molar toxicity calculated by Niboer and Richardson [35].

It is proposed that the term "heavy metals' be abandoned in favor of a classification which separates metal ions into class A (oxygen-seeking), class B (nitrogen/sulfur-seeking) and borderline (or intermediate). A survey of the co-ordination chemistry of metal ions in biological systems (mostly X-ray crystallographic data) demonstrates the potential for grouping metal ions according to their binding preferences (i.e., whether they seek out O-, N- or S-containing ligands). This classification is related to atomic properties and the solution chemistry of metal ions. A convenient graphical display of the metals in each of the three categories is achieved by a plot of a covalent-bonding index versus an ionic-bonding index. A review of the roles of metal ions in biological systems demonstrates the potential of the proposed classification for interpreting the biochemical basis for metal-ion toxicity and its use in the rational selection of metal ions in toxicity studies [35, 36].

According to these series toxicity one can predict that cations of "soft" metals (e.g., Cd(II); $CH_3Hg(I)$; Hg(II), Ag), are more toxic than cations of "intermediate" metals (e.g., Ni(II); Pb(II)), which show higher toxicity relatively to cations of "hard" metals (e.g., Cr(III), Ni(II)). This is related with the higher possibility of "soft" metals to form strong complexes with bio-molecules (on the base of covalent binding). Lead weakly falls into this classification: from one side it strongly interacts with sulfhydryl groups, i.e., showing "softness"; from other hand it forms anomalously strong hydrocomplexes, i.e., shows "harshness" [35].

Namely the formation of metals complexes with certain types of functionally important types of macromolecules is fundamental in the detection of cellular pathogenesis ways [4–6]. For example, numerous studies have implicated structural alterations in chromatin and epigenetic changes as the primary events in nickel carcinogenesis [6, 37–39]. It was shown that phagocytized particles of nickel sulfide affect selectively on the heterochromatin [40]. These effects of nickel compounds on heterochromatin led to the discovery that nickel compounds could silence genes by inducing

DNA methylation [6]. Various studies have reported that arsenite exposure induces both DNA hypo and hypermethylation. Low dose chronic as treatment of a rat liver epithelial cell line resulted in a reduction of *S*-adenosylmethionine levels, an increase in global DNA hypomethylation levels and decreased DNA methyltransferase activity [41]. Three independent studies have reported cadmium-induced changes in global and gene specific DNA methylation levels [6].

The bonding of the metals with cellular structures affects the state of fermentative systems of lysosomal, microsomal and mitochondrial localization, which can lead to switching off or modification of detoxification paths. For example, the modulation of activity of microsomal fraction ferments should affect not only the metabolism of metals themselves, but also on the metabolism of "true" polycyclic carcinogens. The research showed that disorganization of the structures of lysosomal and endoplasm reticulum, and as result such disorganization – disorders of ferment activity, belong to the most distinct metabolic disturbances under the effect of salts of different metals, in particular lead and chrome [6]. The damage of lysosome membranes of tested organs, disorganization of organelles in gonads, in liver and in kidneys under exposure of heavy metals were also noticed. Biochemical changes in gonads under effective doses of metals were accompanied by gonadotoxic effect, and the embryotoxic and mutagenic effect were detected under the impact of lead [6, 31, 42].

It is impossible to exclude that multiple toxic effects of lead [11, 16, 17, 42] are related to its metabolism, in which cytochrome P450, microsomal ferment of classic carcinogens detoxification, participates [43–45]. Moreover, the direct or mediated impact of lead on the activity of other ferments were also detected: the inhibition of aminolevulinic acid dehydratase, inhibition of synthetase of heme, activation of heme oxygenase; the effect on erythrocytes membranes and their destruction because of stimulation of oxidative processes were also observed [45].

The absence of strict specificity in the manifestation of toxic effects on cells of a wide spectrum of metals can be followed also on other examples. Arsenic, as well as lead, inhibits aminolevulinic acid synthetase and easily binds with SH-groups; in dependence on the valency form it can be bonded with thiol (As(III)) and phosphate groups under oxidative phosphorylation (As(V)) [45]. Probably, namely the arsenic binding with sulfide groups conditions hyperkeratosis because of enrichment of skin cells by SH-groups, and the inhibition of aminolevulinic acid synthesis is probably the common feature in the pathogenesis under the effect of both, arsenic, as well as lead, which could indirectly lead to the disorders of cardiovascular system, nervous system and blood system [22–25].

Effects of toxic compounds and ions of mercury at the cellular level are expressed by broad spectrum of disorders at the body level. The peculiarities of exposure of mercury are manifested by formation of more strong complexes with ligands than under exposure of other metals, as well as wider range of binding with different bio-molecules. Beside of the interaction with SH-groups, characteristic for many metals, the ions of mercury ($Hg^{2+}$) forms inert ties with $C_5$-atoms in pyrimidines, and under high pH-values the surplus of $CH_3Hg^+$ replace hydrogen in $NH_2$-groups of $C_4$-atom in cytidine and of $C_5$-атом in adenine. Moreover, mercury binds with amino acids and other molecules [4, 5, 31, 32].

Multiple disorders at the body level are typical under exposure of cadmium, too (which belongs to the group of "soft" metals, as well as mercury). The breadth of spectrum of pathologies at the body level under exposure of cadmium, as well as of mercury, is displayed in the diversity of types of the cadmium interaction with cell: cadmium reacts with proteins and macromolecules and under circulation in the blood it binds with low-molecular protein metallothionein [46] and with hemoglobin of erythrocytes [2, 4, 6].

Other metals evoking toxic and carcinogenic effect at the body level have also a broad spectrum of binding with cell macromolecules. Nickel is able to interact with cysteine and other amino acids, with proteins, with DNA; it can suppress the transportation from cell membranes; under high doses it can lead to the full disintegration of tissue and cells of kidney [2, 45]. Effects of chrome depend on in which form it is absorbed by tissues: in the solutions Cr(VI) is presented in complex anions and Cr(III) in cationic form. Cr(VI) has higher cytotoxicity; it is accumulated in liver mitochondria and penetrates through erythrocytes membranes. After the penetration into the cells Cr(VI) is reduced to Cr(III) and reacts with DNA, proteins and cellular complexes [6, 30]. Vanadium, in spite of the specificity of its effect, which is similar to the effect of ouabaine (the inhibition of $Na^+$, $K^+$-dependent ATPase), as well as other metals, is able to interact with ligands containing phosphorus and sulfur [2, 18].

Thus, common impacts of toxicity under the effect of metals on cells are determined by numerous disorders in organization and functioning of bio-molecules and cell structures. Probably one of the reasons of nonspecific damages in cellular structures under exposure of metals is associated with their ability to bind with SH-groups. Damages in cellular structures contribute to disorganization of work of harmonious ensembles providing viability of cellular systems. Such disorganization is probably based on the inexact correspondence of conformational structures of mutually complementary molecules participating in the regulation of metabolism. In particular, conformational inconsistencies may lead to the decrease of precision in copying and reading of information under matrix synthesis of DNA, RNA, and proteins, as well as in other synthesis processes. This assumption corresponds well with the theory of cell senescence or "catastrophe of errors" of Leslie Orguell (suggested in 1963), adapted to the oncogenesis theory as "theory of errors accumulation" [47].

This conception is based on the explanation of the rise of tumor and tumor progression as a result of accumulation of damaged protein molecules in the cells, which could initiate a "cascade of errors" with progressively increasing decrease of precision of protein synthesis realization and gradual "erasing" of informational structure of molecules. Cascade mechanism of errors accumulation in protein synthesis leads to inevitable death of cells. The metal penetration into cell and its cytotoxicity are the first steps to carcinogenesis.

### *13.3.2 STRUCTURE FUNCTIONAL DISORDERS CONTRIBUTING TO THE RISE OF MUTAGENIC AND CARCINOGENIC ACTIVITY OF METALS*

It is assumed that carcinogenic activity of metals, in contrast with organic chemical carcinogens, does not require metabolic activation besides of oxidation-reduction or dilution of particles [2, 4, 6]. It is considered that if the metal is able to reach a certain organ and to penetrate into the cells and when a sufficiently high concentration of the metal is reached, this metal could initiates a carcinogenic response. It is evident that if metal ion is as active and cytotoxic as $Hg^{2+}$, the cell death will precede carcinogenic response [8, 9], but if death does not occur, it is possible to expect malignant transformation of cells.

The effects may be different in dependence on the way of metal penetration into the cell (phagocytosis or migration of dissolved ion of the metal). The mechanism of metal containing particles utilization via phagocytosis differs completely from the way of penetration of metal ion: phagocytosis increases the probability of malignant transformation of cells, whereas metal ions dissolved in cytoplasm condition the effects of cytotoxicity [48, 49].

It was shown on the tissue cultures that carcinogenetic effect of nickel is related to the easiness of its phagocytosis. Whereas amorphous NiS is not exposed to phagocytosis and has no toxic activity, crystalline NiS is actively subjected to phagocytosis in multiple cell lines, in particular by cells of kidneys, ovary, lungs, macrophages of peritoneum and is characterized by expressed mutagenic and carcinogenetic effect [37, 39, 49].

One of possible mechanisms of carcinogenesis can be considered at the example of the interaction of crystalline nickel with cellular structures: after phagocytosis Ni particles in phagosomes move in cytoplasm, aggregate around nuclear membrane and evolve ion nickel just in these places, penetrating into the nuclei. In nuclei Ni association with heterochromatin occurs, which can be accompanied by chromosomes fragmentation [40]. Probably the main damage, which arises as a result of nickel interaction with DNA, consists in the formation of ties of DNA macromolecules with protein, which are rather stable and complicate the DNA replication. Probably namely this type of damages is responsible for the fragmentation of chromosomes and other breakages, observed under the impact of Ni and other metals [40, 50, 51].

The ability of carcinogenic ions of metals to come upon the cells and penetrate through nuclear membranes is extremely important for their carcinogenicity. It is related with carcinogenicity namely by the ratio of concentrations of carcinogenic metal ions in the nucleus and cytoplasm: high level of metal ions concentration in cytoplasm becomes the cause of cytotoxicity, whereas high level of metal ions concentration in the nuclei will initiate damages of DNA and mutations, i.e., those changes, which precede oncogenesis. The following condition seems to be very important: at subcellular level the main concentration of metal ions is observed in the nucleus: the concentration of nickel and chrome in the nucleus is 90 and 68 times higher than in cytoplasm, respectively. Within the nucleus itself metal ions are accumulated first of all in nucleolus: in nucleolus there is 18 times more of nickel and 11.3 times more of chrome than in the nucleus [52].

It is hard to overestimate the selective absorption of metals by nucleus, because namely nucleolus is responsible for the synthesis of ribosomal RNA and, consequently ribosomal proteins. The disturbance of the transcription of this RNA type can evoke catastrophic consequences, related with degradation and decomposition of protein synthesis apparatus, which, in its turn, may lead either to cell death or to the distortion of information coded by RNA and, consequently, to the disturbance of the transcribed protein molecules.

Damages of the structure and functions of the nucleus with nucleolus, which control of cell proliferation [53], is the key moment in malignant transformation

Besides of DNA-protein links, Ni and Cr can interact with phosphate groups of DNA, promote breaks of single DNA chains and DNA-DNA links, i.e., induce multiple damages at the level of genetic apparatus leading to the change of genetic composition [49], affect cellular growth, DNA replication and reparation. The ability of certain ions of metals to decrease the correctness of DNA reparation and replication can be also the mechanism of mutation changes during oncogenesis. The effect of metals on the genetic apparatus of cell was also shown for As and Hg. The results obtained for Cd and Pb are conflicting, the mechanism of the effect of these metals probably differs from the mechanism of Ni and Cr effect. As the relationship of such metals with SH-groups was shown, it is assumed that these metals can associate with mitotic spindle filaments, which leads to clastogene effects [49, 51, 52]. It was found that a series of metals blocks mitotic cycle, evoking selective block in S-phase (Ni, Cr, Hg, Cd) [49, 52].

In spite of the fact that the relationship between metals mutagenic activity and carcinogenesis seems to be evident, it remains still unclear if they are prime cause of carcinogenesis or its promoters. It could not be excluded that many metals, in particular lead, can affect indirectly the rise of tumors. The data about the evaluation of metals ability affect the system of chromosomal oxidases of broad spectrum of effects with the participation of cytochrome p450 show that metals can be activators or inhibitors of this system. The activity of microsomal oxidases system, in its turn, determines metabolic activation of true carcinogens, which provide their carcinogenic activity.

Thus, it is not excluded that metals can be not true carcinogens, but promoters of chemical carcinogenesis. However, it is rather hard to determine their true role in the processes of neoplasm formation by common approaches. This is because beside of the metal under study, true carcinogens, which effect will be initiated by metal, can be also presented in environment. Consequently, when studying cause-and-effect relations (presence of metal versus statistics of diseases) it is possible to attribute to the effect of metals the role of neoplasm cause without taking into consideration complex effects of different pollutants in the environment [7, 8].

## 13.4 ABOUT THE MECHANISMS OF DAMAGING AND HEREDITABLE EFFECTS OF HEAVY METALS

From the brief consideration of the ways of metals effect on target cells it is evident that the mechanisms of cellular pathogenesis under contact with metals do not have a specific character. Multiple changes of structure and functioning of such macromolecules as DNA, RNA and proteins, as well as the destruction of sub-molecular complexes, affecting, first of all, cell membranes, are the main common characters under the effect of metals on cells. The transformation of macromolecules and cell structures leads to a broad spectrum of functional disorders at different level of organism organization with the measure of specificity determined by the specific of interaction of metal with particular groups of macromolecules. Thus, if metal binds with DNA- and DNA-protein complexes, it is reasonable to expect that such a metal will have mutagenic and carcinogenic activity. However, if metal destroys cellular structures and macromolecules, it is possible to expect expressed cytotoxicity with further various functional disorders in target organs.

In spite of the fact that metals can initiate serious disorders in the organization of genetic material, the question about direct participation of metals in the carcinogenesis process remains open. The carcinogenesis process is multifactor according to its reasons and multi-step in its evolution. It includes both, process initiation, as well as its promotion. The initiation process assumes that carcinogens should be electrophilic either themselves or by means of metabolic activation, at that the covalent form of adducts with nucleophilic residua in DNA and other cellular macromolecules is formed. As a result of the formation of such complexes, the modification of adenine and adenine occurs (although smaller than of guanine), besides modified phosphate residua skeleton of nuclear DNA also appear. The modified DNA is by weakened matrix activity relatively to both replication, as well as transcription. It is considered that covalent binding of chemical agent with DNA is one of initial stages in the initiation of carcinogenesis phases, but it is still insufficient event for malignant transformation. The second phase of carcinogenesis should be initiated by tumor promoters, which modify the properties of cell surface [54].

It possible to assume that metals potentially (in dependence on the form of their contact with cell) could play the role of both, promoters, as well as initiators of carcinogenesis. Especially, taking into account the multiple disorders, which are induced by certain metals in the cells (in particular leading to the formation of covalent adducts with DNA and to the modification of the structure and functions of the cellular membranes). In the more general form, the metals, by producing non-specific damages at different levels of cellular organization, could be induced the instability of genome, and as consequence, the disorder of the homeostasis.

## 13.5 CONCLUSIONS

The disturbance of homeostasis at different levels of cellular organization, manifested in the structural and functional disorders, is a characteristic of cellular pathogenesis under the exposure of metals. It is necessary to notice that the wider the spectrum of the disorders induced by the metal at the cellular level, the more dysfunctions one can detected at the organism level. Other words, a certain projection of the micro-level to the macro-level and vice versa takes place.

The complexity of the issue of the cause-consequence relations between the phenomena on the molecular, on the cellular, on the body levels and between environment contamination by heavy metals is required complicated research, in which need include the element analysis, the medicine and biological monitoring, including the laboratory, clinic and statistic investigations. Only by using such approach, one could find the true causes of cellular pathogenesis under exposure of the heavy metals.

## ACKNOWLEDGEMENTS

The author expresses his sincere gratitude to Larisa Ilyinichna Weisfeld for the initiation of this article and the full support in its preparation.

## KEYWORDS


- **carcinogenesis**
- **homeostasis**
- **metal ions**
- **mutagenesis**
- **phagocytosis**


## REFERENCES


1. Sposito, G. Distribution of potentially hazardous trace metals. Metal Ions in Biological Systems, 20. Ed. Sigel, H. Marcel Dekker, New York, 1986, 1–20.
2. *Martin,* R. B. *Bioinorganic Chemistry of toxic metal ion*, Chapter 2 in Ref 5. Handbook on Metals in Clinical and Analytical Chemistry. Eds. Seiler, H. G., Sigel, A., Sigel, H. 1994, 755 p.
3. Dixit, R., Wasiullah, D., Malaviya, K. et al. Bioremediation of Heavy Metals from Soil and Aquatic Environment: An Overview of Principles and Criteria of Fundamental Processes. Sustainability, 2015, 7(2), 2189–2212.
4. Tchounwou, P. B., Yedjou, C. G., Patlolla, A. K., Sutton, D. J. Heavy Metals Toxicity and the Environment. Supplement, 2012, 101, 133–164.
5. Lars, J. Hazards of heavy metal contamination. British Medical Bulletin, 2003, 68, 167–182.
6. Arita, A., Costa, M. Epigenetics in metal carcinogenesis: Nickel, Arsenic, Chromium and Cadmium. Metallomics, 2009, 1, 222–228.

7. Manti, L., D'Arco, A. Cooperative biological effects between ionizing radiation and other physical and chemical agents. Mutation Research. 2010, 704, 115–122.
8. Neustadt, J., Pieczenik, S. Toxic-metal contamination: Mercury. Integr. Med., 2007, 6, 36–37.
9. Ainza, C., Trevors, J., Saier, M. Environmental mercury rising. Water, Air, and Soil Pollution, 2010, 205, 47–48.
10. Rice, K. V., Walker, E. M., Miaozong Wu, Gillette, Ch., Blough, E. R. Environmental Mercury and Its Toxic Effects. J. Prev. Med. Public Health. 2014, 47(2), 74–83.
11. Salem, H. M., Eweida, E. A., Farag, A. Heavy metals in drinking water and their environmental impact on human health. In ICEHM 2000, Cairo University: Giza, Egypt, 2000, 542–556.
12. Khan, M. A., Ahmad, I., Rahman, I. Effect of environmental pollution on heavy metals content of Withania somnifera. J. Chin. Chem. Soc., 2007, 54, 339–343.
13. Cameron, K. S., Buchner, V., Tchounwou, P. B. Exploring the Molecular Mechanisms of Nickel-Induced Genotoxicity and Carcinogenicity: A Literature Review. Rev. Environ. Health. 2011, 26(2), 81–92.
14. Duda-Chodak, A., Baszczyk, U. The Impact of Nickel on Human Health. J. Elementology, 2008, 13, 685–696.
15. Degraeve, N. Carcinogenic, teratogenic and mutagenic effects of cadmium. Mutat. Res., 1981,86, 115–135.
16. Wuana, R. A., Okieimen, F. E. Heavy metals in contaminated soils: A review of sources, chemistry, risks and best available strategies for remediation. J. International Scholarly Research Notices Ecol. 2011, 1–20.
17. Padmavathiamma, P. K., Li, L. Y. Phytoremediation technology: Hyper accumulation metals in plants. Water, Air, & Soil Pollution, 2007, 184, 105–126.
18. Goc, A. Biological activity of vanadium compounds. Central European Journal of Biology, 2006, 1(3), 314–332.
19. Wang, S., Shi, X. Molecular mechanisms of metal toxicity and carcinogenesis. Molecular Cell Biochem., 2001, 222, 3–9.
20. Chang, L. W., Magos, L., Suzuki, T. Eds. Toxicology of Metals. Boca Raton. CRC Press, FL, USA, 1996, 1198 p.
21. Beyersmann, D., Hartwig, A. Carcinogenic metal compounds: recent insight into molecular and cellular mechanisms. Archiv Toxicol., 2008, 82(8),493–512.
22. Rousselot, P., Laboume, S., Marolleau, J. P., et al. Arsenic trioxide and melarsoprol induce apoptosis in plasma cell lines and in plasma cells from myeloma patients. Cancer Res., 1999, 59, 1041–1048.
23. Yedjou, C. G., Tchounwou, P. B. Oxidative stress in human leukemia cells (HL-60), human liver carcinoma cells (HepG2) and human Jerkat-T cells exposed to arsenic trioxide. Metal Ions Biol. Med., 2006, 9, 298–303.
24. Yedjou, G. C., Tchounwou, P. B. In vitro cytotoxic and genotoxic effects of arsenic trioxide on human leukemia cells using the MTT and alkaline single cell gel electrophoresis (comet) assays. Mol. Cell. Biochem., 2007, 301, 123–130.
25. Tchounwou, P. B., Centeno, J. A., Patlolla, A. K. Arsenic toxicity, mutagenesis and carcinogenesis – a health risk assessment and management approach. Mol. Cell. Biochem., 2004, 255, 47–55.

26. Degraeve, N. Carcinogenic, teratogenic and mutagenic effects of cadmium. Mutat. Res., 1981, 86, 115–135.
27. Jarup, L., Berglund, M., Elinder, C. G. et al., Health effects of cadmium exposure – a review of the literature and a risk estimate. Scand J Work Environ Health, 1998,24 (Suppl 1), 1–51.
28. Cadmium. Environmental Health Criteria. WHO. Geneva: World Health Organization, 1992, vol. 134.
29. Hotz, P., Buchet, J. P., Bernard, A., Lison, D., Lauwerys, R. 1999, Renal effects of low-level environmental cadmium exposure: 5-year follow-up of a subcohort from the Cadmibel study. Lancet, 354, 1508–1513.
30. Bagchi, D., Bagchi, M., Stohs, S. J. 2001, Chromium (VI)-induced oxidative stress, apoptotic cell death and modulation of p53 tumor suppressor gene. Mol. Cell Biochem. 222(1–2),149–158.
31. Ercal, N., Gurer-Orhan, H., Aykin-Burns, N. Toxic metals and oxidative stress part I: mechanisms involved in metal-induced oxidative damage. Curr. Top Med. Chem. 2001, 1(6), 529–539.
32. Chen, Y. W., Huang, C. F., Tsai, K. S., Yang, R. S., Yen, C. C., Yang, C. Y., Lin-Shiau, S. Y., Liu, S. H. The role of phosphoinositide 3-kinase. Akt signaling in low-dose mercury-induced mouse pancreatic beta-cell dysfunction in vitro and in vivo. Diabetes. 2006, 55(6), 1614–24.
33. Sutton, D., Tchounwou, P. B., Ninashvili, N., Shen, E. Mercury induces cytotoxicity, and transcriptionally activates stress genes in human liver carcinoma cells. International, J. Mol. Sci., 2002, 3(9), 965–984.
34. Hunt, P. R., Olejnik, N., Sprando, R. L. Toxicity ranking of heavy metals with screening method using adult Caenorhabditis elegans and propidium iodide replicates toxicity ranking in rat. Food and Chemical Toxicology, 2012, 50(9), 3280–3290.
35. Nieboer, E., Richardson, D. H. S. The replacement of the nondescript term 'heavy metals' by a biologically and chemically significant classification of metal ions. Environ. Pollut., 1980, 1(B), 3–26.
36. Čoga, L. Interactions of metal ions with DNA. University of Ljubljana, Ljubljana, 2012, 18p. http://www-f1.ijs.si/~rudi/sola/Seminar_Lucija_Coga.pdf
37. Fletcher, G, Rossetto, E., Turnbull, D., Niebor, E. Toxicity, uptake, and mutagenicity, of particulate and soluble nickel compounds. Environ. Health Perspect., 1994, 102, 69–79.
38. Biggart, W., Costa, M. Assessment of the uptake and mutagenicity of nickel chloride in salmonella tester strains. Mutat. Res, 1986, 175, 209–215.
39. Costa, M., Simmonz-Hansen, J., Bedrossian, C. W. M., Bonura, J., Capriolli, R. Phagocytosis, cellular distributions, and carcinogenic activity of particulate nickel compounds in tissue culture. Cancer Research, 1981, 41, 2868–2876.
40. Sen, P., Costa, M. Induction of chromosomal damage in Chinese Hamster ovary cells by soluble and particulate nickel compounds: preferential fragmentation of the heterochromatic long arm of the X-chromosome by carcinogenic crystalline NiS particles. Cancer Research, 1985, 45, 2330–2325.
41. Zhao, C., Young, M., Diwan, B., Coogan, T., Waalkes, M. Association of arsenic-induced malignant transformation with DNA hypomethylation and aberrant gene expression. Proc. National Acad. Sci., 1997, 94, 10907–10912.

42. Alperstein, G., Reznik, R. B., Duggin, G. G.(1991) Lead: subtle forms and new modes of poisoning. Med J Aust. 155(6), 407–409.
43. Merkureva, R. V. Biochemical studies in hygiene. Biomedical research in occupational health. Moscow, Medicine, 1986, 16–74 (in Russian).
44. Heidelberger, C. Chemical carcinogenesis. An. Rev. Bioch., 1975, 44, 79–121.
45. Hammond, P. B., Foulkes, E. K. The toxicity of the metal ion in humans and animals. Some questions toxic metals. Moscow, Mir, 1993, 131–165 (in Russian).
46. Coyle, P., Philcox, J. C., Carey, L. C., Rofe, A. M. Metallothionein: the multipurpose protein. Cell Mol. Life Sci. 2002, 59(4), 627–647.
47. Wheldon, T. E., Kirk, A. An Error Cascade Mechanism for Tumor Progression. J. Theor. Biol., 1973, 42, 107–111.
48. Chervona, Y., Arita, A., Costa, M. Carcinogenic Metals and the Epigenome: Understanding the effect of Nickel, Arsenic, and Chromium. Metallomics, 2012, 4(7), 619–627.
49. Costa, M. and Heck, J. D. Metal Ion Carcinogenesis. Mechanistic Aspects. Metal Ions in Biological Systems, 20th Ed. Sigel, H. Marcel Dekker, New York, 1986, 259–278.
50. Sen, P., Costa, M. Incident of localization of sister chromatid exchanges induced by nickel and chromium compounds. Carcinogenesis, 1986, 7, 527–1533.
51. Leonard, A. Disturbances in the chromosomes under the action of heavy metals. Some questions of metal ions toxicity. Vol. 20, Chapter 8. Sigel, H., Sigel, A. Ed., Moscow, Nauka, 1993, 190–212 (in Russian).
52. Ono, Y., Wada, O., Ono, T. J. Toxicol. Environ. Health, 1981, 8, 947. Cited on: Costa, M., Heck, J. D. Metal Ion Carcinogenesis. Mechanistic Aspects. Metal Ions in Biological Systems, 20th Ed. Sigel, H. Marcel Dekker, New York, 1986, 259–278.
53. Baserga, R., Huang, Ch.-Hs., Rossini, M., Chang, H., Ming, P.-M. L. The Role of Nuclei and Nucleoli in the Control of Cell Proliferation. Cancer Res., 1976, 36, 4297–4300.
54. Weinstein, B., Yamasaki, H., Wigler, M. et al. Molecular and Cellular Events associated with Action of Initiating Carcinogens and Tumor Promoters. In Carcinogens: Identification and Mechanisms of Action. Griffin, A. C., Shaw, C. R. (Ed.). Raven, New York, 1979, 399–418.

# PART III

# EFFECTS OF DIFFERENT POLLUTANTS ON ALGAE, FUNGI, AND SOIL MICROORGANISMS

CHAPTER 14

# STRUCTURAL RECONSTRUCTION OF A MEMBRANE AT ABSORPTION OF $MN^{2+}$, $ZN^{2+}$, $CU^{2+}$, AND $PB^{2+}$ WITH GREEN *ALGAE CHLORELLA VULGARIS* BEIJ.

VASIL V. GRUBINKO,[1] ANGELA I. LUTSIV,[1] and KATHERINA V. KOSTYUK[2]

[1]*V.M. Hnatyuk Ternopil National Pedagogical University, 2, M. Krivonos St., Ternopil, 46027, Ukraine*

[2]*O.O. Bogomolets National Medical University, 13, Taras Shevchenko St., Kiev, 01601, Ukraine, E-mail: v.grubinko@gmail.com*

## CONTENTS

## ABSTRACT

We investigated the structural and functional features of membranes at absorption of $Mn^{2+}$, $Zn^{2+}$, $Cu^{2+}$ and $Pb^{2+}$ by unicellular green algae *Chlorella vulgaris* Beij. The accumulation of metals ions is fluctuating. There are four stages: the stage of protective self-isolation of cells as a result of the primary stress response, the stage of the active accumulation (decrease in resistance and destruction of outer membrane), the stage of inhibition of the accumulation (formation of the secondary concentric membrane), and the stage of uncontrolled accumulation (destruction of secondary concentric membrane). The process of a concentric double membrane system formation is universal and does not depend on the nature of the toxicant (biogenic $Mn^{2+}$, $Zn^{2+}$, and $Cu^{2+}$, toxic $Pb^{2+}$). Kinetic parameters of accumulation ($K_m$, $V_{max}$, $E_a$) show that the process of absorption of ions $Mn^{2+}$, $Zn^{2+}$, $Cu^{2+}$, and $Pb^{2+}$ goes according to mixed type of inhibition and is determined by the affinity of metal-binding proteins to ions, and after saturation of their binding sites the process becomes uncontrolled. During absorption of metals happens structural reconstruction the cell wall chlorella by forming double concentric membrane system, which is an adaptive mechanism for regulating the absorption of metals, which functions to further protect cells from toxic levels of metal.

## 14.1 INTRODUCTION

Evolution of algae was in progress if in their habitat environment there were the ions of various metals, including so-called heavy metals in toxic concentrations. As a result, they formed mechanisms of toxin resistance that support their optimal level in cells. Ions of many metals in some concentrations are effective regulators of metabolism, which is an ecological factor in the regulation of their population [1, 2]. Unicellular algae of different taxonomic groups can accumulate metal ions in concentrations that exceed their content in the water thousand times [3, 4]. However, excess of a certain level causes pathological changes of metabolism and their death [1].

Stability and adaptation of hydrobionts to adverse factors aquatic environment determined by the rate of formation and adequacy of response to current factor of protection systems of cell. Great importance is the membrane complex of cell that are a primary barrier for the penetration of metal ions to cells of algae and isolate the contents of cell from the environment how in optimum conditions as well as in adverse. Stability of cell membranes in adverse conditions depends on their structure, functional state and composition of biomolecules in particular quantity and quality changes of membrane lipids [5].

The intensity of penetration of metal ions is determined by: their concentration in the environment, the interaction with membranes, and affinity of the components of cell membranes and intracellular components [2, 6]. It is believed that the penetration of metal ions into the cells is realized by diffusion and by using active transport [7]. Also it is shown [8], that the accumulation of heavy metals in aquatic organisms is a dynamic process that develops according to time gradient and is characterized by a certain intensity and specificity.

The purpose of this research was studying the process of reconstruction cell membranes induced by phased absorption of ions $Mn^{2+}$, $Zn^{2+}$, $Cu^{2+}$ and $Pb^{2+}$ by cells of *Chlorella vulgaris* Beij., and also analysis the kinetic parameters of penetration.

## 14.2 MATERIALS AND METHODOLOGY

We investigated the unicellular green microalgae *Ch. vulgaris* Beij., which was grown in the climate chamber with illumination at a temperature 20±1°C and lighting 2500 lx, in glass flasks (250 $dm^3$) in the Fitzgerald mineral medium in modification by Zehnder and Gorham [9], that contained among other cations 0.058 mg/$dm^3$ $Mn^{2+}$ and 0.023 mg/$dm^3$ $Zn^{2+}$ but without $Cu^{2+}$ and $Pb^{2+}$. We added aqueous solutions of $MnSO_4$, $ZnSO_4$•$7H_2O$, $CuSO_4$•$5H_2O$, $Pb(NO_3)_2$ at a rate per ion: $Mn^{2+}$ – 0.1, 0.2, 0.5 mg/$dm^3$; $Zn^{2+}$ – 1.0, 2.0, 5.0 mg/$dm^3$; $Cu^{2+}$ – 0.001, 0.002, 0.005 mg/$dm^3$; $Pb^{2+}$ – 0.1, 0.2, 0.5 mg/$dm^3$. The duration of the influence of metals on the algae was: 0.083; 0.25; 0.5; 0.75; 1; 3; 6; 12; 24; 48; 72 and 168 hrs (study the accumulation of metal ions). Structural reconstruction of cell membranes investigated after 24; 72; 168 and 336 hrs under the action of toxic ions. The control cells were the cells, which were grown in nutrient medium without the salts of toxic metals.

The penetration of ions into the cells of chlorella was stopped by 2.5 mM EDTA. After centrifugation of the suspension of algae (2000 rpm) the sediment was washed with solution of nutrient medium for cultivation, then it was burned in the nitrate acid [1]. The content of metals was determined by atomic absorption spectrophotometer Selmi C-115 M. The concentration of proteins in the cells of *Ch. vulgaris* Beij. was determined by Lowry's method.

The values of Michaelis constant ($K_m$) and maximum speed of penetration of metal ions ($V_{max}$) into the cells of algae were calculated by the graphic method of double return values in the Lineweaver–Burk coordinates and energy of activation ($E_a$) was determined by Arrenius graphical method [10].

The cell membranes that were received from homogenates of biomass of chlorella by the method Findley and Evans [11] (separated by centrifugation at 1500 g, 20 min), obtained in a mechanical homogenizer at 7000 rpm with 5 mM Tris-HCl buffer (pH 7.6) that containing 0.5 M sucrose, 0.005 M EDTA, 0.01 M KCl and 0.001 M $MgCl_2$ (wet weight : volume of buffer – 1 : 5), centrifugation at 5000 rpm during 15 min. The resulting suspension (cell membranes and two-phase system of solutions 0.25 M sucrose and 30% solution of polyethylene glycol in 0.2 M sodium phosphate) were divided equally into three parts, into each added 10 mL lower phase mixture of solution, then it was mixed and centrifuged at 2000 g during 15 min in baket-rotor. The membrane material is selected in the place of phase separation using a syringe. All procedures are carried out at 4°C. Membrane changes were fixed microscopically (MBI-15, followed by an integrating digital analysis on the complex "SSTU-camera Manual Vision SSD-color-WOYV00020) after their coloring "chloro-zinc-iodine" reagent [12].

## 14.3 RESULTS AND DISCUSSION

As a result of research it was revealed the fluctuating character of the process of accumulation of the investigated metal ions by cells of *Ch. vulgaris* Beij. in time and depending on the concentration.

The intensity of accumulation of $Mn^{2+}$ by cells of algae under metal concentrations 0.2 and 0.5 mg/dm$^3$ is reduced during 30 min. (Figure 14.1) and under ion concentration of 0.1 mg/dm$^3$ – at first it is increased (to 0.25 hr), and then also is decreased (to 0.75 hr). Later it is observed an active accumulation of metal ions (to 24 hr) under action of all investigated concentrations, what can be explained by a breach of resistibility of the cell membrane. After this the process is inhibited (to 48 hr). With increase of duration of the cultivation of chlorella with $Mn^{2+}$ to 72 and 168 hr it is observed the restoration of ions' accumulation during actions of 0.1, 0.5 and 0.2 mg/dm$^3$, respectively, with further decrease of intensity.

This process is a subject to kinetic laws of Michaelis-Menten only during 0.083–0.5 hr and 12–168 hr (Table 14.1). Thus, the values of $V_{max}$ and $K_m$ of the accumulation of $Mn^{2+}$ are reduced by 22% and 82% respectively (during 0.25 hr), then increased by 3% and 38% (to 0.5 hr), later $V_{max}$ is increased by 11% (from 12 to 24 hr), reduced by 17% (to 48 hr), and again increased by 30% (to 72 hr) and reduced by 12% (to 168 hr). $K_m$ is decreased by 4% during 12–24 hr, increased by 52% (to 72 hr) and reduced by 25% (to 168 hr). The energy activation of binding of $Mn^{2+}$ is reduced by 77% (to 0.25 hr), increased by 36% (to 0.5 hr), reduced by 15% (from 12 to 24 hr), increased by 44% (to 72 hr) and again reduced by 14% (to 168 hr).

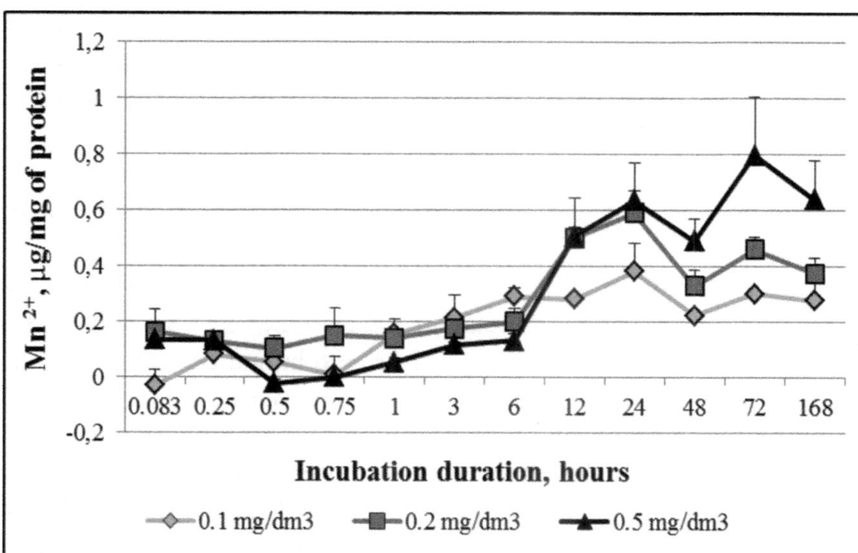


**FIGURE 14.1** Accumulation of $Mn^{2+}$ by *Ch. vulgaris* Beij. cells.

**TABLE 14.1** Kinetic Parameters of $Mn^{2+}$ Accumulation by *Ch.vulgaris* Beij. cells

| Incubation duration, hours | $V_{max}$, μmol/hr • mg of protein | $K_M$, μmol$^{-1}$ | $E_a$, ×10$^{-3}$ kJ/μmol |
|---|---|---|---|
| 0.083 | 34.5 | 0.100 | 2.90 |
| 0.25 | 27.0 | 0.018 | 0.67 |
| 0.5 | 27.8 | 0.029 | 1.05 |
| 0.75 | – | – | – |
| 1 | – | – | – |
| 3 | – | – | – |
| 6 | – | – | – |
| 12 | 54.1 | 0.055 | 1.02 |
| 24 | 60.6 | 0.053 | 0.87 |
| 48 | 50.0 | 0.067 | 1.33 |
| 72 | 71.4 | 0.111 | 1.55 |
| 168 | 62.5 | 0.083 | 1.33 |

*Note*: "–" in Tables 14.1, 14.3, 14.4 – the process is not liable to Michaelis-Menten equation.

The accumulation of $Zn^{2+}$ by the cells of *Ch. vulgaris* Beij. is active (Figure 14.2) (under action of 1 mg/dm$^3$ – to 0.5 hr, under action of 2 and 5 mg/dm$^3$ – to 0.75 hr), which can be explained by their use in the active vital functions of algae; and it is changed by the oppression of the accumulation process (under action of 1 and 2 mg/dm$^3$ – to 3 hr, under action of 5 mg/dm$^3$ – to 1 hr). Then accumulation of metal ions is activated (under action of 1 and 2 mg/dm$^3$ – to 168 hr, under action of 5 mg/dm$^3$ – to 72 hr) with the following decrease of the absorption under concentration of 5 mg/dm$^3$.

The absorption of $Zn^{2+}$ is characterized by such indicators (Table 14.2): $V_{max}$ accumulation of metal ions is increased by 28%, 60%, 38% and 76% during 0,083–0,5 hr, 0,75–1 hr, 3–6 hr and 24–168 hr respectively, and is decreased by 28%, 75% and 50% during 0,5–0,75 hr, 1–3 hr and 6–24 hr respectively, $K_m$ is decreased by 63%, 50%, 62% and 63% during 0,083–0,25 hr, 1–3 hr, 6–24 hr and 48–168 hr, respectively, and increased by 89%, 25% and 25% during 0,25–1 hr, 3–6 hr and 24–48 hr, respectively. $E_a$ binding of $Zn^{2+}$ changes: during the first 0.5 hr it is decreased by 65%, then (to 3 hr) increased by 88%, than (to 12 hr) again

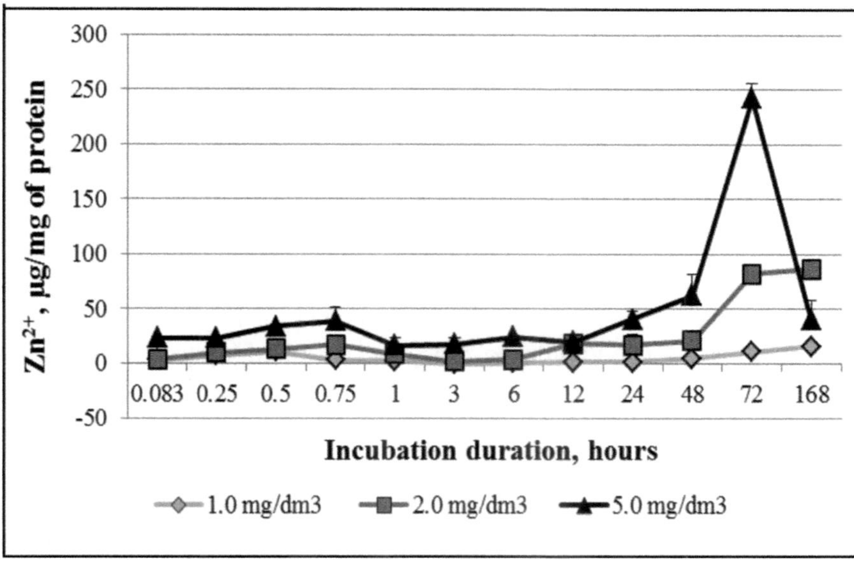


**FIGURE 14.2** Accumulation of $Zn^{2+}$ by *Ch. vulgaris* Beij. cells.

**TABLE 14.2** Kinetic Parameters of $Zn^{2+}$ Accumulation by *Ch.vulgaris* Beij. cells

| Incubation duration, hours | $V_{max}$, μmol/hr • mg of protein | $K_M$, μmol$^{-1}$ | $E_a$, ×$10^{-3}$ kJ/μmol |
|---|---|---|---|
| 0.083 | 2000 | 5.9 | 2.6 |
| 0.25 | 2000 | 2.2 | 1.1 |
| 0.5 | 2778 | 2.5 | 0.9 |
| 0.75 | 2000 | 6.7 | 3.3 |
| 1 | 5000 | 20.0 | 4.0 |
| 3 | 1250 | 10.0 | 8.0 |
| 6 | 2000 | 13.3 | 6.6 |
| 12 | 1250 | 5.7 | 4.5 |
| 24 | 1000 | 5.0 | 5.0 |
| 48 | 2500 | 6.7 | 2.6 |
| 72 | 2941 | 4.3 | 1.5 |
| 168 | 4167 | 2.5 | 0.6 |

decreased by 44%, to 24 hr than increased by 10% and again decreased by 88% to 168 hr.

An active absorption of $Cu^{2+}$ (Figure 14.3) occurs during 0.75 and 1 hr during chlorella cultivation with 0.001 and 0.005 mg/dm$^3$, respectively. During action of 0.002 mg/dm$^3$ it is observed an inhibition of absorption of the ions (to 0.5 hr), which can be explained by self-isolation of cells from metal with following activation (to 0.75 hr). Later the intensity of accumulation of metal ions during action of the investigated concentrations is decreased (to 3 hr), then it is activated (12 hr), inhibited (to 24 hr), activated for the second time (to 72 hr) and inhibited (up to 168 hr). In this case during the action of the investigated concentrations it is observed the accumulation of metal ions to 0.75 and 1 hr and it is changed to inhibition of absorption of ions to 3 hr.

For $Cu^{2+}$ (Table 14.3) value of $V_{max}$ is increased by 20% (to 0.75 hr), decreased by 60% (to 3 hr), again increased by 42% (to 12 hr) and decreased by 23% (for 24 hr); values of $K_m$ and $E_a$ are reduced by 22% and 33%, 92% and 91%, 50% and 33% respectively during 0.083–0.5 hr, 0.75–1 hr, 3–6 hr, and increased by 50% and 45%, 50% and 67%,

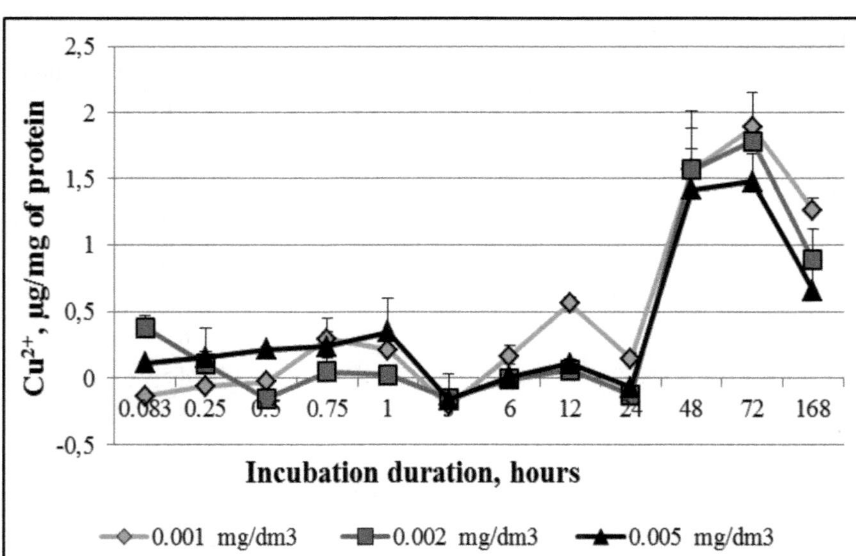


**FIGURE 14.3** Accumulation of $Cu^{2+}$ by *Ch. vulgaris* Beij. cells.

**TABLE 14.3** Kinetic Parameters of $Cu^{2+}$ Accumulation by *Ch. vulgaris* Beij. cells

| Incubation duration, hours | $V_{max}$, µmol/hr•mg of protein | $K_M$, µmol$^{-1}$ | $E_a$, ×$10^{-3}$ kJ/*µmol* |
|---|---|---|---|
| 0.083 | 50.0 | 0.0009 | 0.018 |
| 0.25 | 53.0 | 0.0007 | 0.013 |
| 0.5 | 55.5 | 0.0007 | 0.012 |
| 0.75 | 62.5 | 0.0014 | 0.022 |
| 1 | 48.78 | 0.0001 | 0.002 |
| 3 | 25.0 | 0.0002 | 0.006 |
| 6 | 37.0 | 0.0001 | 0.004 |
| 12 | 43.5 | 0.0003 | 0.006 |
| 24 | 33.3 | 0.0005 | 0.015 |
| 48 | – | – | – |
| 72 | – | – | – |
| 168 | – | – | – |

80% and 73% respectively during 0.5–0.75 hr, 1–3 hr, 6–24 hr. The accumulation of metal ions during 48–168 hr is not a subject to laws of Michaelis-Menten.

The accumulation of $Pb^{2+}$ (Figure 14.4) depends on the concentration: during action of 0.1 mg/dm$^3$ it is observed the decrease of the absorption of the ions of metal (to 0.25 hr), then increase (to 6 hr), decrease (to 24 hr), again increase (to 72 hr) and decrease (168 hr), during the action of 0.2 mg/dm$^3$ – initial accumulation (to 1 hr) is decreased (to 3 hr), then gradually increased (to 168 hr) during the action of 0.5 mg/dm$^3$ – the intensity of the absorption of metal ions is decreased (to 0.25 hr), then activated (6 hr), inhibited (to 12 hr), activated one more time (to 48 hr) and inhibited (to 168 hr).

The accumulation of $Pb^{2+}$ is a subject to laws of Michaelis-Menten only during 0.083 hr and 1–6 hr (Table 14.4).

The value of $V_{max}$ of the absorption of $Pb^{2+}$ is decreased by 79% within 1–3 hr and increased by 60% to 6 hr $K_m$ during this time is increased by 67% and $E_a$ binding of metal ions is increased by 91% within 1–3 hr and decreased by 47% to 6 hr.

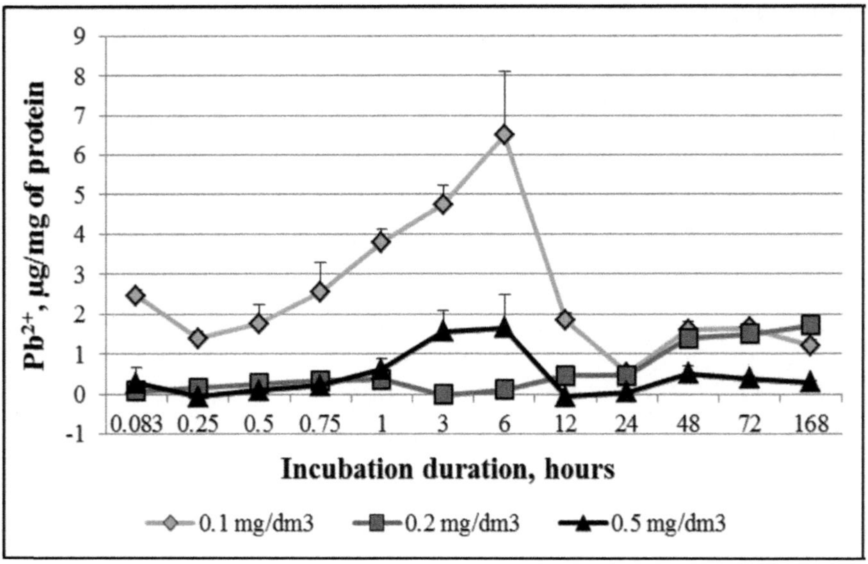


**FIGURE 14.4** Accumulation of $Pb^{2+}$ by *Ch. vulgaris* Beij. cells.

**TABLE 14.4** Kinetic Parameters of $Pb^{2+}$ Accumulation by *Ch. vulgaris* Beij. cells

| Incubation duration, hours | $V_{max}$, *µmol*/hr • mg of protein | $K_M$, *µmol*$^{-1}$ | $E_a$, • $10^{-3}$ kJ/*µmol* |
|---|---|---|---|
| 0.083 | 200 | 0.50 | 2.5 |
| 0.25 | – | – | – |
| 0.5 | – | – | – |
| 0.75 | – | – | – |
| 1 | 238 | 0.22 | 0.9 |
| 3 | 50 | 0.50 | 10.0 |
| 6 | 125 | 0.67 | 5.3 |
| 12 | – | – | – |
| 24 | – | – | – |
| 48 | – | – | – |
| 72 | – | – | – |
| 168 | – | – | – |

The cells of *Ch. vulgaris* Beij. actively accumulate $Mn^{2+}$ to 24 hr of incubation, $Zn^{2+}$ – to 0.5 hr (1.0 mg/dm$^3$) and to 0.75 hr (2 and 5 mg/dm$^3$), $Cu^{2+}$ – to 0.75 hr (0.001 mg/dm$^3$) and to 1 hr (0.005 mg/dm$^3$), $Pb^{2+}$ – to 1 hr (0.2 mg/dm$^3$) and to 6 hr (0.1 and 0.5 mg/dm$^3$), thus controlling the penetration of ions. Later accumulation is inhibited. The mechanism of accumulation of metal ions is fluctuating and has four stages: the stage of self-isolation (stress reaction) of cells, stage of the active accumulation, stage of the inhibition, stage of the second accumulation. The stage of the self-isolation is the response of the cellular organism to the action of stress factor, in our case the ions of metals. The best isolation function of the cells appears during the action of $Mn^{2+}$ (0.2 and 0.5 mg/dm$^3$ – to 0.5 hr), $Cu^{2+}$ (0.002 mg/dm$^3$ – to 0.5 hr), and $Pb^{2+}$ (0.1 and 0.5 mg/dm$^3$ – to 0.25 hr). Decrease of resistance of the primary cellular membrane to the investigated concentrations of metal ions is characterized by the stage of active accumulation of $Mn^{2+}$ (0.1 mg/dm$^3$ – from 0.75 to 24 hr, 0.2 and 0.5 mg/dm$^3$ – from 0.5 to 24 hr), $Zn^{2+}$ (1 mg/dm$^3$ – to 0.5 hr, 2 and 5 mg/dm$^3$ – to 0.75 hr), $Cu^{2+}$ (0.001 mg/dm$^3$ – to 0.75 hr; 0.002 mg/dm$^3$ – from 0.5 to 0.75 hr; 0.005 mg/dm$^3$ – to 1 hr), $Pb^{2+}$ (0.1 and 0.5 mg/dm$^3$ – from 0.25 to 6 hr; 0.2 mg/dm$^3$ – to 1 hr), which is followed by the destruction of the primary membrane [2]. Later the cells of chlorella try to control the absorption of the ions at the stage of secondary inhibition.

The observed fluctuating type of accumulation of the ions corresponds with structural and functional reconstructions of the cellular membrane during the action of metal ions, which consists in formation of double concentric membrane systems (Figure 14.5) as additional protection system,

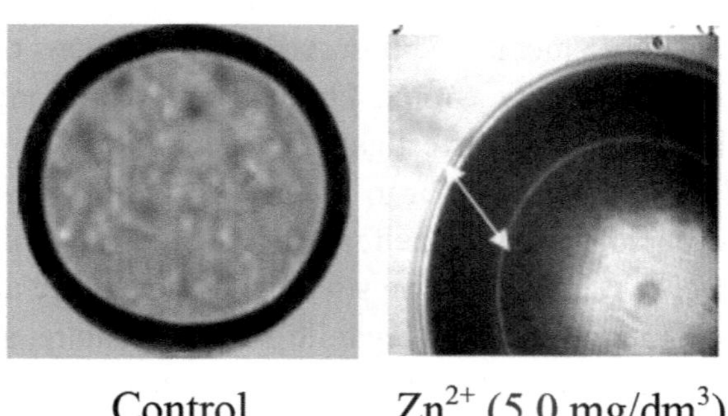


**FIGURE 14.5** The formation of secondary concentric membranes in the cells of *Ch. vulgaris* Beij. under the influence of zinc ions (5.0 mg/dm3), ×900.

which contributes to the normalization of functional and metabolic activity of cells and their survival by unfavorable factors [2].

This process happens already at the first day of stressors action regardless of their nature (biogenic zinc, toxic lead) [2]. In the cells grown in the medium that containing toxic ions detected significant morphological differences related to changes in membrane thickness and size of cells. In particular, the increased granularity of the cytoplasm (24 hr action); in the cells appears the second concentric circle membranes, granularity of cytoplasmic increases, there is increased vacuolization and condensation the substance of white color (48 hr action); concentric membrane thickens, and the area of nuclear and cytoplasmic space is reduced (168 or more hr action).

We believe that a system of concentric membranes is one of the essential parts of a specific cellular response at absorption of ions of toxic metals. We assume that the basis of the membrane adaptation to unfavorable factors lies in hyperplasia of endoplasmic reticulum, since it is proved that in the structures formed by smooth endoplasmic reticulum the content of enzymes responsible for the detoxification is increased [13].

The phenomenon of multiplication of membrane system in the cells of water plants is consistent with the established for some organism cells ability to adapt to the action of stressful factors due to thickening and multiplicative fragmentation of cell membranes. For example, early in a process of *ascosporogenesis* in *Arthroderma vanbreuseghemii* and *Arthroderma simii* [14, 15], dehydrated pollen *Pyrus communis* L. [16], while growing chlorella and micrococcus in radioactive according to deuterium water [17] and testifies about its adaptive value. The process of formation the double concentric membrane observed for the actions of $Mn^{2+}$ (at all investigated concentrations – from 24 hr to 48 hr), $Zn^{2+}$ (5 mg /dm$^3$ – from 0.75 hr to 1 hr), $Cu^{2+}$ (at all investigated concentrations – from 1 hr to 24 hr), $Pb^{2+}$ (0.1 mg/dm$^3$ – from 6 hr to 24 hr; 0.5 mg/dm$^3$ – from 6 hr to 12 hr). Thus, one could argue that the formation of secondary membrane is a universal response of cells to toxic stress, and takes place already during the first hours of action stressors regardless of their nature, and calculated energy of activation at this stage indicates the increase of energy expenditure of maintaining cell activity in chlorella in toxic environment that may be necessary to factor the formation of secondary membrane.

We assume that secondary membrane functions as primary, as evidenced by protein-lipid composition of membranes, mainly activation the synthesis of phospholipids and triacylglycerol during intoxication (Table 14.5).

Content of TAG, DAG and NEFA increases for operating of ions of manganese on 90%, 125% and 91%, for the actions of ions of zinc – on 53%, 6% and 96%, for the actions of ions of copper – on 46%, 76% and 49%, for the actions of ions of lead – on 36%, 71% and 80% accordingly. For the actions of ions of manganese and zinc content of PL diminishes on 12% and 24% accordingly, and for the actions of ions of lead and copper – increases on 1.6% and 7.0% accordingly comparatively with control. At the same time little different correlation of relative content of lipids (of TAG:DAG:PL:NEFA (%): changes in control it was 22:16:47:15; for the actions of $Mn^{2+}$(72 hr) – 27:25:28:20; for the actions of $Zn^{2+}$ (168 hr) – 26:16:33:25; for the actions of $Cu^{2+}$(72 hr) – 22:22:39:17; for the actions of $Pb^{2+}$(168 hr) – 21:21:37:21.

For the actions of ions of manganese relative content of TAG, DAG and NEFA grows on 23%, 56% and 33% accordingly, PL diminishes on 40%. For the actions of ions of zinc relative content of TAG and NEFA increases on 18% and 67% accordingly, content of DAG remains unchanging, PL diminishes on 30%. Relative content of TAG does not change for the actions of ions of copper, content of DAG and NEFA increases on 38% and 13%, PL diminishes on 17%. For the actions of ions of lead part of

**TABLE 14.5** Content of Triacylglycerols (TAG), Diacylglycerols (DAG), Phospholipids (PL), Nonetherified Fatty Acids (NEFA) and Total Lipids in *Ch. vulgaris* Beij. for Actions $Mn^{2+}$, $Zn^{2+}$, $Cu^{2+}$ and $Pb^{2+}$

| Terms of cultivation | Total mass of lipids, mg | Mass of classes lipids, mg | | | |
|---|---|---|---|---|---|
| | | TAG | DAG | PL | NEFA |
| Control | 9.11±1.13 | 1.86±0.17 | 1.50±0.19 | 4.35±0.56 | 1.40±0.15 |
| $Mn^{2+}$ | 13.40±1.31* | 3.54±0.36* | 3.38±0.38* | 3.81±0.31* | 2.67±0.26* |
| $Zn^{2+}$ | 10.47±1.10* | 2.83±0.28* | 1.59±0.19 | 3.32±0.16 | 2.74±0.28* |
| $Cu^{2+}$ | 12.10±1.05* | 2.72±0.37* | 2.64±0.21* | 4.66±0.49 | 2.08±0.24 |
| $Pb^{2+}$ | 12.02±1.73* | 2.52±0.25* | 2.57±0.25* | 4.42±0.47 | 2.52±0.25* |

*Note*. * – P> 99.5 by the criterion of by STUDENT (in relation to control).

TAG and PL diminishes on 4,5% and 21% accordingly, DAG and NEFA increases on 31% and 40%.

Increase of absolute content of TAG in all cases and them relative part for the actions of ions manganese and zinc it takes place as a result of necessity of compression of cellular membranes as a protective mechanism for their toxic action [3, 7], what comports with data about the increase of content of TAG in the cells of chlorella at stress situations up to 80% of their dry biomass [6]. Increase of content of DAG and NEFA at a stress action explained by activating of lipase and phospholipase [18, 19].

Content of PL, as constituents of biological membranes, diminishes for the actions of all investigational ions of metals, that it is possible to explain their participating in binding of metals and their leading-out from a metabolic pool due to high adsorption ability of these lipids [18].

Increase of content of NEFA at the action of the investigated metals is a consequence breaking up of phospholipids, content of that, as marked, diminishes.

Growth content of these classes of lipids is consistent with their participation in compensatory and adaptive activity of cells under stress conditions. Increase of TAG – one of the factors stabilization of cell membranes, which are precursors to the formation of DAG and NEFA [18], and increase their number in a cell corresponds with consolidation and reduced fluidity of membrane [19]; TAG is also reserve energy substrate and increasing their content testifies to energy deficit resulting from toxic effects. As a result, activated lipolysis, order to provide the necessary energetic material for a cell, the level of triacylglycerols after an initial increase their content decreases significantly, that accompanied by an increase in the content of DAG and NEFA. Phospholipids affect not only the fluidity of membranes, but also form a microenvironment for membrane enzymes, ion channels, and regulate the communication of cell with the environment [20]. The increase of their content can be interpreted as a general mechanism of toxic resistance of cells to the action of heavy metal ions, which formed depending on the concentration and duration action of toxicants. It is possible that the intensity of this process explained by the high sorption ability of charged phospholipids [21], or performance of their functions of messengers, which transfer the information into the cell about changes in the environment [22]. Also toxicants stimulate the biosynthesis of proteins, but their number is much less than number of lipids, because the synthesis of protein are longer in time and energy-dependent process. It is proved that the more aggressive toxicant, the greater the weight of accumulating proteins, that indicating of their participation in process of formation of double concentric membranes. As for the indicators of enzymes activity (ATP-ase, alkaline phosphatase), whose activity after its initial decrease by the action of toxicants is resumed synchronously with the formation of a secondary membrane [23, 24]. Reconstructions of the membrane during the action of metal ions are also consistent with violation of the functioning of the membranous ATP-ases [1], particularly $H^+$-ATP-ase, that participates in the regulation of ions [25]. Thus, $Zn^{2+}$ do not practically impact on the membranous ATP-ases, except for high concentrations (5 mg/dm$^3$), because they have high permeability, mobility in the cell, and complexion ability [26]. $Pb^{2+}$ inhibits the activity of ATP-ase [1], because they are characterized by high affinity to proteins and strong restraining of this metal within metallothioneins [26]. A certain number of $Cu^{2+}$ can bind with cellular membranes, and other ones make the complexes with low-molecular organic substances and proteins up to the saturation of their centers of binding [27]. The highest affinity to the proteins among investigated ions is $Cu^{2+}$, and the least – $Zn^{2+}$. $Mn^{2+}$ supplant $Ca^{2+}$ from cellular membranes [28], and therefore their accumulation is limited by duration of antiport.

Next stage of the reactivation process of accumulation, that is observed during the actions of $Mn^{2+}$ (0.1 and 0.5 mg/dm$^3$ – from 48 to 72 hr; 0.2 mg/dm$^3$ – 48 to 168 hr), $Zn^{2+}$ (1 and 2 mg/dm$^3$ – from 3 to 168 hr, 5 mg/dm$^3$ – from 1 to 72 hr), $Cu^{2+}$ (0.001, 0.002 and 0.005 mg/dm$^3$ – from 24 to 72 hr), $Pb^{2+}$ (0.1 mg/dm$^3$ – 24 to 72 hr; 0.2 mg/dm$^3$ – from 3 to 168 hr, 0.5 mg/dm$^3$ – from 12 to 48 hr), is characterized by destruction of the secondary concentric membrane [2], due to exhausting the basic protection of resources in the cell.

The further accumulation of investigated metal ions becomes uncontrolled and can be formed multimembrane system (Figure 14.6). But, the following membranes likely are defective and functioning inefficiently and can lead to the formation of pathological structures, and later occurs a natural process the death of cell by increasing the osmotic pressure inside the cell and cracking of cell.

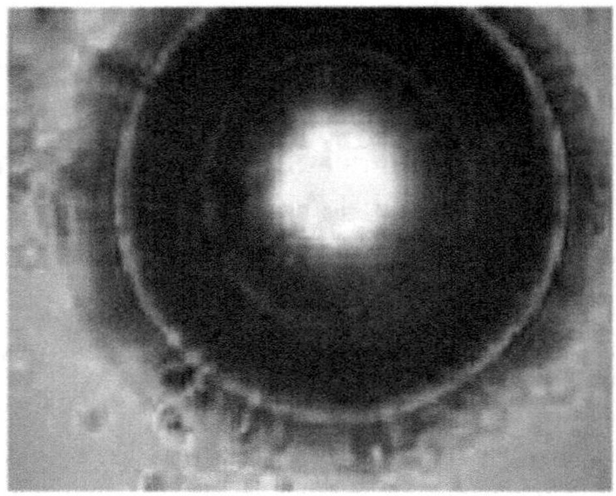

**FIGURE 14.6** The formation of multimembrane system and apoptosis in cells of *Ch. vulgaris* Beij. under the action of the lead ions (0.5 mg/dm3; 336 hr), ×900.

The kinetic indicators conform to the regularity of absorption of metals. The increase of $V_{max}$ accumulation of $Mn^{2+}$ (0.25–0.5; 12–24; 48–72 hr), $Zn^{2+}$ (0.083–0.5; 0.75–1; 3–6; 24–168 hr), $Cu^{2+}$ (0.083–0.75, 3–12 hr), $Pb^{2+}$ (3–6 hr) shows that metal ions are bound with molecules of the cellular walls of alga and molecules-carriers of membranes according to the noncompetitive type [10]. The absorption of ions is also characterized by the decrease of value of $V_{max}$: $Mn^{2+}$ (0.083–0.25; 24–48; 72–168 hr), $Zn^{2+}$ (0.5–0.75; 1–3; 6–24 hr), $Cu^{2+}$ (0.75–3; 12–24 hr), $Pb^{2+}$ (1–3 hr), that indicate the competitive inhibition.

The energy of the activation of binding of $Mn^{2+}$ (0.25–0.5; 24–72 hr), $Zn^{2+}$ (0.5–3; 12–24 hr), $Cu^{2+}$ (0.5–0.75; 1–3; 6–24 hr), $Pb^{2+}$ (1–3 hr) by surface membrane of algae and penetration through the membrane show that the process of accumulation of metals within this period is energy-dependent.

## 14.4 CONCLUSIONS

Thus, accumulation of the ions of $Mn^{2+}$, $Zn^{2+}$, $Cu^{2+}$ and $Pb^{2+}$ by the cells of *Ch. vulgaris* Beij. is fluctuating and is determined by the concentration of ions in the environment and duration of its action on the cells.; noncontrol accumulation is as a result of destruction of secondary membrane. There are four stages: the stage of protective self-isolation of cells as a result of the primary stress response, the stage of the active accumulation as a result of decrease in resistance and destruction of outer membrane, the stage of inhibition of the accumulation as a result of formation of the secondary concentric membrane [2]; the stage of uncontrolled accumulation as a result of destruction of the secondary concentric membrane. The process of a concentric double membrane system formation is universal and does not depend on the nature of the toxicant (biogenic $Mn^{2+}$, $Zn^{2+}$ and $Cu^{2+}$, toxic $Pb^{2+}$).

The absorption of metal ions goes according to mixed mechanism and is determined by the affinity of metal-binding components of membranes, formation of secondary concentric membrane and its resistance to metals, the duration of its structural and functional activities, and after loss of it and after saturation of the centers of binding of metal ions by cytoplasmic components – the process of accumulation becomes passive and uncontrolled.

## KEYWORDS


- **cell wall**
- **copper**
- **double concentric membrane**
- **ions manganese**
- **kinetic parameters absorption**
- **lead**
- **zinc**


## REFERENCES


1. Grubinko, V. V., Gorda, A. I., Bodnar, O. I., Klochenko, P. D. Metabolism of Algae under the Impact of Metal Ions of the Aquatic Medium (a Review). Hydrobiological Journal. 2011, 47(6), 75–88.
2. Kostyuk, K. V., Grubinko, V. V. Change of Composition of the Cellular Membranes of the Aquatic Plants under the Impact of Toxic Substances. Hydrobiological Journal. 2012, 48(4), 75–92.

3. Jain, S. K., Vasudevan, P., Jha, N. K. Taking off some heavy metals from the pollution water helping of water plant: experiences with Azolla. Biological Wastes. 1989, 28(2), 115–126.
4. Whinston, A. I., McAuley, P. J., Smith, V. J. Removal of metals from wastewater by marin microalgae. Journal of Experimental Botany. 1995, 46(1), 1–3.
5. Schmid, K. M., Ohlrogge, J. B. Lipid metabolism in plants. Biochemistry of Lipids, Lipoproteins and Membranes (4th Edn.) [Eds. D. E. Vance, J. E. Vance]. Berlin: Elsevier Science, B. V., 2002, Ch. 4, 93–126.
6. Grubinko, V. V. Features adaptation celled freshwater algae to heavy metals. Actual problem of modern algology. Kyiv, 2012, 83–85. (in Ukrainian).
7. Antonov, V. F. Membrane transport. International Soros Science Education Journal. 1997(6), 6–14. (in Russian).
8. Khomenchuk, V. O., Kurant, V. Z., Konovets, I. M. et. al. Influence of some factors of the aquatic environment on accumulation of heavy metals in the organism of carp. Reports of the National Academy of Sciences of Ukraine. 2000, 5, 97–100. (in Ukrainian).
9. Methods of physiological and biochemical research of algae in hydrobiologycal practic [Ed. A. V. Topachevskiy]. Kyiv: Naukova dumka, 1975, 247 p. (in Ukrainian).
10. Varfolomeev, S. D., Gurevych, K. G. Biokinetics: practical course. Moscow: FAIR – PRESS, 1999, 720 p. (in Russian).
11. Findley, J. B. C., Evans, W. H. Biological membranes: a practical approach. Oxford, Washington: IRL Press, 1987, 304 p.
12. Broda, B. Methods histochemistry of plants. Warsaw: Public organization of medical publishers, 1971, 255 p. (in Polish).
13. Pathoanatomy and physiology, Ultrastructural pathology of cells. URL: http://www.nedug.ru/library/doc.aspx?item=34099 (Accessed 12 May 2015).
14. Ito, H., Hanyaku, H., Harada, T., Tanaka, S. Fine structure in ascosporogenesis of freeze-substituted Arthroderma simii. Revista Iberoamericana de Micología (Bilbao, Spain). 2000.Vol. 699, E-48080, 13–16.
15. Tanaka, S., Fujigaki, T., Watanabe, S. Ultrastructure of the concentric membrane system in asci of Arthroderma vanbreuseghemii. Sabouraudia. 1982, 20(2), 127–136.
16. Tiwari, S. C., Polito, V. S., Webster, B. D. In dry pear (Pyrus communis, L.) pollen, membranes assume a tightly packed multilamellate aspect that disappears rapidly upon hydration. Protoplasma. 1990, 153(3), 157–168.
17. Mosin, O. V. About the phenomenon of cellular adaptation to heavy water. URL: htth://www.gaudeamus.omskcity.com (Accessed 12 May 2015) (in Russian).
18. Lewis, R. N. A. H., McElhaney, R. N. Surface charge markedly attenuates the nonlamellar phase-forming properties of lipid bilayer membranes: calorimetric and 31P-nuclear magnetic resonance studies of mixtures of cationic, anionic, and zwitterionic lipids. Biophysical Journal. 2000, 79(3), 1455–1464.
19. Dyatlovitskaya, E. V., Bezuglov, V. V. Lipids as bioeffectors. Introduction. Biochemistry (Moscow). 1998, 63(1), 3–5 (in Russian).
20. Abbas, C. A., Card, G. L. The relationship between growth temperature, fatty acid composition and the physical state and fluidity of membrane lipids in Yersinia enterocolitica. Biochimica et Biophysica Acta. 1980, 602(3), 469–476.

21. Wang, L., Zhou, Q., Chua, H. Contribution of Cell Outer Membrane and Inner Membrane to Cu2+ Adsorption by Cell Envelope of Pseudomonas putida 5-x. Journal Environmental Science and Health. Part, A. 2004, 39(8), 2071–2080.
22. Vigh, L., Horváth, I., Thompson, G. A. Recovery of Dunaliella salina cells following hydrogenation of lipids in specific membranes by a homogeneous palladium catalyst. Biochimica et Biophysica Acta. 1988, 937(1), 42–50.
23. Kostyuk, K. V., Grubinko, V. V. Role of Membrane ATP-ases in Adaptation of Aquatic Organisms to Environmental Factors (a Review). Hydrobiological Journal. 2010, 46(6), 45–56.
24. Kostyuk, K., Grubinko, V. The effect of zinc, lead ion and diesel fuel on membrane lipid composition of aquatic plants. Bulletin of Lvov University. Biology series. 2010, 54, 257–226 (in Ukrainian).
25. Lionetto, M. G., Giordano, M. E., Vilella, S., Schettino, T. Inhibition of eel enzymatic activities by cadmium. Aquatic Toxicology. 2000, 48(4), 561–571.
26. Dmitrieva, A. G., Kozhanova, O. N., Dronina, N. L. Physiology plant organisms and role of metals. Moscow. Moscow University Press. 2002, 160 p. (in Russian).
27. Kabata-Pendias, A., Pendias, H. Microelements in soils and plants. Moscow. Publishing World ("Mir" in Rus.). 1989, 439 p. (in Russian).
28. Zolotuhina, E. J., Gavrylenko, E. E. Binding of copper, cadmium, iron, zinc and manganese by proteins in water macrophytes. Russian Journal of Plant Physiology. 1990, 37(4), 651–658 (in Russian).

## CHAPTER 15

# ACCUMULATION AND EFFECTS OF SELENIUM AND ZINC ON *CHLORELLA VULGARIS* BEIJ. (CHLOROPHYTA) METABOLISM

HALINA B. VINIARSKA, OKSANA I. BODNAR, OLESYA V. VASILENKO, and ANNA V. STANISLAVCHUK

*Ternopil V. Hnatiuk National Pedagogical University, 2, M. Krivonos St., Ternopil, 46027, Ukraine, Tel. +380977821698, E-mail: viniarska19@gmail.com*

## CONTENTS



## ABSTRACT

In this study we investigated accumulation of selenium and zinc by cells and certain organic compounds, pigments content, activity of energy metabolism enzymes and antioxidant system in *Chlorella vulgaris* Beij. We observed samples after adding sodium selenite in concentration of 10.0 mg/dm$^3$ and $Zn^{2+}$ – 5.0 mg/dm$^3$ during the next 7 days. It was discovered that content of selenium in *Ch. vulgaris* cells increased more significantly after adding sodium selenite in combination with zinc, than in case when we added only selenite, and that the accumulation of $Zn^{2+}$ in the algae cells was very intensive. Lipids in comparison to proteins accumulate selenium in 58% more after adding sodium selenite alone, and in 10% more after simultaneous action of sodium selenite and zinc.

The content of pigments, after adding sodium selenite alone and combination of sodium selenite with zinc ions, increased in comparison to the control, chlorophyll a by 23.1% and 41.2%, chlorophyll *b* in 1.3 and 2.5 times, carotenoids by 46.6%, and 70.6%, respectively. Content of pheopigments decreased by 7.7% after adding selenite and by 28.4% in a result of the combined effect of Se (IV) and Zn (II) in comparison to the control sample.

Sodium selenite when added alone as well as in combination with zinc, stimulated energy metabolism of chlorella by activating cytochrome oxidase and succinate dehydrogenase, NADH- and NADPH-glutamate dehydrogenases as well as the enzyme activity of antioxidant system.

## 15.1 INTRODUCTION

Algae have a high ability to accumulate non-metals and metal ions, due to the high adsorption capacity of their cell membranes for chemical compounds, large assimilating surface, the ability of cells to actively absorb substances against the concentration gradient. Due to these properties, microalgae can accumulate trace elements in quantities that exceed their content in water in many times [1, 2].

Algae cells are able to adapt to metal ions using different mechanisms: membrane and intracellular binding by subcellular structures, binding by exo- and endometabolites [2, 3].

*Chlorella vulgaris* is known as a traditional model object for studying the biochemistry of unicellular green algae and classic object for biotechnologically obtained useful products: proteins, lipids, carotenoids, vitamins, etc. [2].

It is known that between metabolism regulators in freshwater microalgae are, assimilated by them from water, dissolved inorganic selenium compounds (selenites or selenates), which are included in the cell free amino acids, proteins, enzymes, polysaccharides, lipids, and carotenoid pigments. To aquatic organisms selenium is the essential trace element and is directly involved in metabolic, energetic and biophysical processes. Selenium compounds are able to regulate the biosynthesis of fatty acids, carotenoids and pigments [4], thus affecting photosynthesis and energy metabolism.

Besides, selenium is one of the most important trace elements and is a component of antioxidant system of all organisms, including algae. Se (IV) compounds are directly involved in the conversion of methionine to cysteine and glutathione synthesis, which increases antioxidant capacity of cells. It was found that -SeH due to the lower ionization potential and lower binding energy has higher electron-donor activity than -SH group, so that's why the connection with -SeH group is more active and effective than with thiol group [5, 6].

The absorption of selenium by algae and its toxicity greatly vary depending on the morpho-functional characteristics of certain types of algae, concentration and selenium oxidation level, physical and chemical factors of the aquatic environment. It is known that algae growth and development inhibition is higher in samples with selenates than selenites, that's why algae in a processes of their livelihood better absorb Se (IV) compounds from the environment in comparison with Se (VI). In addition, the availability of selenium compounds for microalgae is determined by other underlying factors, such as active ions of biogenic and nonbiogenic metals. Regulatory action of selenite on the metabolism depends on its concentration and duration of its action, which was not studied enough for the algae [7, 8].

Some investigations show that low doses of selenium have the ability to reduce the toxic effects of some metals. Also selenites can interact and adsorb different types of metals, that's why they can be considered as a chemical depot of various microelements that have a physiological importance [7, 9, 10].

Zinc is an important biogenic element, which is contained in the composition of the enzymes involved in the energy and protein metabolism, photosynthesis and regulation of redox processes in cells, activating mainly reduction reactions. It is necessary for the synthesis and the formation of the respiratory enzyme (cytochrome oxidase), cytochromes *a* and *b* as well as chlorophyll [11, 12].

High bioaccumulation of inorganic salts and the formation of their biocomplexes with the algae cells macromolecules *in vitro* can be used to produce dietary supplements that contain microelements that are necessary for the body, such as selenium and ions of biogenic metals [2, 3].

Considering this, we studied the accumulation of selenium and zinc by the cells and main macromolecules of *Chlorella vulgaris* Beij., changes of the pigmentary composition as well as the activity of energy producing and antioxidant enzymes after adding sodium selenite alone and in combination with zinc ions.

## 15.2 MATERIALS AND METHODOLOGY

The experiments were held on crops micropopulations of freshwater green alga *Chlorella vulgaris* Beij. Algae were cultivated under conditions of the accumulating culture on the Fitzgerald's medium N 11 in the modification of A. Zender and P. Gorham under at the illumination of 2500 Lx (16:8 hrs.) at 23–25°C [13].

In experimental conditions, into the culture medium of algae was added aqueous sodium selenite solution at a rate of 10.0 mg Se (IV)/$dm^3$ and an aqueous solution of zinc sulfate at the rate of 5.0 mg Zn (II)/$dm^3$. The biomass of alive cells was collected after 7 days of action. As a control sample was a culture of algae that grew without adding compounds of selenium and zinc ions.

The content of selenium in *Ch. vulgaris* cells, was determined after centrifugation at 3000 rounds per minute during 10 min, washed 3 times by 0.005 M Tris-HCl, then burned with residue nitric acid ($HNO_3$) in sealed cups at 120°C for 2 hrs, later determined spectrophotometrically using o'-phenylendyamin at a wave length of 335 nm [14].

The content of zinc ions in the cells of algae was determined by atomic adsorption method on the spectrophotometer Selmi C-115 M and calculated in μg/g [15].

To determine the total protein content in the biomass of algae, they were precipitated by 5% solution of trichloroacetic acid. After centrifugation the precipitate was dissolved in ethanol and centrifuged again, then washed with ethanol: diethyl ether (3:1) and dried with ether. Proteins solubilized by 5 M KOH at 70°C within 24 hrs, neutralized, dried and weighed, then burned by nitric acid ($HNO_3$) in sealed sample bottle at 120°C for 2 hrs to determine the amount of selenium [14, 16].

To establish the amount of lipids, the algae biomass was extracted by chloroform-methanol mixture in the ratio 2:1 using the Folch method in a modification. Thus to one mass particle of the tissue we added 20 mass particles of extractable mixture and leave for 12 hrs for extraction. Non-lipid impurities from the extract were washed with a solution of 1% KCl. The total lipids amount was determined gravimetrically after stripping the extractable mixture [17, 18].

The separation of lipids into separate lipid fractions was performed by bottom-up dimensional thin layer chromatography in sealed chambers on the plates with a mixture of silica and LS 5/40 μ and L 5/40 μ on a glass base. Before start, plates were activated for 30 min at 105°C at a drying cabinet, processed by the 10% alcohol solution of phosphomolybdic acid and dried in a stream of warm air for 10–15 min. Chloroform solution of lipid probe was applied on the sample plate using microdosator in an amount that does not exceed 200 μg of lipid and the plates were slowly placed into the chromatographic chambers. The mobile phase was a mixture of hexane, diethyl ether and glacial acetic acid in a ratio of 70:30:1 [18]. Obtained chromatograms were developed in a camera which was saturated with iodine vapor, to identify separate lipid fractions: phospholipids (PL), diacylglycerols (DAG), unesterified fatty acids (UFA), lysophospholipids (LPL) and triacylglycerols (TAG), after this they were burned by nitric acid ($HNO_3$) in sealed sample bottles at 120°C for 2 hrs. and determined the amount of selenium [14, 17].

The content of pigments was determined spectrophotometrically. Extraction was provided using the 90% solution of acetone. The intensity of color was measured spectrophotometrically in a wavelengths corresponding to the absorption maxima of carotenoids and chlorophylls *a* and *b*: 430 nm, 480 nm, 630 nm, 645 nm, 663 nm, and 750 nm. All calculations were conducted according to regularly used methods. To determine pheopigments we used spectrophotometric method based on the measurement of the difference of pigment extract optical density on the wave of 665 nm before acidification of the sample by 0.1 N hydrochloric acid and 5 min after acidification [13].

To study the activity of energy metabolism enzymes of Ch. *vulgaris,* cells were separated from the environment using membrane filters "Synpor" with pore diameter of 0.4 microns and prepared homogenates of their biomass in chilled buffer solution (0.066 M $K^+$-$Na^+$ phosphate buffer, pH=7.4), that contained 0.5 M of sucrose, 0.005 M EDTA, 0.01 M KCl, and 0.001 M MgCl, at a ratio of 1:5 (raw weight:buffer). Then homogenate was centrifuged for 3 min at 2000 r/min, after that, in order to highlight the mitochondria, the supernatant was centrifuged again for 3 min at 20,000 r/min. Later the precipitate was resuspended into a medium containing 0.5 M of sucrose and 0.05 M Tris-HCl buffer (pH=7.4).

The activity of succinate dehydrogenase – SDH (E.C. 1.3.99.1) was determined by the ferricyanate method based on the oxidation of succinate to fumarate by potassium ferricyanide, after adding succinate [18]. In this case, the incubation mixture contained 0.1 M phosphate buffer (pH=7.8), 0.1 M succinic acid, 0.025 M EDTA, and 25 mM $K_3Fe(CN)_6$. The reaction was terminated by the addition of 20% tri-chlorine acetic acid. Spectrophotometry was carried out at the length of the wave of 420 nm. The enzymatic activity was expressed in nMol of succinate per 1 mg of protein per 1 min.

The activity of cytochrome oxidase – CO (E.C. 1.9.3.1) – determined by Straus method by the condensation of α-naphthol and paraphenylenediamine hydrochloride with formation of indophenol blue was determined according to [19]. The incubation mixture contained 0.2 M phosphate buffer (pH 7.8), 0.1% solution of α-naphthole, 0.1% solution of paraphenilen diamine hydrochloride, and 0.02% solution of cytochrome c. The reaction was terminated by the addition of the ether alcohol mixture (9:1). Spectrophotometry was carried out at the length of the wave of 540 nm. It was expressed in μg of indophenole blue per mg of protein per 20 min.

The glutamate dehydrogenase activity (E.C. 1.4.1.3) – was determined spectrophotometrically by the rate of oxidation of NADH or NADPH at 340 nm in the reaction mixture composed of 0.05 M of Tris-HCl buffer (pH 7.2 for NADH-GDG and pH=8.3 for NADPH-GDG), 0.01 M of α-ketoglutarate, 0.025 μM of NADH or NADPH and 0.2 M of $(NH_4)_2HPO_4$. Enzymatic activity was expressed in μmole NADH (NADPH)/mg of protein per min [20].

Catalase activity – KT (E.C. 1.11.1.6) was determined by the method [21]. The principle of this method is based on the hydrogen peroxide ability to form stable colored complexes with ammonium molybdate. We prepared 10.0% homogenates in 0.05 M Tris-HCl buffer (pH=7.8) from algae in special cold environment. The reaction was run by adding 0.1 mL of homogenate to 2 mL 0.03% hydrogen peroxide solution. At the same time we prepared the control sample, in which instead of the test material we added 0.1 mL of distilled water. After 10 min the reaction was stopped by adding 1 mL of 4% ammonium molybdate. The color intensity was measured at 410 nm in comparison to the control sample, in which instead of hydrogen peroxide we added 2 mL of water. The catalase activity was expressed in μmol·$H_2O_2$/mg protein per min.

Superoxide dismutase – SOD (E.C. 1.15.1.1) was determined by the level of enzyme inhibition of recovered nitrotetrazolium blue involving NADH and phenazine methosulfate [22]. To determine the activity of SOD we prepared 10.0% homogenate of investigated algae in phosphate buffer (pH=7.4). Then, to the homogenate we added 0.15 mL of chloroform, 0.3 mL of ethanol and 300 mg $KH_2PO_4$ followed by centrifugation at 12000 r/min for 15 min at 4°C. To 0.2 mL of the supernatant we added 1.3 mL of 0.1 M phosphate buffer (pH=8.3), 1 mL of nitrotetrazolium blue 0.3 mL of phenazine methosulfate solution and 2 mL of 0.2 M solution of NADH. Samples were kept in the dark place for 10 min and then photometrically scanned at a wavelength of 540 nm and later compared with samples into which we did not add NADH. Control samples were those, in which instead of homogenate was 0.2 mL phosphate buffer. The enzyme activity was determined by its ability to inhibit the recovery of nitrotetrazolium blue to 50% and this was considered as a 1 conversion unit of activity.

The activity of glutathione peroxidase – GPx (E.C. 1.11.1.9) was determined by the method [23]. In a basis of the color reaction is the interaction of SH-groups with Ellman's reagent (0.01 M solution of 5.5-dithiobis-2 nitrobenzoic acid in methanol) with the formation of colored product – thionytrophenyl anion. To 0.3 mL of algae homogenate we added 0.5 mL

0.25 M Tris-HCl buffer (pH=7.4), 0.1 mL of 25 mM EDTA, 0.1 mL of 0.4 M sodium azide, 70 μL of tertiary butyl hydrogen peroxide and incubated for 5 min at 37°C. The reaction was stopped by 0.4 mL 10% TCA. Samples were centrifuged at 8000 r/min during 10 min. In test tubes that contained 5 mL of Tris-HCl buffer we added 0.1 mL of the supernatant and 100 mL of Ellman's reagent. After 5 min optical density was measured using spectrophotometer at a wavelength of 412 nm. The control sample differed from the investigated one by the fact that the homogenate was added before proteins subsidence.

The amount of protein in algae cells was determined by the method of Lowry [24].

Statistical analysis of data was performed using the application package Statistica 5.5 and Microsoft Office Excel 2007.

## 15.3 RESULTS AND DISCUSSION

### *15.3.1 THE ACCUMULATION OF SELENIUM AND ZINC IN CELLS AND CERTAIN ORGANIC COMPOUNDS OF CH. VULGARIS*

It is known that microalgae are able to assimilate and dissolved inorganic compounds from water accumulate them in their cells, then introduce to free amino acids, proteins, enzymes, polysaccharides, carotenoid pigments and lipids [1, 3, 25]. Therefore, it was important to determine the peculiarities of introduction of studied trace substances Se (IV) and Zn (II) to the basic organic macromolecules.

The results showed that during the whole period of algae incubation with selenite and zinc ions (7 days) there was a significant accumulation of selenium by the *Ch. vulgaris* cells (Table 15.1).

After adding sodium selenite, selenium content in the biomass increased by 15.6%, and in samples, in which it was added with zinc by 62.9% in comparison to the control sample. The accumulation of selenite by proteins occurred in all variants of the experiment – in samples with sodium selenite, selenium content increased by 54.4%, while in samples in which it was added with zinc – by 124.7% compared to the benchmark. Lipids as well as proteins accumulate trace element in all variants of the

**TABLE 15.1** Selenium Accumulation by Cells and Certain Organic Compounds of *Ch. vulgaris* Cells, mg/g in Dry Weight, M±m, n=5

| Compound | Control | Se (IV) | Se (IV)+ Zn (II) |
|---|---|---|---|
| Cells biomass | 6.4±0.2 | 7.4±0.4* | 10.4±0.4* |
| Proteins | 29.5±0.1 | 45.6±1.3* | 66.3±2.4* |
| Lipids | 34.7±1.5 | 73.7±4.3* | 80.3±3.0* |

*Note*: * – $P > 99.5$ by the criterion of by STUDENT (in relation to control).

experiment. Thus, in samples with selenite, selenium content increased by 112.1%, and in a case with its joint action with zinc ions – by 131.3% compared to control sample.

As a result of our study, it was found that algae intensively accumulate zinc ions (Table 15.2). This is because zinc is one of the major trace elements in the body that has an important regulatory functions in many metabolic processes, because it is as able to bind with almost 300 different proteins [12].

Chlorella cells accumulate significant amount of zinc that's why its content has increased in 16 times comparing to the control sample. Zinc amount in proteins increased only by 1.3 times, while in lipids – almost by 5 times comparing to the values of the control sample (see Table 15.2).

The largest amount of selenium and zinc accumulated lipids, because in algae they play an important role in the process of growth, reproduction and photosynthetic activity and also perform energy function. In particular, in hydrophytes lipid usage is significantly enhanced to sustain vital activity when they are under the influence of extreme environmental factors, and their quantity and quality in the cells, especially in cell membranes, reflects the synthesis and degradation of lipids, as well as peculiarities of

**TABLE 15.2** Zinc Accumulation by Cells and Certain Organic Compounds of *Ch. vulgaris* Cells, µg/g in dry weight, M±m, n=5

| Compound | Control | Se (IV) + Zn (II) |
|---|---|---|
| Cells biomass | 810.6±20,4 | 12958.1±631.1* |
| Proteins | 442.0±28.5 | 573.1±41.1* |
| Lipids | 423.3±24.4 | 4202.1±254.1* |

*Note*: * – $P > 99.5$ by the criterion of by STUDENT (in relation to control).

their exchange with the living environment [26]. Therefore, it was appropriate to determine the characteristics of inclusion of the studied micronutrients to the lipids of different classes (Figures 15.1 and 15.2).

We can see in the Figure 15.1, selenium content in phospholipids after adding sodium selenite increased by 49.6%, and after adding selenite in combination with zinc – by 57.6% relative to the benchmarks. Selenium amount in the DAG after adding sodium selenite increased by 39.8%, and in samples with combined effect of selenite with $Zn^{2+}$ – by 63.4% in comparison with control sample. In UFA selenium content increased by 76% in samples with sodium selenite and in samples with combination of selenite with $Zn^{2+}$ – by 99.1%.

Accumulation of selenium in LPL in samples with sodium selenite increased by 68.9%, while in samples with simultaneous adding of selenite with $Zn^{2+}$ – by 81.1% compared to control sample. In TAG selenium amount increased by 60.3% for the samples with selenite and in samples with selenite and $Zn^{2+}$ – by 80.2% compared to control values.

The results showed that the selenium amount in different classes of lipids increases significantly in samples in with selenite was added in combination with zinc than in the samples where selenite was added separately. This may be due to the biological needs of the studied micronutrients, as well as the physiological role of selenium in lipids as antioxidant.

The results of the study of zinc accumulation by lipids of different classes showed (Figure 15.2) that the content of trace elements in PL

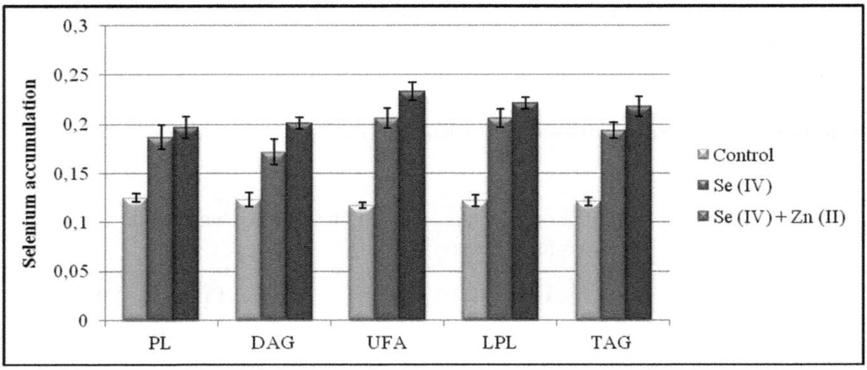


**FIGURE 15.1** Content of selenium in the different classes of lipids, mg/g dry weight, M±m, n=5 (Note: PL – phospholipids, DAG – diacylglycerols, UFA – unesterified fatty acids, LPL – lysophospholipids, TAG – triacylglycerols).

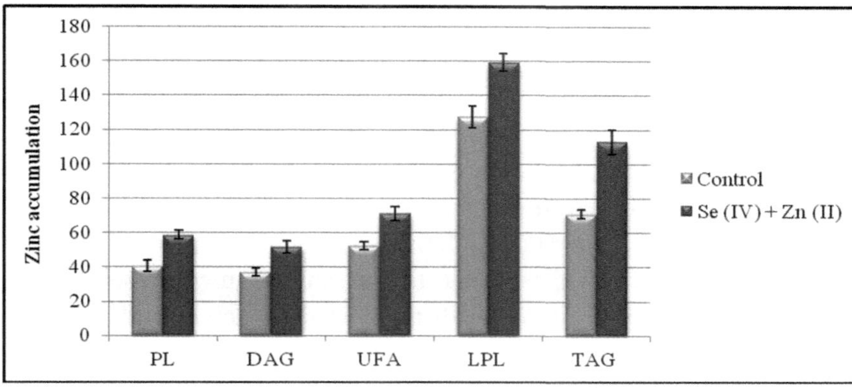


**FIGURE 15.2** Content of zinc in the different classes of lipids, μg/g dry weight, M±m, n=5, (see Figure 15.1) (Note: PL – phospholipids, DAG – diacylglycerols, UFA – unesterified fatty acids, LPL – lysophospholipids, TAG – triacylglycerols).

increased by 44.6%, in DAG – by 39.3%, in UFA – by 35.9%, in LPL – by 25.1% and in TAG – by 59.3% compared to the control sample.

The results that we received in the process of our investigation about the accumulation of zinc were caused mainly by high adsorption capacity of algae cell walls for selenite and zinc ions, substantial assimilation surface and possibly due to the poorly formed mechanisms of micronutrients metabolism regulation. Ions of selenit and zinc in used concentrations could cause some violation of physiological functions and structural changes in the cells, including functional disorders of cell membranes, which in turn, might cause uncontrolled penetration of microelements inside *Ch. vulgaris* cells [1, 27, 28]. It is also possible that in regard to selenium and zinc, that are among the essential micronutrients to the metabolism of cells, cells can have better formed and developed mechanisms of absorption and deposition of the excessive contents of these elements in the environment.

### *15.3.2 PHOTOSYNTHETIC SYSTEM FUNCTIONING AFTER ADDING SELENIUM COMPOUNDS AND ZINC IONS*

Photosynthesis in algae is one of the adaptive factors to possible influence of organic and inorganic ions [29]. Therefore, when studying the effect of sodium selenite in the experiment with its combined action with zinc ions on physiological and biochemical mechanisms, that ensure the livelihoods of chlorella cells, we investigated parameters of the photosynthetic mechanism (Table 15.3).

In samples where we added sodium selenite, chlorophyll a content increased by 23.1%, and in samples with the combination of sodium selenite and zinc ions – by 41.2% relatively to control sample. The content of chlorophyll *b* increased as well as fin the samples in which selenite was added separately and in those with the combined action with zinc ions – at 1.3 and 2.5 times, respectively. Carotenoids increased by 46.6% in samples with sodium selenite and by 70.6% in samples with the combined effect of Se (IV) and Zn (II). Considering the pheophytins, their number decreased comparing with the control sample under the influence of Se (IV) + Zn (II) by 7.8%, and in samples with sodium selenite as the only factor, we observed only minor changes.

Results of our study indicate that the maximum increase of pigments amount was in the samples with the combined action of sodium selenite and zinc ions, while the content of pheophytins under these conditions reduced significantly. The ratio of chlorophyll *a* to chlorophyll *b* also decreased by 1.8 times in comparison to control. The physiological condition and photosynthetic activity of algae was also described by the pigment index – a ratio of carotenoids amount to chlorophyll *a*. In samples with selenite this index grew by 20.0%.

A similar tendency was observed in samples with the simultaneous action of sodium selenite and zinc ions. These changes in correlation of

**TABLE 15.3** Content of Photosynthetic Pigments in *Ch. vulgaris* Cells Under the Impact of Se (IV) and Zn (II), M±m, n=5

| Pigment | Control | Se (IV) | Se (IV) + Zn (II) |
|---|---|---|---|
| Chlorophyll *a*, μg/dm³ | 116.1±1.4 | 143.0±3.0* | 164.0±4.6* |
| Chlorophyll *b*, μg/dm³ | 38.6±1.6 | 49.8±3.1* | 95.8±2.3* |
| Chlorophyll *a*/chlorophyll *b* | 3.0 | 2.9 | 1.7 |
| Carotenoids, μg SPU/dm³ | 46.5±1.7 | 68.1±2.6* | 79.2±1.7* |
| Pigment index | 0.4 | 0.5 | 0.5 |
| Pheophytins, μg/dm³ | 99.8±5.4 | 92.2±3.5 | 71.5±3.7* |

*Note*: * – $P > 99.5$ by the criterion of by STUDENT (in relation to control).

photosynthetic pigments can be explained by high permeability, mobility inside the cell and complex producing ability of zinc ions that can influence directly on chloroplast ultrastructure, penetrating due to Zn/Cd carrier-proteins that are located in the chloroplasts membranes [30, 31]. It is believed that due to its electronegativity Zn (II) inhibits transport of electrons in the cell [32]. Chlorophyll *a* is a part of the reaction centers and peripheral antenna complexes of photosystems I and II, and chlorophyll *b* – is a component of a light-harvesting complex of photosystem II.

Therefore, changes in the ratio of chlorophyll *a*/chlorophyll *b* may indicate on the shift in the stoichiometric balance between complexes of reaction centers of both photosystems and light-harvesting complex of photosystem II. It is believed that the main causes of decreasing in the activity of photosystem II, in the presence of higher concentrations of metals, is the change of the structure of the reaction center proteins and interactions of ions of certain metals, including $Zn^{2+}$, that comprise the reaction center ($Mn^{2+}$, $Ca^{2+}$, $Cl^{-}$). The target of the metal ions, as usual is a primary electron donor of the reaction center of photosystem II – P-680, which is a reducing agent of pheophytin.

It is possible that zinc ions, damaging P-680, cause reduction of pheophytin content in algae cells (see Table 15.3). Pheophytins are the main carriers of electrons in photosystem II. Since it is known that the amount of these pigments is directly proportional to the number of reaction centers in photosystem II, then, such changes of their content indicate a decrease in functional activity of photosynthetic reaction centers. Metal ions can directly affect the process of electron transfer in photochemical reactions [31]. Possible cause that metal ions might inhibit photosynthetic electron transport – changes in chloroplast ultrastructure, including damage of thylakoids [33].

Increasing of pigment index might be explained by the situation that under the unfavorable conditions, primarily chlorophyll a is destroyed which is accompanied by an increase of more sustainable carotenoids. Last are the main protectors for green pigments and nonenzyme antioxidant components of cells. In addition, in the biotransformation process of these pigments are formed physiologically important metabolites that are involved in the regulation of adaptation to the environment, as well as to the impact of sodium selenite and metal ions.

It should be noted that the combined effect of Se (IV) and Zn (II) caused more significant change in the amount of pigments in comparison with the effect of sodium selenite separately. For chlorophyll a this difference was 14.7%, for chlorophyll *b* – 92.4%, for carotenoids – 16.4%, for pheophytins – 77.5%.

Changes in the amount of chlorophylls can occur due to the increasing amount of carotenoids. They, due to their antioxidant properties are involved in the protection of photosynthetic membranes from photooxidation and in the disposal of peroxide radicals. This prevents lipids on the membranes of chloroplasts from oxidation and chlorophyll triplets destruction [34] and therefore increases the amount of green pigment inside cells. While it is known that the content of photosynthetic pigments forms a new adaptive physiological state of cells [35], then changes of the amount and ratio of photosynthetic pigments may indicate an increase in energy needs of chlorella, especially in samples with sodium selenite added together with zinc ions.

### *15.3.3 EFFECT OF SELENIUM COMPOUNDS AND ZINC IONS ON THE ACTIVITY OF ENERGY METABOLISM ENZYMES*

Effectiveness of energy systems functioning in aquatic organism is one of the most indicative success criteria in formulation of survival strategies in an unfavorable environment. Taking into account the extremely large number of enzymes that are responsible for energy generating in the cell, in this study we identified a key enzyme of the citric acid cycle – succinate dehydrogenase and electron transport chain – cytochrome oxidase as well as glutamate dehydrogenase – an enzyme that binds energetic and biosynthetic processes in the cell [28].

It was determined that sodium selenite alone and in combination with zinc ions, increased general activity of these enzymes (Tables 15.4 and 15.5). Thus, in samples with Se (IV) alone we observed increasing activity of cytochrome oxidase by 35.1%, and in samples with combination of sodium selenite and zinc ions – by 54.9%.

Elevation of cytochrome oxidase activity may be associated with an increase in energy demands of chlorella cells for its adaptation to non-specific

**TABLE 15.4** Energy Metabolism Enzymes Activities of in *Ch. vulgaris* Cells Under the Impact of Se (IV) and Zn (II), M±m, n=5

| Enzyme | Control | Se (IV) | Se (IV) + Zn (II) |
|---|---|---|---|
| Cytochrome oxidase, μg indophenol blue/mg protein for 20 min | 0.328±0.008 | 0.443±0.036* | 0.508±0.019* |
| Succinate dehydrogenase, nmol succinate/mg protein per min | 1.055±0.065 | 2.751±0.232* | 4.352±0.323* |

*Note*: * – $P > 99.5$ by the criterion of by STUDENT (in relation to control).

effects. It should be noted that in case of active absorption of studied trace elements by phytoplankton, we can observe an intense zinc accumulation by algae biomass [1, 27], as for selenium, there is only intracellular binding of it by macromolecules and redistribution between proteins, lipids and carbohydrates [8, 36]. Besides, divalent cations, which are the part of cytochrome oxidase are capable to intensify its catalytic properties [37] and in accordance with this increase the amount of reaction product yield.

Regarding succinate dehydrogenase activity, in samples with selenite it increased by 1.6 times in comparison with control sample, and in samples with combined influence of trace elements we observed its growth in more than 3 times (see Table 15.4).

SDH has a high catalytic potential, which can be implemented at different physiological conditions of the organism. This enzyme is involved in the regulation and interconnection not only of the oxidation, but also of constructive metabolism [18, 28]. Therefore, increasing of the succinate dehydrogenase activity is likely to be a compensatory response of energy metabolism to the elevated concentrations of selenite and zinc ions in the environment and is consistent with the increased activity of cytochrome oxidase section of ETC.

NADH- and NADPH-glutamate dehydrogenase activity appealed similarly as in samples with sodium selenite alone and in samples with combination of sodium selenite ions and zinc (Table 15.5). Thus, in samples with sodium selenite we observed a slight inhibition of both enzymes – by 24% and 39%, respectively, and in samples with combined effect of Se (IV) and Zn (II), activity of NADH-GDH and NADPH-GDH was restored and its values were higher than in the control samples by 2.5 and 1.4 times, respectively.

**TABLE 15.5** NADH- and NADPH-Glutamate Dehydrogenase Activities in *Ch. vulgaris* Cells Under the Impact of Sodium Selenite and Zinc Ions, M±m, n=5

| Enzyme | Control | Se (IV) | Se (IV) + Zn (II) |
|---|---|---|---|
| NADH-GDH, μmol×$10^{-3}$ NADH/mg protein min. | 7.00±0.55 | 5.30±0.53 | 23.80±0.55* |
| NADPH-GDH, μmol×$10^{-3}$ NADPH/mg protein min. | 11.94±1.02 | 7.2±0.33* | 28.94±1.11* |
| NADH-GDH/NADPH-GDH | 0.59 | 0.74 | 0.82 |

*Note*: * – $P > 99.5$ by the criterion of by STUDENT (in relation to control).

Orientation of glutamate dehydrogenase reaction which is reversible, and its direction is determined by the presence of coenzyme (NADH-dependent – straight, NADPH-dependent – reverse) causes the direction of metabolism [28, 38, 39]. In direct reaction glutamate deamination occurs with formation of α-ketoglutarate followed by its use in Krebs cycle or other metabolic systems, playing energetic function. In the reverse reaction observed bounding of excess ammonia enzyme, that performs detoxification function, and directs nitrogen for amino acids biosynthesis – synthetase way, thereby α-ketoglutarate is being removed from the Krebs cycle, reducing the activity of aerobic way of energy formation.

In samples with selenite the ratio of NADH-GDH/NADPH-GDH increased by 25%, and in samples with simultaneous adding of Se (IV) and Zn (II) – by 40%, indicating on the activation of the catabolic part of nitrogen metabolism.

It was found that an important physiological role of zinc in plant cells is its direct involvement and activation of redox reactions [11], so that's why additional zinc restored and stimulated the activity of both glutamate dehydrogenases.

Taking into account the metabolic connection of studied enzymes, intensive glutamate deamination by NADH-GDH, caused possibly by the supply of α-ketoglutarate into the Krebs cycle, which was greatly intensified after adding Se (IV) and Zn (II). Herewith, consequently, the formation of glutamate in the Ch. *vulgaris* cells involving NADPH-GDH decreased. Because that, glutamate can be actively used as additional energy substrate in the state of nonspecific action [38, 39].

### 15.3.4 ANTIOXIDANT STATUS OF CH. VULGARIS CELLS AFTER ADDING SELENIUM COMPOUNDS AND ZINC IONS

Enzymes of the antioxidant system are characterized by their specificity toward inclusion of some metals and metalloids, including zinc and selenium, into its composition [25].

Adding sodium selenite into the culture medium altered antioxidant system enzyme activity (Table 15.6). Adding to the culture medium Se (IV) stimulated the increased activity GPx by 16 times compared with the control sample. At the same time, the decrease in catalase activity by 2 times and an increase of superoxide dismutase activity by 1.3 times indicates on the activation of antioxidant protection in Ch. *vulgaris* cells. And that selenium is involved in this process.

In samples with added selenite and zinc ions we observed the increased activity in the GPx by 14.2 and decrease of the SOD and CAT (catalase) activity in 1.2 and 2.9 times comparing to the control.

Studies have shown that GPx activity increased while comparing with the control both in samples with selenite alone as well as in samples with selenite combined with zinc ions, and correlated with selenium and zinc accumulation in the cells (see Table 15.6). SOD activity, which has the opposite to GPx effect, stated that smaller amounts superoxide ions were produced in the cells due to the high activity of GPx. We assume that the increase in GPx activity that is an acceptor of $H_2O_2$ and hydroperoxides, led to a decrease in the formation of superoxide radicals through dynamic transformation of various oxygen forms [10]. This confirms that the

**TABLE 6** Antioxidant enzyme activities in *Ch. vulgaris* cells under the impact of sodium selenite and zinc ions, Mm, n=5

| Enzyme | Control | Se (IV) | Se (IV) + Zn (II) |
|---|---|---|---|
| Glutathione peroxidase, μmol GSH/100 mg protein per min | 2.29±0.02 | 36.89±0.07* | 32.48±0.07* |
| Catalase, μmol·$H_2O_2$/mg protein per min | 1.08±0.08 | 0.47±0.02* | 0.37±0.01* |
| Superoxide dismutase, 1 standard unit/ mg protein | 10.43±0.05 | 13.81±0.01 | 8.73±0.03* |

*Note*: * – P> 99.5 by the criterion of by STUDENT (in relation to control)

increased Se contributed to the raising GPx activity and by reducing the content of produced hydrogen peroxide and superoxide radicals decreased the need in their absorber – SOD. It should be noted that the presence of selenium in cells alone and in combination with zinc increased antioxidant status of chlorella, as evidenced by a significant decrease in CAT activity.

In addition, it is known that protection from reactive oxygen species in plants other than antioxidant enzymes is implemented by nonenzymatic components, including carotenoids [34]. Participation of carotenoids in antioxidant protection of cells in this experiment is confirmed by the growth of their content after adding selenite, and then selenite in combination with zinc ions (see Table 15.3). Activation of the components of two antioxidant protection systems might be appropriate due to the different mechanisms of peroxide products neutralization: enzymes – by the destruction of peroxides and hydroperoxides; carotenoids – by neutralizing singlet oxygen. Elevated levels of chlorophyll content in samples with higher sodium selenite and zinc amounts compared to the benchmarks is connected with the effective disposal of peroxide radicals by both parts of antioxidant defense mechanisms, thereby preventing the degradation of chlorophyll triplets and oxidation of lipids that are located on the membranes of chloroplasts. As the process of lipid peroxidation and the degree of oxidative stress are considered as the main organisms growth factors [40], it can be assumed that increasing activity of CO (cytochrome oxidase) and SDH has also contributed to the activation of energetic and biosynthetic processes.

## 15.4 CONCLUSIONS

The results of this study showed that the selenium content in *Ch. vulgaris* cells increases significantly after adding sodium selenite ions in combination with zinc, than in the samples in which selenite was added alone. The presence of sodium selenite in the culture medium at a concentration of 10.0 mg/dm$^3$ contributed to the accumulation of zinc ions by the algae cells. The largest amount of Se (IV) and Zn (II) was accumulated by lipids, proteins – less, all of this may be associated with different structural and functional relationship of these macromolecules.

The biological effect of accumulated selenium and zinc was detected in: activation of pigment systems as a whole, but there has been a redistribution of value to individual carotenoid pigments in favor of that, except photosynthesis functions involved in ensuring nonenzymatic way antioxidant protection.

Considering zinc ions we can assume that they have a dual effect on the photosynthetic system of *Ch. vulgaris*: direct – on the content of pigments and regulation of electron transport speed, and indirectly – by influencing on other enzyme systems.

These changes in the functioning of the photosynthetic system of the *Ch. vulgaris* are reflected on the whole complex of metabolic changes, including the energy supply of the cell. Sodium selenite alone and in combination with zinc stimulated energetic links of chlorella metabolism by activating cytochrome oxidase and succinate dehydrogenase as well as NADPH- and NADH-glutamate dehydrogenases that actively involve aminoacids to the oxidation as an energy substrates.

In general, increased content of selenium and zinc in *Chlorella* cells by adding them as selenite 10.0 $mg/dm^3$ and zinc ions 5.0 $mg/dm^3$ within 7 days modified cell metabolism through activation of photosynthetic systems and adaptive adjustment of energy metabolism and antioxidant protection that increased physiological and biochemical status of cells with simultaneous accumulation of selenium and zinc in macromolecules, mostly – in lipids.

Received effect can be regarded as a basis for the development of technologies for selenium, zinc-lipid/protein biologically active medications.

## ACKNOWLEDGEMENT

Authors would like to thank to our supervisor Professor V.V. Grubinko for help in the discussion of the results of this study.

## KEYWORDS


- **antioxidant enzymes**
- **classes of lipids**
- **pigments**
- **proteins**
- **selenates**
- **selenites**


## REFERENCES


1. Bodnar, O. I. Adaptive properties of algae for the actions of metal ions. Thesis for the Scholarly of PhD of Biological Sciences, specialty 03.00.17 "Hydrobiology." Kyiv. 2009, 22 p. (in Ukrainian).
2. Zolotareva, O. K., Shnyukova, E. I., Sivash, O. O., Mikhaylenko, N. F. Prospects of the use of microalgae in biotechnology. Kyiv. Alterpres. 2008, 234 p. (in Ukrainian).
3. Holtvyansky, A. V. Bioaccumulation of metal ions by green algae cells and production of biomass enriched with microelements. Abstract of dissertation for the degree of candidate of biological sciences, specialty 03.00.20 "Biotechnology." Kyiv. 2002, 17 p. (in Ukrainian).
4. *Zhou, Z., Li, P., Liu, Z.,* et al. Study on the accumulation of selenium and its binding to the proteins, polysaccharides and lipids from *Spirulina maxima, S. platensis* and, *S. subsalsa.* Oceanol. et Limnol. Sin. = Haiyang yu huzhao. 1997, 28(4), 363–370.
5. Whanger, P. D. Selenocompounds in plants and animals and their biological significance. J. Amer. College Nutr. 2002, 21(3), 223–232.
6. Zhi-Yong Li, Si-Yuan Guo, Lin Li. Bioeffects of selenite on the growth of Spirulina platensis and its biotransformation. Bioresource Technology, 2003, 89, 171–176.
7. Araie, H., Shiraiwa, Y. Selenium Utilization Strategy by Microalgae: Review. Molecules. 2009, 14, 4880–4891.
8. Umysová, D., Vítová, M., Doušková, I., et al. Bioaccumulation and toxicity of selenium compounds in the green alga *Scenedesmus quadricauda.* BMC Plant Biology. 2009, 9(58), 58–74.
9. Davydova, O. E., Veshytski, V. A., Yavorivski, P. P. Physiological-biochemical and stress protect functions of selenium in plants. Physiology and Biochemistry of Cultivated Plants. 2009, 41(2), 109–122 (in Ukrainian).
10. Hartikainen, H., Xue, T., Piironen, V. Selenium as an anti-oxidant and pro-oxidant in ryegrass. Plant Soil. 2000, 225(1–2), 193–200.
11. Taiz, L., Zeiger, E. Plant Physiology. Fourth Edition. Sinauer Associates. Sunderland, MA, 2006, 764 p.

12. Vallee, B. L., Falchuk, K. H. The Biochemical Basis of Zinc Physiology. Physiol. Rev. 1993, 73, 79–118.
13. Romanenko, V. D. (ed.). Methods of hydroecological investigation of surface waters. Kyiv. Logos. 2006, 408 p. (in Ukrainian).
14. Dedkov, J. M., Musatov, A. V. Selenium: biological role, chemical properties and methods of determination. All-Russian Institute of Scientific and Technical Information. [VINITI by RJ.- in Rus.] Chemistry. №1688, 2002 (in Russian).
15. Havezov, I., Tsalev, D. Atomic absorption analysis. Leningrad. Chemistry. 1983, 144 p. (in Russian).
16. Ermakov, A. I., Arasimovich, V. V., Jarosch, N. P. Methods of biochemical investigation in plant. Leningrad. Agropromizdat. 1987, 430 p. (in Russian).
17. Kates, M. Techniques of lipidology: isolation, analysis and identification of lipids. Moscow. World. 1975, 322 p. (in Russian).
18. Prohorova, M. P. Methods of biochemical research: lipid and energy metabolism. Leningrad. 1982, 273 p. (in Russian).
19. Straus, W. Colorimetric microdetermination of cytochrome c oxidase. J. Biol. Chem. 1954, 207(2), 733–743.
20. Sofin, A. V., Shatilov, V. R., Kretovich, V. L. Glutamate dehydrogenases of unicellular green alga Ankistrodesmus braunii. Kinetic properties. Biochemistry. 1984, T. 49(2), 334–343 (in Russian).
21. Korolyuk, M. A. Method of catalase activity determination. Laboratory Science. 1988, 1, 16−19 (in Russian).
22. Beauchamp, C., Fridovich, I. Superoxide dismutase: improved assays and an assay applicable to acrylamide gels. Anal Biochem. 1971, 44, 276–87.
23. Moin, V. M. A simple and specific method of glutathione peroxidase activity determination in erythrocytes. Laboratory science. 1986, 12, 724–727 (in Russian).
24. Lowry, O. H., Rosenbroug, N. I., Farr, A. L., Randall, R. I. Protein measurement with the folin phenol reagent. J. Biol. Chem. 1951, 193(1), 265–275.
25. Sors, T. G., Ellis, D. R., Salt, D. E. Selenium uptake, translocation, assimilation and metabolic fate in plants. Photosynthesis Research. 2005, 86(3), 373–389.
26. Chirkova, T. V. Cell membranes and plant resistance to stress effects. Soros Educational Journal. 1997, 9, 12–17 (in Russian).
27. Lutsiv, A. I., Grubinko, V. V. Characteristics of the absorption of Mg2+, Zn2+, Cu2+, and Pb2+ by cells of Chlorella vulgaris Beijer. Reports of the National Academy of Sciences of Ukraine. 2013, 7, 138−145 (in Ukrainian).
28. Metzler, D. E. Biochemistry: The Chemical Reactions of Living Cells. 2nd edition. New York. London. Academic Press. 2003, 1973 p.
29. Prasad, M. N. V., Strzalka, K. Impact of heavy metals on photosynthesis. Heavy Metal Stress in Plants. Springer Verlag. Berlin. 1999, 117–138.
30. Arevalo-Ferro, C., Hentzer, M., Reil, G., et al. Identification of quorum-sensing regulated proteins in the opportunistic pathogen Pseudomonas aeruginosa by proteomics. Environmental Microbiology. 2003, 5(12), 1350–1369.
31. Vassilev, L. T., Vu, B. T., Graves, B. et al. In vivo activation of the p53 pathway by small-molecule antagonists of MDM2. Science. 2004, 303(6), 844–848.

32. Polishchuk, A. V., Topchiy, N. N., Sytnik, K. M. Effect of heavy metal ions on electron transfer on the acceptor side of photosystem II. Reports of the National Academy of Sciences of Ukraine. 2009, 6, 203–210. (in Ukrainian).
33. Maksymiec, W., Russa, R., Urbanik-Sypniewska, T., Baszyński, T. Changes in acyl lipid and fatty acid composition in thylakoids of copper non-tolerant spinach exposed to excess copper. J. Plant Physiol. 1992, 140, 52–55.
34. Demmig, A. Carotenoids and photoprotection in plants: a role for the xanthophyll zeaxanthin. Biochim Biophys Acta. 1990, 1020, 1–24.
35. Mutygullina, Y. R. The contents dynamic and the role of the pigments of photosynthesis in species of the genus Dianthus, L. of Caucasus flora. Bulletin of the Moscow State Regional University series "Science." 2009, 1, 52–55 (in Russian).
36. Vinyarska, H. B., Bodnar, O. I., Stanislavchuk, A. V., Grubinko, V. V. The binding of selenium in the culture of Chlorella vulgaris. Ukrainian Biochemical Journal. Kyiv. 2014, 86(5) (Suppl. 2), 50–51 (in Ukrainian).
37. Koepke, J., Olkhova, E., Angerer, H. et al. High resolution crystal structure of Paracoccus denitrificans cytochrome c oxidase: new insights into the active site and the proton transfer pathways. – Biochim Biophys Acta. 2009, Jun. 1787(6), 635–645.
38. Grubinko, V. V., Bodnar, O. I., Vasilenko, O. V. et al. Function of glutamate dehydrogenase of ammonium assimilation pathway in freshwater algae. Scientific Issues of Ternopil Volodymyr Hnatiuk Pedagogical University. Section: Biology before: 2014, 60(3), 31–37 (in Ukrainian).
39. Bodnar, O. I., Grubinko, V. V., Gorda, A. I., Klochenko, P. D. Metabolism of algae under the impact of metal ions of the aquatic medium (a review). Hydrobiological Journal. 2011, 47(2), 75–88 (in Ukrainian).
40. Blokhina, O., Virolainen, E., Fagerstedt, K. V. Antioxidants, oxidative damage and oxygen deprivation stress: a review. Ann Bot. 2003, 91(2), 179–194.

CHAPTER 16

# DEVELOPMENT OF *RHIZOBIUM GALEGAE* LAM. UNDER EXTREME CONDITIONS

LILIYA E. KARTYZHOVA and ZINAIDA M. ALESCHENKOVA

*Institute of Microbiology, National Academy of Sciences, Kuprevich Str., 2, 220141, Minsk, Belarus, Tel. +37529638-54-16, E-mail: Liliya_Kartyzhova@mail.ru*

## CONTENTS



## ABSTRACT

The chapter presents results of studies on selection of *Rhizobium galegae* strains resistant to the action of disinfectants, herbicides and petroleum contamination. The authors performed screening of mutant *R. galegae* strains retaining nodulating properties upon direct contact with applied disinfectants and herbicides and ensuring efficient symbiosis with host plant under extreme conditions.

## 16.1 INTRODUCTION

Permanent use of mineral fertilizers and pesticides in agriculture, soil pollution with crude oil and derived refining products, accumulation of heavy metals are adverse environmental conditions affecting proper development of agricultural crops and their capacity to produce quality yields and to avoid build-up of hazardous ingredients in green biomass and seeds. To reduce unfavorable influence of plant chemical control agents advanced strategy to ecologies agricultural production envisages their simultaneous introduction with biological preparations, promoting thereby adaptation of cultivars to extreme environment. The efficient approach to secure stable growth and development of plants exposed to extreme factors is generation of artificial sustained plant–microbial association.

*Galega orientalis* Lam. is one of most attractive legume crops in Belarus distinguished by huge nutritive value, perennial type of development, long farming record and 2 harvests per season yielding 250 c/ha green mass and 3–4 c/ha seeds. The main challenge in *Galega* cultivation technology is handling of seedlings during the first year marked by their slow growth rate and extensive weed proliferation. Stimulation of *Galega* development in the initial period is achieved by seed inoculation with specific bacteria *R. galegae* not available in local soils, whereas field colonization by microbial pathogens and weeds is controlled with disinfectants and herbicides. Application of laboratory-selected *R. galegae* strains displaying maximum tolerance to disinfectants, herbicides, oil–refining products and possessing a whole spectrum of agronomically valuable properties will enable to enhance crop resistance owing to plant–microbial symbiosis and to minimize negative impact of chemicals used to counter pathogen infection, weed invasion and oil pollution.

Cultivation of legume grasses capable to withstand extreme natural background is grounded on formulation of plant–microbial association where strains of nodulating bacteria set up efficient symbiosis with host plant. Vital criteria for their selection and application are the ability to adapt rapidly to changing environmental conditions, to propagate actively and maintain the major physiological-biochemical characteristics.

Aim of this study was investigation of individual aspects governing adaptation and survival of specific nodulating bacteria *Rizobium galegae* on disinfected seeds of *Galega orientalis* Lam., in herbicide-treated and oil-polluted soils.

## 16.2 MATERIALS AND METHODOLOGY

Nodulating bacterial strains *R. galegae* 1 (BIM B–436), *R. galegae* 5 (BIM B–437), *R. galegae* 8 (BIM B–438) deposited at collection of Institute of Microbiology, National Academy of Sciences, Belarus served as microbial cultures for further studies.

In experiments evaluating pesticide resistance of *R. galegae* strains contact and soil herbicides (Basagran M, Treflan) and seed disinfectants (Fundasol, Dividend-star, Raxil) were supplied in doses practiced in agrotechnology of *Galega* cultivation. Tolerance of *R. galegae* to pesticides was assessed by wells technique [1], survival rate – during mixed culture of nodulating bacteria with test chemicals in liquid leguminous medium and on disinfected seeds. The number of viable bacterial cells was estimated in the course of 48 h submerged fermentation. To examine hereditary herbicide resistance of *R. galegae* strains and to choose the most resistant variants gradient adaptation method was preferred [2]. Large–scale herbicide doses used for field weed elimination are: 6 L/ha for treflan (24% concentrated emulsion), 4 L/ha for basagran M (aqueous solution). The amount of herbicides per 100 mL of agar legume medium equaled: 1.5 mL of treflan containing 0.36 g of trifluramine, 0.75 mL of basagran M comprising 0.19 g of bentasone and 0.002 g 2M-4X. The working herbicide solutions were fed to the molted legume medium in the following manner: the medium was dispensed onto Petri plates located on the smooth surface at 45° angle until agar setting. The plates with solid agar medium were supplemented with liquid agar containing herbicide solution – either one chemical or both and left to solidify. The grown colonies were screened in the zone with maximal herbicide concentration. Resistance of *R. galegae* strains to herbicide action was tested (Figure 16.1) according to the method of Simarov and Fyodorov [2] using the Eq. (1):

$$H_A = (l : d) \times H_{in.}, \qquad (1)$$

where $H_A$ – calculated concentration; $l$ – distance from plate edge (zero herbicide level) to the colony grown at definite herbicide concentration, mm; $d$ – diameter of Petri plate, mm; $H_{in.}$ – initial herbicide level.

Drug-resistant and adapted to herbicides *R. galegae* strains were used to check survival of nodulating bacteria on seeds. The number of viable rhizobial cells was counted upon inoculation of aqueous suspension (10 g seeds +100 mL of sterile water) on agar medium with antibiotic rifampicin at concentration 150 mg/mL. Ability of *R. galegae* strains to degrade hydrocarbons of crude oil and diesel fuel in 0.1% (100 mg/L medium), 0.5% (500 mg/L medium), 1% concentrations (1000 mg/L) was analyzed on agar leguminous medium.

Effect of oil hydrocarbons on survival rate of nodulating bacterial strains *R. galegae* [3] was studied by wells technique [1]. Capacity of most resistant *R. galegae* strain to utilize hydrocarbons as the sole source of carbon (crude oil and diesel fuel in 0.5 and 1.0% concentrations (v/v) was examined in medium E-8 (g/L): NaCl – 0.5; $MgSO_4$ – 0.8; $KH_2PO_4$ – 0.7; $(NH_4)_2HPO_4$ – 1.5 [4]. Submerged fermentation of *R. galegae* was carried out in 500 mL Erlenmeyer flasks containing 250 mL of above-mentioned nutrient medium on laboratory shaker (180 rpm) at 28°C. The inoculum was added in 5% ratio of media volume. Amount of oil in medium E-8

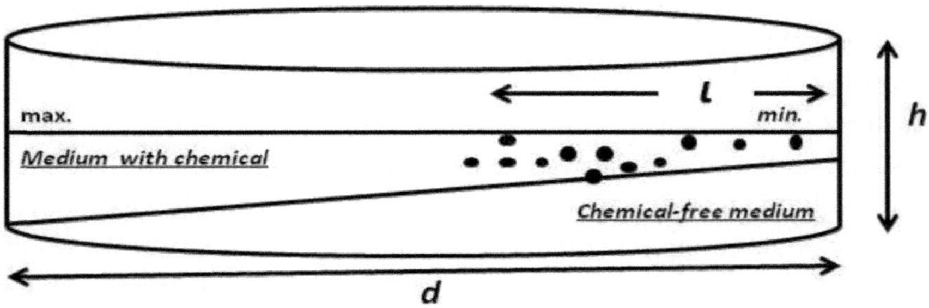


**FIGURE 16.1** Determination of herbicide tolerance concentration at the specific location of bacterial colony resistant to this herbicide value.

throughout fermentation process (120 h) with or without Rhizobia was determined gravimetrically [5].

To estimate the impact of tested parameters, mono- and bifactor dispersion analysis was conducted according to descriptive statistical methods reported in manuals [6–8]. To construct relationship model, define its accuracy and predict results we resorted to correlation-regression analyzes with computed determination coefficients.

## 16.3 RESULTS AND DISCUSSION

### *16.3.1 HERBICIDES*

Studies on effect of herbicides treflan and basagran M on adaptation and survival of Galega nodulating bacteria have revealed high degree of resistance to treflan in strains *R. galegae* 5 and 8. The chemical agent diffusing into agar (wells technique) did not suppress development of both microbial cultures so that their even growth was observed across the whole surface of agar plate, whereas only separate colonies were recorded for *R. galegae* 1.

Submerged culture of strain *R. galegae* 5 with herbicide treflan proved its high adaptive potential similar to *R. galegae* 1 culture where treflan presence stimulated rather than inhibited propagation of microbial cells (Figure 16.2).

It was found in laboratory experiments that resistance of strains *R. galegae* 1 and 5 to herbicide treflan was correlated with interaction type: direct contact on plate or mixed culture. With respect to strain *R. galegae* 8 inhibitory effect of tested herbicides was established (plate method, mixed culture). On day 3 of fermentation in the presence of treflan the cell titer of strain *R. galegae* 8 tended to decline down to $6.0\times10^9$ CFU/mL. The obtained findings agree well with data of other researchers [9], evidencing higher biocidal parameters in plate test as compared to mixed submerged culture.

Investigation of plant–microbial interaction and adaptation capacity of *R. galegae* strains exposed to herbicide action under lab conditions (vegetation experiments) showed that herbicide application caused unfavorable influence on symbiotic properties of nodulating bacteria: the number of

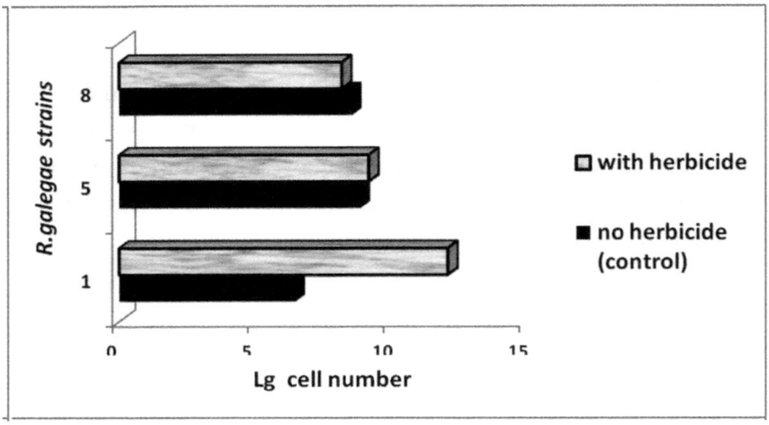


**FIGURE 16.2** Effect of herbicide treflan on population of *R. galegae* during submerged fermentation in liquid medium (72 h).

nodules on roots of *Galega orientalis* Lam. fell by 40% on average and the respective nitrogen-fixing activity dropped more than 3-fold. Reduction of rhizobial symbiotic potential affected considerably green mass yields (Figure 16.3).

**FIGURE 16.3** Effect of herbicides on plant growth and development (Note: "I" – no herbicides (control), "II" – with herbicides).

Average crude weight of phytomass was 30% lower than the control value (Table 16.1).

Monofactor dispersion analysis of results confirmer significance of this factor: herbicides decrease biometric parameters and symbiotic properties of rhizospheic cultures. The data obtained in microvegetation experiment unequivocally indicate that introduction of herbicides at pre-and post-germitation stages in conventional doses induced biocidal impact on tested *R. galegae* strains and inhibited legume-rhizobial symbiotic relationship during early vegetative phases (see Table 16.1).

It appears natural therefore that our further research was focused on production of herbicide–resistant mutants of *R. galegae* strains 1, 5, 8 and further evaluation of their symbiotic stability. The studies resulted in 9 *R. galegae* forms resistant to diverse doses of soil herbicide treflan (in the range 0.08 to 0.28 g/100 mL medium) and 5 variants with dual herbicide resistance – treflan + basagran M.

Three *R. galegae* strains were distinguished by the maximal symbiotic activity: $1^{Rif}$ TB-4 showed resistance to treflan (T) and basagran M (B), $5^{Rif}$ T-0 and $8^{Rif}$ T-1 – to treflan. Herbicide – resistant mutant

**TABLE 16.1** Effect of Herbicides on Symbiotic Efficiency of *Galega orientalis* Lam.: *R. galegae* (Microvegetation Experiment)

| ***R. galegae* strains** | **Plant height, cm** | **Phytomass crude weight, g per plant** | **Nodulating capacity, number of nodules per 1 plant** | **Nitrogen-fixing activity, μg $N_2$/ plant in 30 min** |
|---|---|---|---|---|
| No herbicides | | | | |
| *R. galegae* $1^{Rif}$ | 15.9±0.01 | 220.0±0.7 | 19±2.6 | 21.3±2.8 |
| *R. galegae* $5^{Rif}$ | 15.5±0.008 | 245.0±1.5 | 12±1.2 | 39.1±3.2 |
| *R. galegae* $8^{Rif}$ | 17.1±0.03 | 320.0±2.1 | 23±2.8 | 35.5±3.4 |
| Average value | 16.1 | 261.0 | 18 | 32.0 |
| Herbicide presence | | | | |
| *R. galegae* $1^{Rif}$ | 14.0±0.56 | 120.0±1.8 | 5±0.3 | 7.1±0.3 |
| *R. galegae* $5^{Rif}$ | 18.5±0.05 | 225.0±16 | 6±0.2 | 14.6±0.6 |
| *R. galegae* $8^{Rif}$ | 13.2±0.04 | 220.0±1.2 | 10±0.8 | 9.5±0.5 |
| Average value | 15.2 | 1883 | 7 | 10.4 |

*Note*: "Rif" – resistance to rifampicin.

forms of *R. galegae* strains retained stability of symbiotic potential and exceeded the parent strains in nodulating capacity and nitrogen-fixing activity (Table 16.2).

### 16.3.2 DISINFECTANTS

Experimental data resulting from contact interaction of *R. galegae* strains 1, 5, 8 with disinfecting agents (raxil, fundasol, divident-star) demonstrated that the latter 2 chemicals inhibited growth of bacteria. Inhibition zone around wells filled with disinfectant averaged 0.57–0.60 cm (Figure 16.4). Beyond it dense abundant growth of nodulating cultures was recorded at the entire area of agar medium. As the exception Raxil did not suppress development of *R. galegae*.

About 48 h fermentation of nodulating bacteria *R. galegae* in liquid medium with disinfectants pointed out significant distinctions in resistance and adaptation of selected strains to applied chemical agents (Table 16.3).

Fermentation of strain *R. galegae* 8 with disinfectants caused stimulating effect by the first 2 hrs, with microbial population growing 9-fold on the average. Cell titer of strain *R. galegae* 1 during initial 24 h of culture with fundasol decreased from $4.1\times10^7$ to $2.4\times10^7$ CFU/mL. After 48 h of mixed fermentation with this chemical the number of rhizobial cells rose by 14% over the control value. The period of *R. galegae* 1 adaptation to raxil reached 24 h. Upon 2h of submerged culture with dividend-star cell density of strain *R. galegae* 1 fell 1.7 times below the control parameter.

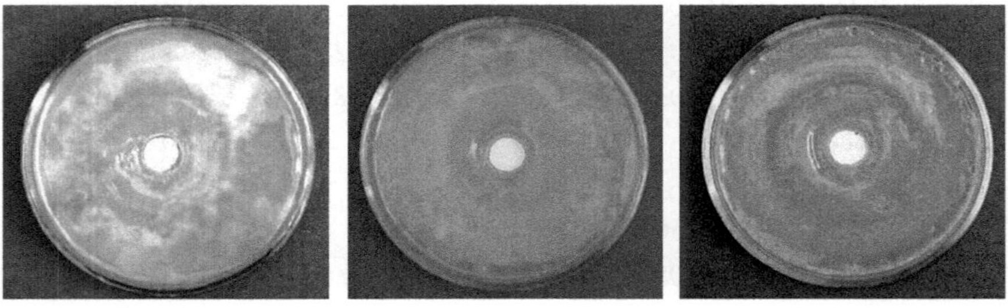

1 2 3

1- *R. galegae* 1, 2 - *R. galegae* 5, 3 - *R. galegae* 8

**FIGURE 16.4** Response of *R. galegae* strains to fundasol action (wells technique).

**TABLE 16.2** Symbiotic Properties of Parent and Herbicide – Resistant *R. galegae* Strains (in Artificial Light Culture)

| ***R. galegae* strains** | **Plant height, cm** | **Phytomass crude dry weight, mg 1 per plant** | **Nodulating capacity, number of nodules per 1 plant** | **Nitrogen-fixing activity, µg $N_2$/plant in 30 min** |
|---|---|---|---|---|
| *R. galegae* 1$^{Rif}$ (control) | 10.5±0.15 | 18.5 /14.0 | 6±0.8 | 1.7±0.2 |
| *R. galegae* 1$^{Rif}$T-5 | 8.0±2.5 | 19.0/12.0 | 14±0.8 | 2.6±0.2 |
| *R. galegae* 1$^{Rif}$TB-4 | *8.5±0.12* | *20.0/14.0* | *9*±0.4 | *2.6*±0.1 |
| *R. galegae* 5$^{Rif}$ (control) | 10.5±1.01 | 23.0/15.0 | 8±0.6 | 2.6±0.3 |
| *R. galegae* 5$^{Rif}$T-0 | *10.0*±1.3 | *21.0/11.0* | *14*±1.2 | *2.6*±0.2 |
| *R. galegae* 5$^{Rif}$TB-2 | 9.0±0.60 | 21.0/11.0 | 6±0.4 | 2.9±0.4 |
| *R. galegae* 8$^{Rif}$ (control) | 7.0±0.6 | 25.0/15.0 | 10±1.2 | 3.8±0.6 |
| *R. galegae* 8$^{Rif}$T-1 | *7.0*±0.20 | *25.0/14.0* | *16*±2.4 | *4.2*±0.4 |
| *R. galegae* 8$^{Rif}$T-2 | 7.5±0.15 | 21.0/12.0 | 6±0.6 | 3.3±0.1 |
| *R. galegae* 8$^{Rif}$TB-4 | 8.5±0.10 | 20.0/8.0 | 3±0.08 | 2.9±0.2 |

*Note*: "Rif" – resistance to rifampicin. Italics and semi-bold symbols denote mutant forms of nodulating bacteria *R. galegae* $^{Rif}$ showing herbicide tolerance and preserving symbiotic characteristics.

By 24 h the culture succeeded to adapt to new environment and propagated intensively to attain concentration $3.4\times10^9$ CFU/mL, exceeding 4.2 times the control value. By the end of fermentation (48 h) population density remained at relatively high level ($1.2\times10^9$ CFU/mL), yet, it was almost twice inferior to the control.

Raxil recorded the outstanding impact on development of *R. galegae* 5 among other disinfectants. Within 2 hrs of culture cell titer surged up virtually by 2 hours if compared to the control. Inhibiting effect was stated in mixed culture with dividend star – in the course of the first day bacterial population spiraled down by an order in comparison with the control. By 48 h of submerged fermentation with disinfectants fundasol, raxil, divi-

**TABLE 16.3** Growth Dynamics of *R. galegae* Strains During Submerged Fermentation with Disinfectants in Liquid Medium

| | *R. galegae* **cell titer, CFU/mL** | | | **% of the control** |
|---|---|---|---|---|
| ***R. galegae* strains** | **2 h** | **24 h** | **48 h** | |
| *R. galegae* 1 (control) | $(4.1 \pm 0.06) \times 10^7$ | $(8.2 \pm 0.12) \times 10^8$ | $(2.1 \pm 0.6) \times 10^9$ | 100 |
| *R. galegae* 1 + fundasol | $(2.4 \pm 0.09) \times 10^7$ | $(6.6 \pm 0.3) \times 10^8$ | $(2.4 \pm 0.8) \times 10^9$ | 114 |
| *R. galegae* 1 + raxil | $(2.9 \pm 1.8) \times 10^7$ | $(10.7 \pm 03) \times 10^8$ | $(2.0 \pm 0.5) \times 10^9$ | 95 |
| *R. galegae* 1 + divident-star | $(2.4 \pm 0.07) \times 10^7$ | $(3.4 \pm 0.2) \times 10^9$ | $(1.2 \pm 0.1) \times 10^9$ | 57 |
| *R. galegae* 5 (control) | $(2.6 \pm 0.7) \times 10^7$ | $(2.4 \pm 0.4) \times 10^9$ | $(4.7 \pm 0.03) \times 10^9$ | 100 |
| *R. galegae* 5 + fundasol | $(2.5 \pm 0.09) \times 10^7$ | $(2.2 \pm 0.2) \times 10^8$ | $(3.2 \pm 0.03) \times 10^9$ | 68 |
| *R. galegae* 5 + raxil | $(2.2 \pm 0.3) \times 10^9$ | $(1.2 \pm 0.8) \times 10^9$ | $(3.1 \pm 0.06) \times 10^9$ | 66 |
| *R. galegae* 5 + divident-star | $(0.7 \pm 0.03) \times 10^6$ | $(2.2 \pm 0.5) \times 10^8$ | $(1.8 \pm 0.4) \times 10^9$ | 38 |
| *R. galegae* 8 (control) | $(2.5 \pm 0.5) \times 10^7$ | $(4.4 \pm 0.03) \times 10^8$ | $(8.6 \pm 0.09) \times 10^8$ | 100 |
| *R. galegae* 8 + fundasol | $(2.7 \pm 0.2) \times 10^8$ | $(1.6 \pm 0.6) \times 10^8$ | $(2{,}5 \pm 0{,}08) \times 10^9$ | 277 |
| *R. galegae* 8 + raxil | $(2.7 \pm 0.09) \times 10^8$ | $(1.8 \pm 0.06) \times 10^8$ | $(4.0 \pm 0.09) \times 10^9$ | 444 |
| *R. galegae* 8 + divident-star | $(1.2 \pm 0.2) \times 10^8$ | $(0.8 \pm 0.2) \times 10^8$ | $(1.7 \pm 0.07) \times 10^9$ | 189 |

*Note*: Semi-bold type states maximal cell titers of *R. galegae* strains determined by selective specificity to disinfectants in the course of submerged fermentation.

dend-star the level of *R. galegae* 5 biomass dropped by 32, 34 and 62%, respectively.

*R. galegae* 8 culture on nutrient media with the tested disinfectants demonstrated rise of cell titer from $10^8$ to $10^9$ during 48 h period in contrast to the corresponding downward slope for the control parameter. Considerable promotion of bacterial life activities was registered in the presence of all examined disinfectants. 48 h culture with raxil stimulated *R. galegae*

8 cell titer to the level $4.0 \times 10^9$ CFU/mL, which is 4.7-times above the control.

Optimization of conditions favoring introduction of microbial species into ecosystem implies analysis of all factors governing stability of physiological-biochemical properties of biopopulation in the course of its adaptation to local agrocenoses.

Survival rate of nodulating bacterial strains inoculated on disinfected seeds of *Galega orientalis* Lam. depended on the applied chemical, making the following cell titer, (percentage of the control):

- control 1 (not sterile untreated seeds – 100%), sterile seeds – 171%; disinfected seeds – 112%: treated with fundasol – 118%, raxil – 92%, divident-star – 125% (average strain values);
- control 2 (inoculum – 100%), not sterile seeds – 49%, sterilized seeds – 84%, disinfected seeds – 54% (average strain values);
- *R. galegae* 1 – 138%, *R. galegae* 5 – 80%, *R. galegae* 8 – 116% (average data for three disinfectants).

Evaluation of *R. galegae* viability on treated seeds of *Galega orientalis* indicated that seed sterilization secured tight attachment and endurance of nodulating bacterial cells on the seed surface. Survival rate of *R. galegae* surpassed 50% on non–sterile seeds treated with fundasol, raxil, and dividend-star, being 12% lower on not-disinfected seeds. The cell number of strains *R. galegae* 1 и *R. galegae* 8 on the surface of disinfected seeds averaged 127% of the control. It was found that maximal survival of nodulating bacteria *R. galegae* 1, 5, 8 was achieved on seeds treated with dividend-star and fundasol.

Specificity of tested rhizobial cultures to disinfectants accounting for their survival on Galega seeds was established: strain *R. galegae* 1 – to fundasol and dividend–star, strain *R. galegae* 5 – to divident-star, strain *R. galegae* 8 – to raxil and fundasol.

Adaptive seed tolerance of *R. galegae* cannot ultimately guarantee generation of effective symbiosis with host plant. It triggered studies to investigate influence of disinfected seed inoculation on symbiotic plant-microbial relationship. The laboratory artificial light culture confirmed that inoculation of *Galega orientalis* seeds with strain *R. galegae* 1 preceded by fundasol treatment produced beneficial effect on growth and development of host plant. Accumulation of green mass solids equaled 118% of the control. Inoculation of nodulating bacteria *R. galegae* 5 and 8 onto *Galega orientalis* seeds pretreated with dividend-star and raxil resulted in stimulation of cultivar development and significant build-up of phytomass (dry weight). Considerable growth-promoting effect was observed following application of *R. galegae* 5 on divident-star exposed seeds. In the latter case the yields of green mass solids doubled those of the control variant. Application of raxil for disinfection of *Galega orientalis* seeds provided for enhanced survival rate of nodulating strain *R. galegae* 8. Accumulation of green mass solids due to inoculation of raxil–treated *Galega* seeds increased by 5% versus control as a clear evidence of effective symbiosis.

Superiority in phytomass productivity to the control parameters indicates efficiency of symbiotic interaction between legume crop *Galega orientalis* and nodulating bacteria *R. galegae* promoted by the afore-tested disinfectants.

### 16.3.3 CRUDE OIL AND PETROLEUM HYDROCARBONS

Taking into account that *Galega orientalis* is an excellent phytoremediating agent for soils polluted with crude oil [3], studies on resistance and survival of microsymbionts under extreme conditions appear relevant and indispensable. Screening of nodulating bacteria *R. galegae* for hype resistance to oil hydrocarbons, characterization of strain stability subject to impact of stress factors and subsequent application of microorganisms for inoculation of *Galega orientalis* seeds is a prerequisite for formulation of effective plant–microbial association used for remediation of soils contaminated with petroleum or oil–refining products.

Tests to estimate viability of local *R. galegae* strains exposed to hydrocarbon pollution (crude oil, toluene, benzene, hexane, hexadecane, diesel fuel) have revealed different extent of resistance in nodulating bacterial cultures. Effect of diverse hydrocarbons on growth of rhizobia on the surface of nutrient agar medium was investigated. Strain *R. galegae* 5 showing the highest resistance was chosen for further studies (Table 16.4).

Resistance to oil hydrocarbons and degrading activity of bacteria are known to be correlated with synthesis of exopolysaccharides [10]. It may be deduced therefore that enhanced exopolysaccharide production pro-

**TABLE 16.4** Growth Evaluation of *R. galegae* Strains on Agar Medium Supplemented with Hydrocarbons in 1% (v/v) Concentration

| | *R. galegae* strains | | |
|---|---|---|---|
| **Hydrocarbons** | **1** | **5** | **8** |
| Crude oil | ++ | +++ | ++ |
| Benzene | + | ++++ | +++ |
| Toluene | ++ | ++ | ++ |
| Xylene | – | ++ | ++ |
| Hexane | ++ | +++ | ++ |
| Hexadecane | ++ | +++ | +++ |
| Diesel fuel | ++ | +++ | ++ |

*Note*: "–" – no growth; "+" – poor growth; "++" – good growth; "+++" – abundant growth; "++++" – profuse growth.

vides for elevated resistance and degradation capacity of strain *R. galegae* 5 withstanding increased levels of hydrocarbons in the environment [11]. Survival of selected bacterial strain *R. galegae* 5 was examined during 72 h submerged fermentation in liquid mineral medium E-8 with the sole carbon source sucrose, crude oil or diesel fuel).

It may be seen from Table 16.5 that hydrocarbon constituents of the medium directly affect vital functions of nodulating bacteria *R. galegae* 5. By 24 h of submerged culture retarded growth and cell proliferation of strain *R. galegae* 5 was recorded in variants with 0.5% crude oil or diesel fuel. Cell titer in both cases decreased by 2.8 and и 1.2 times, respectively.

Upon 48 h of submerged fermentation with tested substrates the strain *R. galegae* 5 readily adapted to severe conditions and demonstrated active growth and cell reproduction in all experimental variants. 1% oil concentration accelerated 1.5-fold propagation of microbial culture but by 72 h the promoting effect faded leading to reduction of population density (Figure 16.5).

It was found that strain *R. galegae* 5 actively consumed crude oil and diesel fuel as the sole carbon sources. The maximal titer of viable *R. galegae* 5 cells equaled $1.7\times10^{9}$ CFU/mL after 48 h culture with 1% crude oil. 72 h fermentation in liquid medium E-8 with 0.5% oil or diesel fuel resulted in maximal titer values $1.3\times10^{9}$ and $1.4\times10^{9}$ CFU/mL, respectively.

**TABLE 16.5** Dynamics of *R. galegae* 5 Growth in Medium E-8 With Hydrocarbons During Submerged Flask Culture

| Experimental variant | Fermentation time, h | | | |
|---|---|---|---|---|
| | 0 | 24 | 48 | 72 |
| 0.5% sucrose (control) | $(3.5\pm0.56)\times10^8$ | $(5.2\pm0.64)\times10^8$ | $(1.09\pm0.05)\times10^9$ | $(3.9\pm0.12)\times10^8$ |
| 0.5% diesel fuel | $(6.97\pm1.16)\times10^8$ | $(2.5\pm0.08)\times10^8$ | $(1.2\pm0.06)\times10^9$ | $(1.3\pm0.03)\times10^9$ |
| 1.0% diesel fuel | $(2.9\pm0.05)\times10^8$ | $(4.8\pm0.32)\times10^8$ | $(1.03\pm0.08)\times10^9$ | $(3.3\pm0.05)\times10^8$ |
| 0.5% crude oil | $(7.9\pm0.24)\times10^8$ | $(6.6\pm0.08)\times10^8$ | $(1.1\pm1.0)\times10^9$ | $1.4\pm0.08\times10^9$ |
| 1.0% crude oil | $(6.0\pm1.08)\times10^8$ | $(7.7\pm1.2)\times10^8$ | $(1.7\pm0.02)\times10^9$ | $(1.5\pm0.96)\times10^8$ |

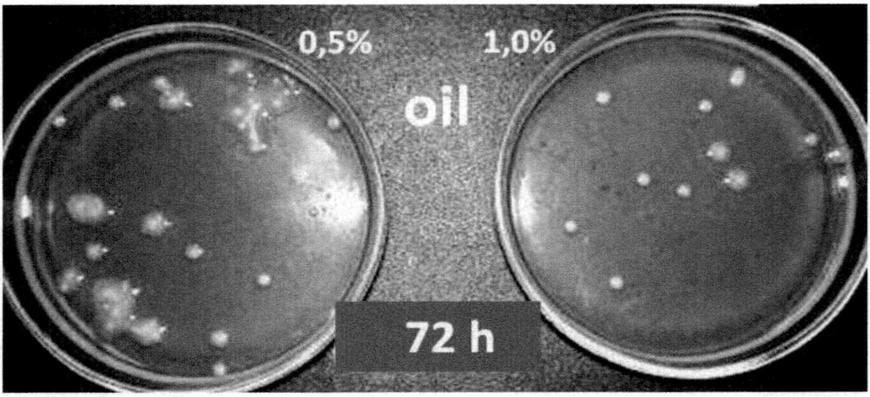


**FIGURE 16.5** Effect of crude oil concentration (0.5% and 1.0%) on cell proliferation of strain *R. galegae* 5 in the course of 72 h culture.

Quantitative analyzes of oil concentrations in medium E-8 prior to and after 3-month submerged fermentation of strain *R. galegae* 5 pointed out dramatic reduction of pollution levels in 0.5 and 1% variants – by 97.4 and 97.8%, respectively (Table 16.6). In the course of microvegetation experiment in soil massive oil decomposition was recorded owing to selected plant – microbial association (*R. galegae* 5+*Galega orientalis* Lam.) (see Table 16.6).

**TABLE 16.6** Degree of Crude Oil Degradation by *R. galegae* 5 Culture

| Experimental variants | | Cultural media | | | |
|---|---|---|---|---|---|
| | | **Liquid** | | **Soil** | |
| | | **Time of culture** | | | |
| | | ***3 months*** | | ***5 months*** | |
| **Oil concentration** | **Scheme of experiment** | **Residual oil amount, mg/L** | **Oil degradation degree, % of the control** | **Residual oil amount, mg/100 g soil** | **Oil degradation degree, % of the control** |
| 0.5% oil | Control | 30.1±0.02 | 97.4 | $0.212 \pm 3.1 \times 10^{-5}$ | 86 |
| | *R. galegae* | 0.78± 0.0003 | | $0.03 \pm 3.1 \times 10^{-5}$ | |
| 1.0% oil | Control | 86.7±0.3 | 97.8 | $1.265 \pm 7.9 \times 10^{-6}$ | 61 |
| | *R. galegae* | 1.9±0.005 | | 0.499±0.0001 | |

Note: control: cultural medium + oil, *R. galegae*: cultural medium + *R. galegae* 5 + oil.

Successful performance of plant–microbial consortium *R. galegae* – *G. orientalis* Lam. in variants with 0.5% and 1% crude oil concentration guaranteed soil decontamination within 5 months by 86% and 61%, respectively.

## 16.4 CONCLUSIONS

1. *R. galegae* strains show specificity to Galega orientalis seed disinfectants: fundasol and dividend-star (*R. galegae* 1, *R. galegae* 5), raxil and fundasol (*R. galegae* 8).
2. Herbicides treflan and basagran generally display bacteriostatic action on *R. galegae* strains, however, resistant microbial varieties retain symbiotic potential so that in nodulating capacity and nitrogen-fixing activity they surpass parent strains by 62% and 15%, respectively.
3. Strain *R. galegae* 5 is not susceptible to adverse petroleum impact since it is able to utilize oil as the sole carbon source. Submerged fermentation of the tested strain during 3 months in liquid medium with 0.5 and 1.0% concentrations of oil hydrocarbons resulted in almost complete pollutant decomposition – by 97.4 and 97.8%, respectively.
4. Introduction of plant-microbial association G. orientalis Lam.-*R. galegae* 5 into oil-polluted (0.5–1.0% v/v) soil caused 86 and 61% decontamination of the media within 5 month period, supporting thereby attractive application prospects of legume-rhizobial system for phytoremediation of hydrocarbon–deteriorated environment.

## KEYWORDS

- **disinfectants**
- **herbicides**
- **nodulating bacteria**
- **oil contaminated soil**

## REFERENCES

1. Segi, J. Methods of soil microbiology, Minsk, Kolos, 1983, 296 p. (in Russian).
2. Bromfield, E. S. P., Michele, S., White, R. P. Identification of Rhizobium strains on antibiotic concentration gradients, Annals of Applied Biology, 1982, 101(2), 261–267, 269–277.
3. Lindström, K. Potential of the Galega – Rhizobium galegae System for Bioremediation of Oil Contaminated Soil, Food Technol. Biotechnol., 2003, 41(1), 11–16.
4. Eroshin, V. K. Growth of Mucorales fungi on paraffin, Microbiology, 1965, 34(5), 883–887 (in Russian).
5. Lurie, Y. Yu. Analytical Chemistry of Industrial Effluents, Moscow, 1984, 306–308 (in Russian).
6. Rokitsky, P. F. Biological Statistics, Minsk, 1973, 221 p. (in Russian).
7. Zinchenko, A. P. Agricultural statistics with principles of social-economic statistics, Moscow, 1998, 427 p. (in Russian).
8. Golberg, A. M. (Ed.). General Theory of Statistics. Finansy i statistica, Moscow, 1985, 320 p. (in Russian).
9. Milto, N. I. Nodulating bacteria and productivity of legume crops. Minsk, Science and Statistics, 1982, 175–176 (in Russian).
10. Bezborodov, A. M. Enzymatic processes in biotechnology. Bezborodov, A. M., Zagustina, V. O., Popov, V. O. (Eds.). Moscow, Nauka, 2008, 334 p. (in Russian).
11. Kartyzhova, L. E. Study on the capacity of local *R. galegae* strains to synthesize exopolysaccharides. Grodno, 2010, 182–184 (in Russian).

## CHAPTER 17

# BIOSORPTION OF HEXAVALENT CHROMIUM AND URANIUM BY BACTERIA, MICROALGA, AND FUNGI

INGA ZINICOVSCAIA,[1,2] ALEXEY V. SAFONOV,[3] and TATIANA V. KHIJNIAK[4]

[1]*Joint Institute for Nuclear Research, 6, Joliot-Curie Str., Dubna, 141980, Russia, E-mail: zinikovskaia@mail.ru*

[2]*The Institute of Chemistry of the Academy of Sciences of Moldova, 3, Academiei Str., Chisinau, 2028, Republic of Moldova*

[3]*A.N. Frumkin Institute of Physical Chemistry and Electrochemistry, Academy of Sciences, 31, Leninsky Prospect, Moscow, 199071, Russia, E-mail: alexeysafonov@gmail.com*

[4]*Winogradsky Institute of Microbiology ,Research Center for Biotechnology, 33/2 Leninskiy Pr., Moscow, 119071, Russia, E-mail: khijniaktv@yandex.ru*

## CONTENTS

## ABSTRACT

Due to widespread industrial use, high solubility and migration capacity heavy metals in several oxidation states (chromium, uranium) are considered as serious pollutants of the environment. The main source of metal pollution is industry: galvanic plants, tailing mills and energy stations with coal and peat burning technologies. Nowadays, it is a great need to develop new environmentally friendly methods of metals removal from ecosystems. Biosorption is the most promising technology involved in the removal of toxic metals from industrial waste streams and natural waters. Biological methods of environment remediation outstand with high effectiveness, low cost of the sorbent materials and prevention of secondary environment pollution. The present review revealed hexavalent chromium and uranium biosorption using different types of microorganisms: bacteria, microalgae and fungi, as well as adsorption isotherms (Langmuir, Freundlich, Langmuir-Freundlich) used in the evaluation of the adsorptive capacity of the biomaterials.

## 17.1 INTRODUCTION

Rapid industry development has led to large-scale release of toxic metals and radionuclides in the environment [1]. High toxicity is characteristic for metals with several states of oxidation because of their high migration activity, carcinogenicity and mutagenicity.

Chromium, the seventh most abundant transition metal element on Earth, is located in group VI-B of the periodic table [2]. Chromium exists in several valence states ranging from −2 to +6. Hexavalent chromium Cr(VI) and trivalent chromium Cr(III) are the most common forms in the environment, which possess different chemical proprieties [2–4]. Cr(VI) is known to be highly toxic for all living organisms and is usually present in the form of chromate ($CrO_4^{2-}$) or dichromate ($Cr_2O_7^{2-}$) ions [2]. Its toxicity is associated with chromium penetration into the cell via surface anion transport systems ($SO_4^{2-}$ channels) with further formation of various reactive intermediates [5]. Cr(VI) release in the environment is explained by its application in different industries such as: leather tanning, metal plating and alloying, wood preservation [5–7], electroplating, paint and pigment manufacturing, textile and fertilizer industries [3]. Cr(III), on the other hand, is more stable and is approximately 100 times less toxic and 1000 times less mutagenic than Cr(VI) [2, 3].

Uranium is the heaviest naturally occurring element to be found ubiquitous in rock, soil, and water. It has two main stable states of oxidation +6 and +4. Uranium(VI) is a common part of nuclear industry wastes, tailing mills and liquid radioactive wastes. In nature, uranium(VI) forms highly soluble carbonate complexes at alkaline pH that leads to an increase of its mobility in soil and groundwater. Uranyl ion $UO_2^{2-}$, which is similar to other actinyl ions ($NpO_2^{2-}$, $PuO_2^{2-}$), is its most mobile and toxic form [8]. All uranium isotopes are unstable with half-lives varying between 69 years and 4.5 billion years. Three items determine the toxicity of radioactive materials: radiological effect, chemical effect and particle size. Regarding radiological effect 1 g of uranium releases 13,000 alpha particles per second chemically is a very toxic heavy metal, and from the point of view of size, uranium particles within the air fit in the nanometer range (aerodynamic diameter of 0.1 microns or less). Emission of alpha radiation causes not only chemical but also physical toxic effects by damaging DNA and others biomolecules. Uranium is a reactive element that is able to combine with, and affect the metabolisms of: lactate, citrate, pyruvate, carbonate and phosphate. Uranyl cations bind to protein, nucleotides, and they can be absorbed by phosphate or carbonate compounds. Thus, different uranium forms have singular biological activities and thus, different toxicity [9].

A wide range of physicochemical methods used for metal removal from wastewater such as: reduction, precipitation, ion-exchange, reverse-osmosis, electro-dialysis [10]. However, the conventional methods have several disadvantages, which include high operating costs, the necessity of preliminary treatment, the difficulty of treating the solid waste subsequently generated, and the requirement of large quantities of chemical adsorbents

[11, 12]. In recent years, great attention is paid to biological methods of metal removal, which become a hot topic in metal pollution control practices because of their potential application [13]. Various chemical and biological methods such as ion exchange, biosorption and bioreduction have been reported to successfully remove metals [14]. Among them biosorption has been considered environmentally friendly and greener alternative for environment bioremediation, due to low cost, high efficiency, easy reclamation of heavy metal, selectivity for absorbing heavy metals in low concentrations and easy recycling of the biosorbent [15, 16].

The main aim of the present work is to summarize the results of chromium and uranium biosorption by microorganisms and to demonstrate the possibility of their use for solution of industrial tasks.

## 17.2 REVIEW OF THE LITERATURE ON BIOSORPTION OF CHROMIUM(VI) AND URANIUM(VI) BY MICROORGANISMS

### *17.2.1 SCIENTIFIC BASIS OF METAL BIOSORPTION*

There are three main mechanism of metal removal by microorganisms: bioreduction (e.g., $U^{+6}$ to $U^{+4}$, $Cr^{+6}$ to $Cr^{+3}$) in anaerobic conditions, biosorption, bioaccumulation and bioprecipitation in phosphates, carbonates and other chemical forms [17, 18].

Biosorption can be defined as the microbial uptake of organic and inorganic metal species, both soluble and insoluble by physical sorption or chemisorption. Chemisorption is the process of complexation between metal cations and negatively charged extracellular components. A detailed investigation carried out by Strandberg et al. [19] provided evidence that surface associated reactive groups chemically similar to high molecular weight polyphosphates were responsible for uranium uptake by *Saccharomyces cerevisiae.*

The biosorption capacity of microorganisms is determined by their large surface area, ease biomass/liquid separation, metal tolerance (for living cells), and complex cell wall composition [20, 21]. Biosorption is possible with both living and non-living biomass. In dead biomass biosorption involves the surface binding of metal ions to cell wall and extracellular material. In living cell is a metabolism-dependent process, which includes metal intracellular uptake or bioaccumulation in some cases *via* specific membrane transport systems. The bioaccumulation is influenced by pH, Eh, organic and inorganic nutrients, and metabolites [22].

In metal binding participate different specific and nonspecific compounds produced by microorganisms. Nonspecific metal-binding compounds include different classes of chemical compounds ranging from simple organic acids and alcohols to macromolecules, such as polysaccharides, humic and fulvic acids [23]. An important role in metal binding belongs to extracellular polymeric substances, a mixture of polysaccharides, mucopolysaccarides and proteins, which are produced by bacteria, algae and fungi [24]. Large polysaccharides of floating cyanobacterial mats participated in metal removal from wastewater [9]. Chitin, an important structural component of fungal cell walls, and its derivatives are considered as effective biosorbents for radionuclides [25]. Fact confirmed by Tsezos and Volesky [26] study, who showed that uranium biosorption by *Rhizopus arrhizus*, takes place through its coordination to the amine of chitin and further precipitation of hydroxylated derivatives. Beside chitin, fungal phenolic polymers and melanins contain different oxygen-containing groups, which are potential metal binding sites [27]. Another biomolecules involved in biosorption processes are low molecular weight coordination molecules – siderophores. Siderophores possess the ability to bind iron, magnesium, manganese, chromium(III), gallium(III) and radionuclides such as plutonium(IV) (see, [27]). Peptidoglycan's carboxyl groups are the main binding sites for cations on Gram-positive bacterial cell walls, while phosphate groups contribute mainly in Gram-negative species [27, 28].

Cervantes and co-authors in their work [2] have shown that Cr(VI) biosorption process included three steps:

1. the binding of anionic Cr(VI) ions with the positively charged groups present on the biomass surface;
2. the reduction of Cr(VI) to Cr(III) by adjacent electron-donor groups;
3. the release of the Cr(III) ions into the aqueous phase due to electronic repulsion between the positively charged groups and the Cr(III) ions, or the formation of complexes of the Cr(III) with adjacent groups capable of Cr-binding.

### *17.2.2 BIOSORPTION BY BACTERIAL BIOMASS*

Bacteria are excellent biosorbents because of their small size, ubiquity, ability to grow under controlled conditions, resilience to a wide range of environmental conditions, high surface-to-volume ratios and a high content of functional groups on the cell wall [15]. Bacteria sorption capacity is determined by:

1. Bacterial surface charge. Usually bacterial cell walls are uncharged at pH below 2, but they become increasingly negatively charged at higher pH value and the magnitude of the negative charge varies from strain to strain [29, 30]. From Mamo [31] study is known that in gram-positive bacteria, teichoic and teichuronic acids, acidic polypeptides and polysaccharides evidently contribute to a negative surface charge. For example, the destruction of the cell-wall ribitol teichoic acid in *Staphylococcus aureus* led to reduction of the surface negative charge [32].
2. Bacterial cell walls composition. Bacterial surface contains different functional groups, which display electrostatic and chemical affinities for positively charged metals. The most common organic functional groups present on the bacterial cell walls are carboxyl, hydroxyl, and phosphoryl sites, with amino groups present to a lesser extent [27, 29]. The cell wall of Gram-positive bacteria is composed primarily of peptidoglycan with lesser amounts of other polymers such as teichoic acids; the most likely metal binding sites are the carboxyl groups in peptidoglycan and the phosphoryl groups in the teichoic acids [33]. Gram-negative bacteria, the cell wall architecture is more complex, as the cells have an additional outer membrane (as well as the cytoplasmic cellular membrane found in all bacteria with lipopolysaccharide (LPS) groups outside a thin peptidoglycan cell wall. Phosphoryl groups in the LPS are the most probable sites of metal complexation in these organisms [27, 29]. In spite of complex Gram-negative bacteria cell composition, the sorptive capacity of Gram-positive bacteria is higher due peptidoglycan's, numerous sorptive sites [33].
3. Existence of bacterial extracellular capsule. Extracellular capsule contains exopolysaccharides or other polymer (polyglutamic acids, hyaluronic acid) components, which cover bacterial cells. Bacte-

rial exopolysaccharides are polymers of different type of sugar residue, and they often contain uronic acids and/or pyruvyl ketal groups, which confer a negative charge on the polymers [31].

4. S-layers. Almost all bacteria and archea are covered by so-called S-layer proteins, which constitute approximately 15% of total bacterial proteins. S-layer proteins possess unique property of recrystallization or self-assembly in suspension, in air-liquid or liquid-solid interphase and execute protection function. S-layer proteins with the molecular weight in the range of 40–200 kDa are synthesized in a very short period of time. S-layer proteins are weakly acidic (pH 4–6) and contain a high proportion of uncharged hydrophobic amino acids [34].
5. Biofilms. Biofilms are defined as microbial aggregates and floccules within the pore spaces of porous media [35], being at least 500 times more resistant to antibacterial agents. In biofilms, the main role in metal sorption belongs to exopolysaccharides, which contains cationic groups (in amino sugars and proteins) and anionic groups (in uronic acids and proteins). Generally, exopolysaccharides are anionic, thus the process of metal cations removal takes place [33].

A large number of studies reflect the role of selected biomolecules in metal biosorption, however the full picture of metal interaction with microorganisms is not always fully understood.

#### 17.2.2.1 Chromium

Biosorption of Cr(VI) have been reported in a variety of aerobic, facultative, and anaerobic bacterial strains. Sorption studies with four bacterial strains *Pseudomonas aeruginosa* CA207Ni, *Burkholderia cepacia* AL96Co, *Corynebacterium kutscheri* FL108Hg, and *Rhodococcus* sp. AL03Ni were performed to determine the ability to remove chromium from solution with various concentrations, pH and temperatures. The optimal temperature values for biosorption of chromate ions by the bacterial strains were in the range 30–35°C. The process of chromate ions removal showed to be pH dependent, with the maximum sorption at pH 2. The ion removal was highly concentration-dependent, maximum Cr(VI) sorption by the bacterial strains were attained at concentrations ranging from 350 to 450 mg/L. Bacteria chromium Cr(VI) affinity was described by the following line *Rhodococcus* sp. AL03Ni > *Burkholderia cepacia* AL96Co > *Corynebacterium kutscheri* FL108Hg > *Pseudomonas aeruginosa* CA207Ni. *Rhodococcus* sp. AL03Ni showed to be better biosorbent with a maximum uptake of 107 mg/g (dry weight) of biomass [36].

Biosorption capacity of 71 morphologically distinct Cr(VI) resistant bacterial strains was studied by Srinath et al. [37]. The experiments were performed with living and dead cells. Biosorption by dead cells showed to be 5–20% higher than by living biomass. The higher efficiency in chromium removal showed two bacterial strains *Bacillus coagulans* and *Bacillus megaterium.* Living and dead cells of *Bacillus coagulans* biosorbed 23.8 and 39.9 mgCr/g dry weight, respectively, whereas, 15.7 and 30.7 mgCr/g dry weight was biosorbed by living and dead cells of *Bacillus megaterium*, respectively.

Alam and Ahmad [13] conducted an experiment for Cr(VI) biosorption using *Exiguobacterium* sp. ZM-2, *Stenotrophomonas maltophilia* ZA-6, *Pantoea sp.* KS-2 and *Aeromonas sp.* KS-14 and found that maximum biosorption occured at pH 2.5 during the first 15 min. Electron micrographs confirmed the bioaccumulation of chromium in the test bacterial isolates. Chromate sensitive isolates *Pantoea* sp. KS-2 and *Aeromonas sp.* KS-14 were not efficient chromate reducers. *Pseudomonas aeruginosa* and *Bacillus subtilis* adsorb Cr(VI) at pH 2 and temperature 32°C.

Biosorption of hexavalent chromium by non-living biomass of *Aeromonas caviae* at pH 2.5 and interaction time 120 min was investigated by Loukidou et al. [38] in a well-stirred batch reactor. It was shown that the major part of adsorption took place during 30 min of interaction. *Chroococcus* sp. HH-11 was found to be suitable for the development of an efficient biosorbent for the removal of Cr(VI) from wastewater [39].

The greatest capacity of biosorption for Cr(VI) ions by *Pantoea* sp. TEM18 was obtained at pH 3.0. The biosorption capacity of the biomass increased with the increase of chromium concentration in the solution. At the chromium (VI) concentration in the solution 28.9–245 mg/L the loading capacity of *Pantoea* sp. TEM18 constituted 7.8–53 mgCr/g dry weight [40]. *Micrococcus* sp. showed maximum removal of Cr(VI) (90%) at pH 7.0 [28].

Biosorption of the chromium Cr(VI) ions by *Pseudomonas aeruginosa*, a gram-negative aerobic species, commonly found in near-surface systems was investigated in batch experiments. It was found that biosorption of Cr(VI) includes two processes. The first process was the reduction of Cr(VI) to Cr(III) by reductive functional groups and the second one chromium ions removal from wastewater using the adsorptive functional groups. FT-IR study showed that –NH groups were mainly involved in Cr(VI) ions removal [12].

Cr(VI) can be efficiently removed by *Azotobacter chroococcum, Bacillus* sp. and *Pseudomonas fluorescens* [41], *Staphylococcus saprophyticus* [42], *Bacillus licheniformis* [43], *E. coli* ASU 7[44], *Pseudomonas* species [45], and *Acinetobacter haemolyticus* [46].

### 17.2.2.2 Uranium

A number of studies have been conducted to investigate uranium(VI) sorption by different bacterial strain. The obtained results show the precipitation of uranium in the insoluble form by bacterial strains: *Cellulomonas flavigena A* [47], *Clostridium* sp. [18], *Desulfomicrobium norvegicum* and *Desulfovibrio sulfodismutans* DSM 3696 [48], *Geobacter daltonii* sp. nov. [49].

Merroun et al. [50] studied the sorption of U(VI) to *Bacillus sphaericus* JG-A12, a Gram-positive bacterium isolated from a uranium mining waste pile. Obtained results demonstrated than carboxyl and phosphate groups participated in metal binding. The biosorption of uranium by aerobic *Bacillus* sp. dwc-2, isolated from a potential disposal site for (ultra-) low uraniferous radioactive waste in Southwest China, was studied by different techniques. The received results implied that the biosorption of uranium on *Bacillus* sp. is complex mechanisms, which involves at least bioaccumulation, ion exchange and complexation processes [51].

Uranyl adsorption onto the Gram-positive soil bacterium *Bacillus subtilis* was conducted at 25°C as a function of pH, solid/solute ratio, and equilibration time. The rate of uranium bound to the bacterium was strongly dependent on the solid:solute ratio and the solution pH. The adsorption of uranium reached equilibrium within 30 min and it did not changed further. A maximum uranium adsorption of 90% was observed at pH 4.9 (1.5 g of bacteria/L). It was suggested that uranium adsorption data require two separate adsorption reactions: with the neutral phosphate functional groups and with the deprotonated carboxyl functional groups of the bacterial cell wall [52].

The process of uranium sorption by *Mycobacterium smegmatis* showed to be three steps process, which include a rapid initial phase, deceleration, and a plateau corresponding to equilibrium conditions (after 120–180 min of interaction). For 5% bacterial suspensions at pH 1, maximum $UO_2^{2+}$ biosorption per gram of dry biomass reached the value of 187 μmol [53]. Results on arabinogalactan-peptidoglycans (composed of polypeptides with carboxyl groups) isolation from the cell wall of bacterium, *Mycobacterium smegmatis* (Actinobacteria) indicated that about 100 mg of uranium could be potentially adsorbed per gram of cell dry weight [54].

The kinetics of uranium sorption by lyophilized Pseudomonas biomass revealed the rapid uranium uptake (>90%) within 10 min of reaction and saturation after 2 h at pH 4.0. pH plays a critical role in metal sorption by influencing both the bacterial surface chemistry as well as the solution chemistry. The results of uranium sorption in the presence of equimolar amounts of competing ions showed the significant antagonism to U sorption for $Th^{4+}$, $Fe^{2+}$ and $Fe^{3+}$, $Al^{3+}$ and $Cu^{2+}$ while metals like cadmium(II), lead(II), silver(II) and anions like chloride(I), phosphate(II) and sulfate(II) had no effect biosorption process [55].

Living, heat-killed, permeabilized *Pseudomonas aeruginosa* strain was characterized with respect to its sorptive activity. Obtained results suggested that uranium removal by *Pseudomonas aerugino*sa occurs independently of oxygenative respiration or other apparent metabolic activity. Uranium binding by *Pseudomonas aerugino*sa was clearly pH dependent, increasing with increasing pH (4–8). The uranium biosorption equilibrium was described by the Langmuir isotherm [56].

The adsorption of uranium from by *Streptomyces levoris* in dependence of pH, uranium concentration and interaction time was studied. *Streptomyces levoris* cells were able to adsorb uranium from a solution over a wide pH range from 3.5 to 6. The amount of uranium adsorbed by *Streptomyces levoris* cells increased with the external uranium concentration increase. At the external uranium concentration 100 μM, the maximum uranium adsorption was about 380 μmol of uranium per gram dry weight. In dependence of interaction time the amount of uranium adsorbed by the *Streptomyces levoris* cells from a solution containing uranium increased very rapidly during the first 5 min of interaction [57].

*Arthrobacter* sp. G975 can effectively remove soluble U(VI) ions from aqueous solution (carbonate-free and carbonate amended synthetic groundwater, SGW). The presence of calcium and bicarbonate ions in SGW affects the sorption behavior of U(VI) due to the formation of highly soluble and stable uranyl-carbonate and calcium uranyl carbonate complexes that reduce adsorption of U(VI) [58].

Among the microorganisms that can sorb uranium *Citrobacte*r sp. stand out (Table 17.1) due to overproduction of phosphates and specific enzyme – phosphatase [59–61]. Also, it was shown that *Sulfolobus acido-*

**TABLE 17.1** Uranium Biosorption by Several Microbial Strains

| Biomass | Microorganism | Uranium uptake, % mg of dry weight |
|---|---|---|
| Fungi | *Aspergillus niger* | 21 |
| | *Aspergillus terreus* | 0.1 |
| | *Penicillium chrysogenum* | 8–9 |
| | *Mucor hiemalis* | 7.1 |
| | *Rhizopus oryzae* | 3.4 |
| Yeasts | *Saccharomyces cerevisiae* | 10–15 |
| | *Candida albicans* | 1.6 |
| | *Rhodotorula glutinus* | 3.6 |
| Alga | *Chlorella regularis* | 15 |
| | *Dunaliella* sp. | 0.02 |
| Actinomycetes | *Actinomyces flavoviridus* | 7.8 |
| | *Streptomyces albus* | 8.7 |
| Bacteria | *Escherichia coli* | 2.3 |
| | *Citrobacter* sp.* | 900 |
| | *Bacillus subtilis* | 8.5 |
| | *Pseudomonas aeruginosa* | 15 |
| | *Micrococcus luteus* | 7.5 |

*Highest uranium uptake.

*caldarius* are able to precipitate uranium as U(VI)-phosphates [62]. The biosorption of uranium by *Citrobacter freundii* could be described well by the Langmuir or Freundlich isotherm, and the latter was better.

### *17.2.3 BIOSORPTION BY MICROALGAE*

Microalgal biomass wide application for metal removal can be explained by possibility of its cultivation in open ponds or in large-scale laboratory culture, providing a reliable and consistent supply of biomass for scale-up work and use of light as an energy source, facilitating the maintenance of metabolism in the absence of organic carbon sources [63].

#### 17.2.3.1 Chromium

Cr(VI) adsorption by filamentous algae *Spirogyra s*pecies in batch experiments shows that adsorption capacity of the biomass strongly depends on equilibrium pH. The percent adsorption of Cr(VI) increases within crease in pH from pH 1.0 to 2.0 and thereafter decreases with further increase in pH. The concentration of chromium in biomass increased with time from 0 to 120 min and after that becomes almost constant up to the end of the experiment (180 min). The chromium sorption isotherms followed the Langmuir model and the maximum chromium adsorption constituted $14.7\times10^3$ mg metal/kg of dry weight biomass [64].

Chromium biosorption by microalga *Phormedium bohneri, Oscillatoria tenuis, Chlamydomonas angulosa, Ulothrix tenuissima* was investigated. The maximum accumulation of Cr was shown by *Phormedium bohneri* (8550 μg/g) followed by *Oscillatoria tenuis* (7354 μg/g), *Chlamydomonas angulosa* (5325 μg/g), *Ulothrix tenuissima* (4564 μg/g), and Oscillatoria nigra (1862 μg/g) [5].

Chromium(VI) removal onto nonviable freshwater cyanobacterium *Nostoc muscorum* biomass was studied by Gupta and Rastogi [63]. The maximum Cr(VI) biosorption capacity for *Nostoc muscorum* has been found to be 23 mg/g at a dose of 1.0 g/L with initial Cr(VI) concentration of 100 mg/L and optimum pH of 3.0. Results of application of Langmuir and Freundlich mathematical models indicated the applicability of *Nostoc muscorum* for Cr(VI) removal in both monolayer biosorption and heterogeneous surface conditions.

Adsorption of Cr(VI) by heat-dried biomass of the cyanobacterium *Phormidium laminosum* has been reported at pH 2.0 [65]. The biosorption of chromium(VI) ions by *Scenedesmus obliquus* and *Chlorella vulgaris*, procaryotic, green algae were investigated as a function of initial pH, initial metal ion concentration and cell concentration. The optimum biosorption pH was 2.0 for both microalgae. Maximum equilibrium uptakes of chromium(VI) ions were also determined as 33.8 mg/g and 30.2 mg/g for *Chlorella vulgaris* and *Scenedesmus obliquus*, respectively, at chromium(VI) ion initial concentration 250 mg/L. Obtained adsorption equilibrium data fitted very well to Freundlich and Langmuir adsorption models [66].

Biosorption of chromium by residual *Nannochloris oculata* after lipid extraction was investigated. Increased surface area of *Nannochloris oculata* was observed after lipid extraction. Cr(VI) removal was highest at pH 2 and it decreased with the increase in pH [67].

The biosorption of chromium(VI) from saline solutions on two strains of living *Dunaliella* algae were tested under laboratory conditions in a batch system. Maximum chromium(VI) sorption capacities of both sorbents were obtained at pH 2.0. The uptake of chromium(VI) by two strains reached a plateau at 250–300 mg/L showing the saturation of binding sites at higher concentration levels. However, chromium uptake by *Dunaliella* sp. 1 (53%) appeared to be slightly higher than that of *Dunaliella* sp. 2 (45%). Both the Freundlich and Langmuir adsorption models were suitable for describing the biosorption of chromium(VI) [68]. Travieso et al. [69] have demonstrated better chromium removal efficiencies by *Scenedesmus acutus* than *Chlorella vulgaris*. Microalga *Chlorella pyrenoidosa, Spirulina maxima, Spirulina platennis, Selenastrum capriornutum* and *Scenedesmus quadricauda* [70] can be successfully used for chromium biosorption.

#### 17.2.3.2 Uranium

*Chlorella vulgaris* and *Dunaliella salina* was efficiently applied for purification of wastewater contaminated with $UO_2^{2+}$ ions that are by-products of the technology of nuclear fuel elements preparation. Uptake of $UO_2^{2+}$

ions was higher for *Chlorella vulgaris* than for *Dunaliella salina* due to biomass biochemical and physiological properties. The authors suggested that in $UO_2^{2+}$ ions binding participated polysaccharides and nucleic acids of the biosorbents [71]. Another three types of alga *Nostok linckia, Spirulina platensis, Porphyridium cruentum* were tested for their ability of uranium biosorption. According to obtained result the retention degree on alga decreased in the series: *Spirulina platensis* > *Porphyridium cruentum* ≥ *Nostok linckia* [14].

*Cystoseria indica* biomass exhibited the highest uranium uptake capacity at 15°C, uranium ion concentration 500 mg/L and pH 4. The obtained data fitted well with Langmuir model [72]. Uranium biosorption by powdered biomass of lake-harvested cyanobacterium water-bloom, which consisted predominantly of *Microcystis aeruginosa* was studied by Li et al. [73]. Optimum uptake of uranium was at pH 4.0–8.0 during one-hour experiment. The biosorption data fitted the Freundlich model.

### 17.2.4 BIOSORPTION BY FUNGI

Wide fungi application as biosorbents is explained by the production of high yields of biomass, ease grow under wide range of environmental conditions, resistance to high metal ion concentrations, and possibility of enzymes (reductase, DNA polymerase etc.) production. In general, the fungal organisms are resistant to higher metal ion concentrations [74]. It is known that the cell wall of fungi consists of large quantity of functional groups. Among these groups are carboxyl (–COOH), amide ($–NH_2$), thiol (–SH), phosphate ($PO_4^{3-}$) and hydroxide (–OH), which are believed to play an important role in metal chelation [75].

#### 17.2.4.1 Chromium

A wide range of fungal species under nonliving condition have been studied by different researchers for the removal of Cr(VI) from the wastewaters [74]. Dead fungal biomass of *Aspergillus niger*, *Aspergillus sydoni* and *Penicillium janthinellum* was used to investigate biosorption of Cr(VI) from aqueous solution as well as from electroplating effluent.

For batch solutions the maximum percent removal of chromium(VI) was 89±2%, 81±2.2% and 78±2.3% for *Aspergillus niger*, *Aspergillus sydoni* and *Penicillium janthinellum,* respectively. As a function of adsorbent dose the removal of Cr(VI) was 91±2.2% by *Aspergillus niger* at biosorbent dose 0.6 g/50 mL, whereas, 88±1.6% and 86±1.6% by *Aspergillus sydoni* and *Penicillium janthinellum* at 0.8 g/50 mL and after that no appreciable amount of Cr(VI) ions removed from the solution. A major fraction of Cr(VI) was adsorbed onto biomass after 60 min of interaction and remained nearly constant afterwards. Adsorption data fitted well with Freundlich and Langmuir models. In case of wastewater Cr(VI) sorption was lower in comparison with as compared to synthetic sample, that can be explained by presence of other metal ions in wastewater [76].

Chen et al. [77] investigated the effects of pH, initial concentration, and sorption time on Cr(VI) removal by polyethylenimine (PEI)-modified *Phanerochaet chrysosporium.* The optimum pH was approximately 3.0. The maximum removal for Cr(VI) was 344.8 mg/g.

Biosorption of Cr(VI) ion was investigated using biomass of *Agaricus bisporus.* Optimum biosorption conditions were determined to be pH 1, $C_0$ = 50 mg/L and t = 20°C. Beside the set of traditional parameters, the optimal rotation speed was assessed. The maximum uptake yield of Cr(VI) was obtained at 150 rpm. Application of mathematical models showed that biosorption of Cr(VI) ions onto biomass was better suitable to Freundlich adsorption model [78].

The use of the Cr-resistant fungus *Paecilomyces lilacinus* to remove Cr(VI) from two physicochemically different undiluted tannery industry effluents was studied by Sharma et al. [79]. The fungus has broad pH tolerance range and was able to reduce Cr(VI) content both in acidic (pH 5.5) and alkaline (pH 8.0) conditions. *Saccharomyces cerevisiae* has the ability to sorb Cr(VI) and the sorption capacity of dehydrated cells is considerably higher than that of intact cells [80]. The removal rate of Cr(VI) by biomass of marine *Aspergillus niger* was increased with a decrease in pH and an increase in Cr(VI) and biomass concentration. Chromium can be also successfully removed by *Aspergillus* niger [81] and *Aspergillus* sp [28].

#### 17.2.3.2 Uranium

The reports on fungal biomass use as a potential biosorbent of uranium are very scanty. The ability of non-living biomass of *Penicillium citrinum* has been explored for the removal and recovery of uranium from aqueous solutions. The *Penicillium citrinum* exhibited the highest uranium sorption capacity at an initial pH of 6.0, concentration of 50 g/mL, and at 5 h. Obtained data fitted well with both Langmuir and Freundlich models [82].

Uranium(VI) uptake by *Aspergillus fumigatus* showed to be a rapid process. During one hour 86% of uranium was removed from solution at pH 5. Biosorption data fitted to Langmuir model of isotherm and a maximum loading capacity of 423 mg U/g was obtained. Presence of iron, calcium and zinc cations did not affect the process of uranium sorption, while aluminum showed an inhibitory effect [83].

The effect of aluminum on uranium biosorption by another fungi *Rhizopus arrhizus* was investigated by Tsezos et al. [84]. In comparison to *Aspergillus fumigatus* in *Rhizopus arrhizus* aluminum did not interfered with the kinetic of uranium uptake. The authors suggested the interference has physical not chemical character as uranium-chitin complexes were formed.

The optimum biosorption of uranium by immobilized *Aspergillus fumigatus* beads in Wang et al. [85] study was observed to occur at pH 5.0, biosorbent dose 2.5%, and initial U(VI) concentration 60 mg/L during 120 min. The adsorption process conformed to the Freunlich and Temkin isothermal adsorption models. The ability of uranium biosorption by *Aspergillus niger* at different parameters was examined in Solat and co-authors work [86]

## 17.3 SORPTION MODELS

The adsorption isotherm is the initial experimental test step to determine feasibility of adsorption treatment and whether further test work should be conducted [76]. Although the Langmuir and Freundlich isotherms were first introduced about 90 years ago, they still remain the two most commonly used adsorption isotherm equation. Their success undoubtedly reflect their ability to fit a wide variety of adsorption data quite well, but it may also partly reflect the appealing simplicity of the isotherm equations and the ease with which their adjustable parameters can be estimated [39, 87].

The Langmuir equation is the most widely used isotherm equation for modeling adsorption equilibrium data [88, 89]. This model is based on the assumption that the sorption energy is uniform and homogeneously distributed on the sorbent surfaces, and the energy of adsorption is constant [76].

The Langmuir isotherm model is expressed as:

$$q = (q_{\max} bc)/(1+bc)$$

where $c$ – is the concentration of metal ions; $q_{max}$ – represents the maximum metal accumulation; $b$ – is the affinity parameter of the isotherm reflecting the high affinity of the biosorbent for the sorbate [56].

The constants $q_{max}$ and b are evaluated from the linear plot of logarithmic equation:

$$(1/q) = (1/q_{\max}) + (1/bq_{\max})\,(1/c)$$

The Freundlich equation is basically empirical and was developed for heterogeneous surfaces nature [73, 90, 91]. The Freundlich adsorption isotherm is mathematically expressed as

$$x/m = Kc^{(1/n)}$$

It is also can be written in logarithmic form:

$$\log(x/m) = \log k + (1/n)\log c$$

where $x$ = mass of adsorbate; $m$ = mass of adsorbent; $c$ = equilibrium concentration of adsorbate in solution; $K$ and $n$ are constants for a given adsorbate and adsorbent at a particular temperature [92].

In some studies to describe adsorption process in heterogeneous materials Langmuir- Freundlich (LF) model was applied [93–96].

$$q = [q_{\max}\,(bc)^n]/[1\text{–}(bc)^n]$$

where $c$ is the concentration of metal ions; $q_{max}$ represents the maximum metal accumulation; $b$ is the affinity parameter of the isotherm reflecting the high affinity of the biosorbent for the sorbate, and $n$ is the empirical parameter that varies with the degree of heterogeneity.

Other types of adsorption isotherms are well described elsewhere [21, 97, 98].

## 17.4 CONCLUSIONS

The present review shows that microorganisms (bacteria, microalga and fungi) represent an efficient and potential class of biosorbents for the removal of hexavalent chromium and uranium from industrial wastewater. The microorganisms contain a variety of functional groups responsible for metal adsorption, which allows producing cheap and effective biosorbents for large-scale application. In a large part of study on metal biosorption the data are processed using Langmuir and Freundlich isotherms.

## ACKNOWLEDGEMENTS

This work was supported grant of Russian Foundation for Basic Research #15-05-08919 and 15-33-20069.

## KEYWORDS

- **biofilms**
- **cell wall composition**
- **metal-binding compounds**
- **metals**
- **microorganisms**
- **pollutants**
- **polymeric substances**

## REFERENCES

1. Lloyd, J. R., Lovley, D. R. Microbial detoxification of metals and radionuclides, Current Opinion in Biotechnology, 2001, 12(3), 248–253.
2. Cervantes, C., Campos-Garcia, J., Devars, S., Gutierrez-Corona, F., Loza-Tavera, H., Torres-Guzman, J. C., Moreno-Sanchez, R. Interactions of chromium with microorganisms and plants, FEMS Microbiology Reviews, 2001, 25(3), 335–347.
3. Kaushik, S., Juwarkar, A., Malik, A., Satya, S. Biological removal of Cr (VI) by bacterial isolates obtained from metal contaminated sites, Journal of Environmental Science and Health Part A Toxic/Hazardous Substances and Environmental Engineering, 2008, 43, 419–423.
4. Deng, L., Zhang, Y., Qin, J., Wang, X., Zhu, X. Biosorption of Cr(VI) from aqueous solutions by nonliving green algae Cladophora albida, Minerals Engineering, 2009, 22, 372–37.
5. Baldi, F., Vaughan, A. M., Olson, G. J. Chromium(VI)-resistant yeast isolated from a sewage treatment plant receiving tannery wastes, Applied and Environmental Microbiology, 1990, 56(4), 913–918.
6. Mabbett, A. N., Macaskie, L. E. A novel isolate of Desulfovibrio sp. with enhanced ability to reduce Cr(VI), Biotechnology Letters, 2001, 23(9), 683–687.
7. Dwivedi, S., Srivastava, S., Mishra, S., Kumar, A., Tripathi, R. D., Rai, U. N., Dave, R., Tripathi, P., Charkrabarty, D., Trivedi, P. K. Characterization of native microalgal strains for their chromium bioaccumulation potential: Phytoplankton response in polluted habitats, Journal of Hazardous Materials, 2010, 173, 95–101.
8. Renshaw, J. C., Lloyd, J. R., Livens, F. R. Microbial interactions with actinides and long-lived fission products, Comptes Rendus Chimie, 2007, 10, 1067–1077.
9. Gadd, G. M. Metal interaction with metals/radionuclides the basis of bioremediation. Interaction of microorganisms with radionuclides. Keith-Roach M. J., Livens F. R. (Eds.). Elsevier Science Ltd, UK, 2002, 179–203.
10. Singh, B., Kumar, Das S. Adsorptive removal of Cu (II) from aqueous solution and industrial effluent using natural/agricultural wastes, Colloids and Surfaces B: Biointerfaces, 2011, 1, 221–232.
11. Liu, Y., Feng, B., Fan, T., Zhou, H., Li, X. Tolerance and removal of chromium(VI) by Bacillus sp. strain YB-1 isolated from electroplating sludge, Transactions of Nonferrous Metals Society of China, 2008, 18, 480–487.
12. Kang, S. Y., Lee, J. U., Kim, K. W. Biosorption of Cr (III) and Cr (VI) onto the cell surface of Pseudomonas aeruginosa, Biochemical Engineering Journal, 2007, 36, 54–58.
13. Alam, M. Z., Ahmad, S. Chromium removal through biosorption and bioaccumulation by bacteria from tannery effluents contaminated soil, Clean – Soil, Air, Water, 2011, 39, 226–237.
14. Cecal, A., Humelnicu, D., Rudic, V., Cepoi, L., Ganju, D., Cojocari, A. Uptake of uranyl ions from uranium ores and sludges by means of Spirulina platensis, Porphyridium cruentum and Nostok linckia alga, Bioresource Technology, 2012, 118, 19–23.
15. Dhankhar, R., Guriyan, R. B. Bacterial biosorbents for detoxification of heavy metals from aqueous solution: A review. International Journal of Advances in Science and Technology, 2011, 103–128.

16. Mishra S., Doble M. Novel chromium tolerant microorganisms: Isolation, characterization and their biosorption capacity, Ecotoxicology and Environmental Safety, 2008, 71, 874–879.
17. Francisб A. J. 2006, Biotransformation of uranium complexes with organic ligands, Merkel, B. J., Hasche-Berger A., Uranium in the Environment. Berlin; Heidelberg: Springer, 191–197.
18. Francis, A. J., Dodge, C. J. Bioreduction of uranium (VI) complexed with citric acid by Clostridia sp. affects its structure and solubility, Environmental Science and Technology, 2008, 42, 8277–8282.
19. Strandberg, G. W., Shumate, S. E., Parrott Jr., J. R., North S. E. Microbial accumulation of uranium, radium, and cesium. Chemical Technology Division Oak Ridge National Laboratory Oak Ridge, Tennessee, 1981, 1–12.
20. Akhtar, N., Iqbal, M., Zafar, S. I., Iqbal, J. Biosorption characteristics of unicellular green alga Chlorella sorokiniana immobilized in loofa sponge for removal of Cr (III), Journal of Environmental Sciences, 2008, 20, 231–239.
21. Vijayaraghavan, K., Yun, Y. S. Bacterial biosorbents and biosorption, Biotechnology Advances, 2008, 3, 266–291.
22. Malik, A., Grohmann, E., Alves, M. Management of microbial resources in the environment, Springer Science & Business Media, 2013, 530 p.
23. Gadd, G. M. Microbial influence on metal mobility and application for bioremediation, Geoderma, 2004, 122, 109–119.
24. Sasek, V., Glaser, J. A., Baveye, P. The utilization of bioremediation to reduce soil contamination: Problems and solutions. Springer Science + Business Media Dordrech, 2003, 417 p.
25. Juwarkar, A. A., Kumar, G. P., Nair A. Biomolecules in environmental application, Mishra, C. S. K., Juwarkar A. A., Environmental biotechnology, New Delhi, APH Publishing Corporation, 2007, 19–54.
26. Tsezos, M., Volesky, B. The mechanism of uranium biosorption by Rhizopus arrhizus. Biotechnology and Bioengineering, 1982, 24(2), 385–401.
27. Gadd, G. M. Transformation and mobilization of metals, metalloids, and radionuclides by microorganisms, Violante, A., Huang, P. M., Gadd G. M., Biophysicochemical process of heavy metals, metalloids in soil environments, New Jersey, John Wiley & Sons Inc., 2008, 53–97.
28. Congeevaram, S., Dhanarani, S., Park J., Dexilin, M., Thamaraiselvi, K. Biosorption of chromium and nickel by heavy metal resistant fungal and bacterial isolates, Journal of Hazardous Materials, 2007, 146, 270–277.
29. Fein, J. B. Thermodynamic Modeling of Metal Adsorption onto Bacterial Cell Walls: Current Challenges, Advances in Agronomy, 2006, 55, 179–202.
30. Dickson, J. S., Koohmaraie, M. Cell surface charge characteristics and their relationship to bacterial attachment to meat surfaces, Applied and Environmental Microbiology, 1989, 832–836.
31. Mamo, W. Physical and biochemical surface properties of gram-positive bacteria in relation to adhesion to bovine mammary cells and tissues. A review of the literature, Scientific and Technical Review of the Office International des Epizooties, 1989, 8, 163–176.

32. Brown, S., Santa, Maria, Jr. J. P., Walker, S. Wall Teichoic Acids of Gram-Positive Bacteria. Annual Review of Microbiology, 2013, 67, 10.1146/annurev-micro-092412–155620.
33. van Hullebusch, E. D., Zandvoort, M. H., Lens P. N. L. Metal immobilization by biofilms: mechanisms and analytical tools, Reviews in Environmental Science and Biotechnology, 2003, 2, 9–33.
34. Mulani, M. S., Majumder, D. R. Review S-layer protein: tailor – made nanoparticles, International Journal of Innovative Research in Science, Engineering and Technology, 2013, 4693–4706.
35. Salcedo, F., Pereyra, C. M., Di Palma, A. A., Lamattina, L., Creus, C. M. Methods for studying biofilms in Azospirillum and other PGPRs, Cassan, F. D., Okon, Ya., & Creus C. M., Handbook for Azospirillum. Technical issues and protocols, Switzerland, Springer International Publishing, 2015, 199–321.
36. Oyetibo, G. O., Ilor, I. M. O., Obayori, O. S., Amund, O. O. Chromium (VI) biosorption properties of multiple resistant bacteria isolated from industrial sewerage, Environmental Monitoring and Assessment, 2013, 185, 6809–6818.
37. Srinath, T., Verma, T., Ramteke, P. W., Garg, S. K. Chromium (VI) biosorption and bioaccumulation by chromate resistant bacteria. Chemosphere, 2002, 48, 427–435.
38. Loukidou, M. X., Zouboulis, A. I., Karapantsios, T. D., Matis, K. A. Equilibrium and kinetic modeling of chromium (VI) biosorption by Aeromonas caviae, Colloids and Surfaces A: Physicochemical and Engineering Aspects, 2004, 242, 93–104.
39. Anjana, K., Kaushik, A., Kiran, B., Nisha, R. Biosorption of Cr(VI) by immobilized biomass of two indigenous strains of cyanobacteria isolated from metal contaminated soil, Journal of Hazardous Materials, 2007, 148, 383–386.
40. Ozdemir, G., Ceyhan, N., Ozturk, T., Akirmak, F., Cosar, T. Biosorption of chromium (VI), cadmium (II) and copper (II) by Pantoea sp. TEM18, Chemical Engineering Journal, 2004, 102, 249–253.
41. Parameswari, E., Lakshmanan, A., Thilagavat, T. Bacterial biosorption: an alternative treatment option for heavy metal management, Asian Journal of Microbiology, Biotechnology & Environmental Sciences Paper, 2010, 12, 681–686.
42. Ilhan, S., Nourbacksh, M. N., Kilicarslan, S., Ozdag, H. Removal of chromium, lead and copper ions from industrial waste waters by Staphylococcus saprophyticus, Turkish Electronic Journal of Biotechnology, 2004, 146, 50–57.
43. Zhou, M., Liu, Y., Zeng, G., Li, X., Xu, W., Fan, T. Kinetic and equilibrium studies of Cr(VI) biosorption by dead Bacillus licheniformis biomass, World Journal of Microbiology and Biotechnology, 2007, 23, 43–48.
44. Gabr, R. M., Gad-Elrab, S. M. F., Abskharon, R. N. N., Hassan, S. H. A., Shorei, A. A. M. Biosorption of hexavalent chromium using biofilm of E. coli supported on granulated activated carbon, World Journal of Microbiology and Biotechnology, 2009, 25, 1695–1703.
45. Aravindhan, R., J. Sreeram, K., Rao, J. R., Nair, B. U. Biological removal of carcinogenic chromium(VI) using mixed Pseudomonas strains, The Journal of General and Applied Microbiology, 2001, 53, 71–79.
46. Pei, Q. H., Shahir, S., Santhana Raj, A. S., Zakaria, Z. A., Ahmad, W. A. Chromium (VI) resistance and removal by Acinetobacter haemolyticus, World Journal of Microbiology and Biotechnology, 2009, 25, 1085–1093.

47. Sani, R. K., Peyton, B. M., Smith, W. A., Apel, W. A., Petersen, J. N. Dissimilatory reduction of Cr(VI), Fe(III), and U(VI) by Cellulomonas isolates, Applied Microbiology and Biotechnology, 2002, 60, 192–199.
48. Lovley, D. R., Roden, E. E., Phillips, E. J., Woodward J. C. Enzymatic iron and uranium reduction by sulfate-reducing bacteria, Marine Geology, 1993, 113, 41–51.
49. Prakash, O., Gihring, T. M., Dalton, D. D., Chin, K. J., Green, S. J., Akob, D. M., Wanger, G., Kostka, J. E. Geobacter daltonii sp. nov., an iron(III)- and uranium(VI)-reducing bacterium isolated from the shallow subsurface exposed to mixed heavy metal and hydrocarbon contamination, *International Journal* of Systematic and Evolutionary *Microbiology,* 2010, 60, 546–553.
50. Merroun, M. L., Raff, J., Rossberg, A., Hennig, C., Reich, T., Selenska-Pobell, S. Complexation of uranium by cells and S-Layer sheets of Bacillus sphaericus JG-A12, Applied and Environmental Microbiology, 2005, 71, 5532–5543.
51. Li, X., Ding, C., Liao, J., Lan, T., Li, F., Zhang, D., Yang, J., Yang, Y., Luo, S., Tang, J., Liu, N. Biosorption of uranium on Bacillus sp. dwc-2: preliminary investigation on mechanism, Journal of Environmental Radioactivity, 2014, 135, 6–12.
52. Fowle, D. A., Fein, J. B., Martin, A. M. Experimental study of uranyl adsorption onto Bacillus subtilis, Environmental Science and Technology, 2000, 34, 3737–3741.
53. Andres, Y., MacCordick, H. J., Hubert, J. C. Adsorption of several actinide (Th, U) and lanthanide (La, Eu, Yb) ions by Mycobacterium smegmatis, Applied Microbiology and Biotechnology, 1993, 39, 413–417.
54. Kalin, M., Wheeler, W. N., Meinrath, G. The removal of uranium from mining wastewater using algal/microbial biomass, Journal of Environmental Radioactivity, 2004, 78, 151–177.
55. Sar, P., Kazy, S. K., D'Souza, S. F. Radionuclide remediation using a bacterial biosorbent, International Biodeterioration and Biodegradation, 2004, 54, 193–202.
56. Hu, M. Z., Norman, J. M., Faison, B. D., Reeves, M. E. Biosorption of uranium by Pseudomonas aeruginosa strain CSU: characterization and comparison studies, Biotechnology and Bioengineering, 1996, 51, 237–247.
57. Tsuruta, T. Adsorption of uranium from acidic solution by microbes and effect of thorium on uranium adsorption by Streptomyces levoris, Journal of Bioscience and Bioengineering, 2004, 97, 275–277.
58. Carvajal, D. A., Katsenovich, Y. P., Lagos, L. E. The effects of aqueous bicarbonate and calcium ions on uranium biosorption by Arthrobacter G975 strain, Chemical Geology, 2012, 330–331, 51–59.
59. Macaskie, L. E. The application of biotechnology to the treatment of wastes produced from the nuclear fuel cycle: biodegradation and bioaccumulation as a means of treating radionuclide-containing streams, CRC Critical Reviews in Biotechnology, 1991, 11, 41–112.
60. Macaskie, L. E., Bonthrone, K. M., Rouch, D. A. Phosphatase-mediated heavy metal accumulation by a *Citrobacter* sp. and related enterobacteria, FEMS Microbiology Letters, 1994, 121, 141–146.
61. Macaskie, L. E., Bonthrone, K. M., Yong, P., Goddard, D. T. Enzymically mediated bioprecipitation of uranium by a *Citrobacter* sp.: a concerted role for exocellular lipopolysaccharide and associated phosphatase in biomineral formation, Microbiology (UK), 2000, 146, 1855–1867.

62. Reitz, T., Merroun, M. L., Selenska-Pobell, S. Interactions of Paenibacillus sp. and Sulfolobus acidocaldarius strains with U(VI), Uranium, Mining and Hydrogeology, 2008, 703–710.
63. Gupta, V. K., Rastogi, A. Sorption and desorption studies of chromium(VI) from nonviable cyanobacterium Nostoc muscorum biomass, Journal of Hazardous Materials, 2008, 154, 347–354.
64. Gupta, V. K., Shrivastava, A. K., Jain, N. Biosorption of chromium (VI) from aqueous solutions by green algae spirogyra species, Water Research, 2001, 35, 4079–4085.
65. Sampedro, M. A., Blanco, A., Llama, M. J., Serra, J. L. Sorption of heavy metals to Phormidium laminosum biomass, Biotechnology and Applied Biochemistry, 1995, 22, 355–366.
66. Dönmez, G. Ç., Aksub, Z., Ö ztürk, A., Kutsalb, T. A comparative study on heavy metal biosorption characteristics of some algae, Process Biochemistry, 1999, 34, 885–892.
67. Kim, E. J., Park, S., Hong, H. J., Choi, Y. E., Yang, J. W. Biosorption of chromium (Cr(III)/Cr(VI)) on the residual microalga Nannochloris oculata after lipid extraction for biodiesel production, Bioresource Technology, 2011, 102, 11155–11160.
68. Arun, N., Vidyalaxmi, Singh, D. P. Chromium (VI) induced oxidative stress in halotolerant alga Dunaliella salina and D. tertiolecta isolated from sambhar salt lake of Rajasthan (India), Cellular and Molecular Biology, 2014, 24, 90–96.
69. Travieso, L., Cañizares, R. O., Borja, R., Benítez, F., Domínguez, A. R., Dupeyrón, R., Valiente, V. Heavy Metal Removal by Microalgae, Bulletin of Environmental Contamination and Toxicology, 1999, 25, 144–151.
70. Chen, H., Pan, G., Yan, H., Qin, Y. Toxic effects of hexavalent chromium on the growth of blue-green microalgae, Huan Jing Ke Xue, 2003, 24, 13–18.
71. Cecal, A., Humelnicu, D., Rudic, V., Cepoi, L., Cojocari, A. Removal of uranyl ions from UO2(NO3)2 solution by means of Chlorella vulgaris and Dunaliella salina algae, Open Chemistry, 2012, 10, 1669–1675.
72. Khani, M. H., Keshtkar, A. R., Ghannadi, M., Pahlavanzadeh, H. Equilibrium, kinetic and thermodynamic study of the biosorption of uranium onto Cystoseria indica algae, Journal of Hazardous Materials, 2008, 150, 612–618.
73. Li, P. F., Mao, Z. Y., Rao, X. J., Wang, X. M., Min, M. Z., Qiu, L. W., Liu, Z. L. Biosorption of uranium by lake-harvested biomass from a cyanobacterium bloom, Bioresource Technology, 2004, 94, 193–195.
74. Sen, M., Ghosh Dastidar, M. Chromium removal using various biosorbents. Iranian Journal of Environmental Health Science and Engineering, 2010, 7, 182–190.
75. Yesim, S., Yücel, A. Kinetic studies on sorption of Cr(VI) and Cu(II) ions by chitin, chitosan and Rhizopus arrhizus, Biochemical Engineering Journal, 2002, 12, 143–153.
76. Kumar, R., Bishnoi, N. R., Garima, & Bishnoi, K. Biosorption of chromium (VI) from aqueous solution and electroplating wastewater using fungal biomass, Chemical Engineering Journal, 2008, 135, 202–208.
77. Chen, G. Q., Zhang, W. J., Zeng, G. M., Huang, J. H., Wang, L., Shen, G. L. Surface-modified Phanerochaete chrysosporium as a biosorbent for Cr (VI)-contaminated wastewater, Journal of Hazardous Materials, 2011, 186, 2138–2143.

78. Ertugay, N., Bayhan, Y. K. Biosorption of Cr (VI) from aqueous solutions by biomass of Agaricus bisporus, Journal of Hazardous Materials, 2008, 154, 432–439.
79. Sharma, S., Adholeya, A. Detoxification and accumulation of chromium from tannery effluent and spent chrome effluent by Paecilomyces lilacinus fungi, International Biodeterioration and Biodegradation, 2011, 65, 309–317.
80. Paknikar, K. M., Bhide, J. V. Aerobic reduction and biosorption of chromium by a chromate resistant Aspergillus sp., Biohydrometallurgical Technologies, 1993, 237–244
81. Narasimhulu, K., Setty, Y. P. Studies on biosorption of chromium ions from wastewater using biomass of Aspergillus niger, Open Access Scientific Reports, 2012, doi: 10.4172/scientificreports.146
82. Pang, C., Liu, Y. H., Cao, X. H., Li, M., Huang, G. L., Hua, R., Wang, C. X., Liu, Y. T., An, X. F. Biosorption of uranium(VI) from aqueous solution by dead fungal biomass of Penicillium citrinum, Chemical Engineering Journal, 2011, 170, 1–6.
83. Bhainsa, K. C., D'Souza, S. F. Biosorption of uranium(VI) by Aspergillus fumigatus, Biotechnology Techniques, 1999, 13, 695–699.
84. Tsezos, M., Georgousis, Z., Remoundaki, E. Mechanism of aluminum interference on uranium biosorption by Rhizopus arrhizus, Biotechnology and Bioengineering, 1997, 55, 16–27.
85. Wang, J. S., Hu, X. J., Liu, Y. G., Xie, S. B., Bao, Z. L. Biosorption of uranium (VI) by immobilized Aspergillus fumigatus beads, Journal of Environmental Radioactivity, 2010, 101, 504–508.
86. Solat, S., Reza, R., Soheila, Y. Biosorption of uranium from aqueous solution by live and dead Aspergillus niger, Journal of Hazardous, Toxic, and Radioactive Waste, 2014, 18, 65–74.
87. Kinnlburgh, D. G. General-purpose adsorption isotherms, Environmental Science and Technology, 1986, 20, 895–904.
88. Wilson, D. J., Jones, M. G., Jones, M. M. Chemical models for the lethality curves of toxic metal ions. Chemical Research in Toxicology, 1989, 2, 123–130.
89. Vecchio, A., Finoli, C., Di Simine, D., Andreoni, V. Heavy metal biosorption by bacterial cells, Fresenius' Journal of Analytical Chemistry, 1998, 338–342.
90. Souag, R., Touaibia, D., Benayada, B., Boucenna, A. Adsorption of heavy metals (Cd, Zn and Pb) from water using keratin powder prepared from Algerien sheep hoofs, European Journal of Scientific Research, 2009, 416–425.
91. Hussain, A., Ghafoor, A., Anwar-ul-Haq, M., Nawaz, M. Application of the Langmuir and Freundlich equations for P adsorption phenomenon in saline-sodic soils, International Journal of Agricultural and Biologica, 2003, 5, 449–456.
92. Xie, S., Yang, J., Chen, C., Zhang, X., Wang, Q., Zhang, C. Study on biosorption kinetics and thermodynamics of uranium by Citrobacter freudii, Journal of Environmental Radioactivity, 2008, 99, 126–133.
93. Tsibakhashvili, N., Kalabegishvili, T., Mosulishvili, L., Kirkesali, E., Kerkenjia, S., Murusidze, I., Holman, H.-Y., Frontasyeva, M. V., Gundorina, S. F. Biotechnology of Cr (VI) transformation into Cr (III) complexes, Journal of Radioanalytical and Nuclear Chemistry, 2008, 278, 565–569.
94. Umpleby, R. J., Baxter, S. C., Chen, Y., Shah, R. N., Shimizu, K. D. Characterization of molecularly imprinted polymers with the Langmuir−Freundlich isotherm, Analytical Chemistry, 2001, 73, 4584–4591.

95. Tsai, S. C., Juang, K. W., Jan, Y. L. Sorption of cesium on rocks using heterogeneity-based isotherm models, Journal of Radioanalytical and Nuclear Chemistry, 2005, 266, 101–105.
96. Bennajaha, M., Darmane, Y., Ebn Touhamic, M., Maalmia, M. A variable order kinetic model to predict defluoridation of drinking water by electrocoagulation-electroflotation, International Journal of Engineering, Science and Technology, 2010, 2, 42–52.
97. Dursun, A. Y. A comparative study on determination of the equilibrium, kinetic and thermodynamic parameters of biosorption of copper (II) and lead (II) ions onto pretreated Aspergillus niger, Biochemical Engineering Journal, 2006, 28, 187–195.
98. Akar, T., Anilan, B., Gorgulu, A., Akar, S. T. Assessment of cationic dye biosorption characteristics of untreated and non-conventional biomass: Pyracantha coccinea berries, Journal of Hazardous Materials, 2009, 168, 1302–1309.

## APPENDIX

List of microorganisms (in Latin) listed in the chapter.

| Latin name | Definition |
|---|---|
| *Acinetobacter haemolytic*us | Gram-negative bacterium |
| *Actinomyces flavovirid*us | Gram-positive bacterium |
| *Aeromonas* sp. | Gram-negative, non-spore-forming, rod-shaped, facultative anaerobic bacteria |
| *Arthrobacter sp.* | soil-bacterium |
| *Aspergillus fumigat*us | saprophytic fungus |
| *Aspergillus niger* | mold |
| *Aspergillus* sp | fungus and one of the most common species of the genus Aspergillus |
| *Aspergillus sydon*i | fungus |
| *Aspergillus terreu*s | fungus (mold) found worldwide in soil |
| *Azotobacter chroococcum* | free-living diazotrophic bacterium |
| *Bacillus coagulans* | lactic acid-forming bacterial species |
| *Bacillus licheniformis* | Gram-positive, spore-forming soil bacterium |
| *Bacillus megaterium* | Gram-positive, endospore forming, rod shaped bacterium |
| *Bacillus* sp. | aerobic, sporulating, rod-shaped bacteria |
| *Bacillus sphaericu*s | obligate aerobe bacterium used as a larvicide for mosquito control |
| *Bacillus subtilis* | Gram-positive, rod-shaped bacterium |

| Latin name | Definition |
|---|---|
| *Burkholderia cepacia* | catalase-producing, lactose-nonfermenting, Gram-negative bacterium |
| *Candida albicans* | diploid fungus |
| *Cellulomonas flavigena* | Gram-positive, rod-shaped bacterium |
| *Chlamydomonas angulosa* | genus of green algae consisting of unicellular flagellates |
| *Chlorella pyrenoidosa* | freshwater green algae |
| *Chlorella regularis* | single-cell fresh water microalga |
| *Chlorella vulgaris* | single-cell green algae |
| *Chroococcus* sp. | cyanobacteria |
| *Citrobacter freundii* | Gram-negative, facultative anaerobic bacterium |
| *Citrobacter* sp. | Gram-negative, facultative anaerobic bacterium |
| *Clostridium* sp. | Gram-positive, anaerobic, rod-shaped bacterium |
| *Corynebacterium kutscheri* | Gram-positive, aerobe, rod-shaped bacterium |
| *Desulfomicrobium norvegicum* | Gram-negative, sulfate-reducing bacterium |
| *Desulfovibrio sulfodismutans* | sulfate reducing bacterium |
| *Dunaliella salina* | halophile green microalga |
| *Dunaliella* sp | halophile green microalga |
| *Escherichia coli* | Gram-negative, facultatively anaerobic, rod-shaped bacterium |
| *Exiguobacterium* sp. | thermophilic bacterium |
| *Geobacter daltonii* sp. nov | Fe(III)- and uranium(VI)-reducing bacterium |
| *Micrococcus luteus* | saprotrophic bacterium |
| *Micrococcus* sp. | gram-positive, oxidase-positive, and strictly aerobic cocci |
| *Microcystis aeruginosa* | freshwater cyanobacteria |
| *Mucor hiemalis* | fungal plant pathogen |
| *Mycobacterium smegmatis* | acid-fast bacterial species |
| *Nannochloris oculata* | microalga with high lipid content |
| *Nostoc muscorum* | free-living cyanobacterium |
| *Nostok linckia* | free-living cyanobacterium |
| *Oscillatoria tenuis* | filamentous cyanobacterium which is named for the oscillation in its movement |
| *Paecilomyces lilacinus* | saprobic filamentous fungus |

| Latin name | Definition |
|---|---|
| *Pantoea* sp. | Gram-negative bacterium |
| *Penicillium chrysogenu*m | fungus |
| *Penicillium citrinu*m | anamorph, mesophilic fungus |
| *Penicillium janthinellu*m | mold |
| *Phanerochaet chrysospo-riu*m | model white rot fungus |
| *Phormedium bohne*ri | epilithic filamentous cyanobacterium |
| *Phormidium laminosu*m | thermophilic cyanobacterium |
| *Porphyridium cruent*um | species of red alga |
| *Pseudomonas aeruginosa* | Gram-negative, rod-shaped, asporogenous, and mono-flagellated bacterium |
| *Pseudomonas fluorescen*s | Gram-negative, rod-shaped bacterium |
| *Pseudomonas* sp. | Gram-negative, aerobic gammaproteobacterium |
| *Rhizopus arrhizus* | filamentous fungus |
| *Rhizopus oryz*ae | filamentous fungus |
| *Rhodococcus* sp | aerobic, nonsporulating, nonmotile, Gram-positive bacterium |
| *Rhodotorula glutin*us | unicellular pigmented yeast |
| *Saccharomyces cerevisiae* | Baker's yeast, brewer's yeast |
| *Scenedesmus acutus* | nonmotile colonial green alga |
| *Scenedesmus obliquus* | nonmotile colonial green alga |
| *Scenedesmus quadricaud*a | microalga |
| *Selenastrum capriornutu*m | freshwater microalga |
| *Spirulina maxim*a | microalga rich in protein and other essential nutrients |
| *Spirulina platensi*s | blue-green microalga |
| *Staphylococcus aureus* | Gram-positive bacterium |
| *Staphylococcus saprophyti-cus* | Gram-positive, coagulase-negative bacterium |
| *Stenotrophomonas malto-philia* | aerobic, nonfermentative, Gram-negative bacterium |
| *Streptomyces albu*s | streptomycete strain |
| *Streptomyces levori*s | soil-bacterium |
| *Sulfolobus acidocaldari*us | aerobic thermoacidophilic crenarcheon |
| *Ulothrix tenuissi*ma | filamentous green algae, generally found in fresh and marine water |

# PART IV

# THE RISKS OF ENVIRONMENTAL POLLUTION WITH HEAVY METALS FOR WARM-BLOODED ANIMALS: THE POSSIBILITY OF PROTECTING FOOD PRODUCTS FROM POLLUTANTS

CHAPTER 18

# EFFECT OF THE LEAD NITRATE ON DEVELOPMENT OF THE EARLY AND REMOTE CONSEQUENCES UNDER IRRADIATION OF MICE AT LOW DOSE

LYUDMILA N. SHISHKINA,[1] ALEVTINA G. KUDYASHEVA,[2] OKSANA G. SHEVCHENKO,[2] NADEZHDA G. ZAGORSKAYA,[1] and MIKHAIL V. KOZLOV[1]

[1] *N.M. Emanuel Institute of Biochemical Physics of Russian Academy of Sciences, 4, Kosygin St., Moscow, 119334, Russia, E-mail: shishkina@sky.chph.ras.ru*

[2] *Institute of Biology, Komi Scientific Centre of Ural Branch of Russian Academy of Sciences, 28, Kommunisticheskaya St., Syktyvkar, 167982, Russia, E-mail: kud@ib.komisc.ru*

## CONTENTS

## ABSTRACT

Comparative analysis of the early and long-term biological consequences under the low-intensity γ-radiation action at low dose and combined action of γ-radiation and lead nitrate at different concentrations was performed by using of the morphophysiological parameters and the some indices of the lipid peroxidation (the TBA-reactive substances amount, the activity of catalase, the phospholipid composition) in the functionally distinct tissues of CBA mice. It was found that the direction and scale of changes of studied parameters are complex nonlinear character, depending on the lead nitrate dose and the period after the radiation action, as well as the studied tissue or organ.

The data obtained and analysis of literature allow us to consider parameters of the lipid peroxidation regulatory system in tissues of animals and also the changes of interrelations between them as markers for the estimation of the biological consequences under the action of radiation at the low doses and its combined action with the chemical toxicants at the different concentrations.


## 18.1 INTRODUCTION

The impossibility to predict the combined effect of different factors occurs due to the maximum probability of appearance of synergistic effect from different chemical and physical factors at low doses and intensity [1]. The research of the combined effect of the different factors on organism is a high priority problem due to the magnification of the pollutant and radionuclide content in environment. Lead and its compounds are among the technogenic unfavorable factors, which are recognized as a long-live and persistent environmental toxicant [2, 3]. Lead cause the bonding of SH-groups of proteins, the hematological, gastrointestinal and neurological dysfunction in organism and chronic nephropathy, alter cellular calcium metabolism and cause irreversible injury in isolated hepatocytes [2, 4–6]. In the last new years, the ability of lead ions to induce phospholipidosis and cholesterogenesis in animal tissues is considered as a mechanism of lead toxicity [7, 8]. The problem of the combined action of the ionizing radiation at low doses and heavy metals is important for radioecology because of the necessity to predict effects of anthropogenic influences upon the organism and population levels. Earlier a high sensitivity of the lipid peroxidation (LPO) regulatory system parameters both to increased natural radiation background and to technogenic radioactive contamination of the environment is obtained in tissues of wild rodent inhabiting local radioactively contaminated areas in the Komi Republic or in 30-kilometer exclusion zone of the Chernobyl accident [9, 10]. Besides, a long-time inhabiting of tundra voles under an increased natural radiation background provokes distortions on a morphophysiological level [10]. Lead is known to be in termination of the radioactive decay of many nuclides. At present there is sufficient experimental data on the influence of the combined action of radionuclides and heavy metals in different concentrations for plant objects [11–15]. However, information about biological consequences of the combined action of the ionizing radiation at low doses and lead ions depending on its concentration for organism of animals is scanty.

The aim of present work was to investigate the influence of lead nitrate at different concentrations on the development of the early and remote consequences under the chronic γ-radiation of laboratory mice at low dose using the morphophysiological indices and parameters of the LPO regulatory system in tissues.

## 18.2 MATERIALS AND METHODOLOGY

The chronic γ-irradiation at low dose under the low dose rate and lead nitrate were chosen as one of the most widespread among the technogenic damaging factors. The experiments were carried out on the laboratory 2.5–3 month aged CBA mice (males). 110 mice were divided in 7 groups (per 7–10 species in each) as follows:

A. intact control;
B. irradiation;
I. irradiation + lead nitrate at the dose of 0.003 g/kg;
II. irradiation + lead nitrate at the dose of 0.01 g/kg;
III. irradiation + lead nitrate at the dose of 0.03 g/kg;
IV. irradiation + lead nitrate at the dose of 0.1 g/kg;
V. irradiation + lead nitrate at the dose of 0.3 g/kg.

Two ampoules with $^{226}Ra$ which had the $0.474 \times 10^6$ and $0.451 \times 10^6$ kBq activities and placed in 2.5 m distance used as a source of γ-irradiation. In the irradiation zone mice were within 30 days under the dose rate 2.0–2.2 mR/h. The total absorbed dose measured by DGRZ radiometers in each cell was 1.44–1.6 cGy. Mice of A and B groups received clean drinking water, while the solutions of lead nitrate (analytical pure grade, "Reakhim" USSR) in drinking water were used during in the course of irradiation for the other groups of mice at the calculated concentrations which furnished the uptake of the lead nitrate at the doses of 0.003, 0.01, 0.03, 0.1 and 0.3 g/kg of the body weight of mouse. The calculations were performed taking into account the weight of animals and the volume of consumed liquid.

Decapitation of mice in experimental groups was done within 1 day and 30 days after irradiation and decapitation of mice in A and B groups was simultaneously performed. Following decapitation of mice their organs (spleen, liver and brain) were placed on ice. The blood was collected in test tubes treated by 5% solution of sodium citrate. The blood plasma was separated from the blood corpuscles by centrifugation. Body weights and relative weights (index, ‰), i.e., the ratio of the organ weight (in mg) to the bogy weight of species (in g), were recorded as morphophysiological parameters.

The content of the LPO secondary products reacting with 2-thiobarbituric acid (TBA-reactive substances, TBA-RS) was determined using the method described in Ref. [16]. Protein content was estimated according to Ref. [17]. The catalase activity in liver was measured spectrophotometrically at a wavelength 410 nm according to the formation of a colored complex of ammonium molybdate in the presence of hydrogen peroxide [18].

Lipids were isolated by the method of Blay and Dyer in the Kates modification [19]. The qualitative and quantitative composition of phospholipids (PL) was determined by thin-layer chromatography as it describes in Ref. [20]. In was used type G or H silica gel (Sigma, USA), glass plates 9×12 cm and a mixture of solvents chloroform:methanol:glacial acetic acid:distilled water (25:15:4:2) were used as a mobile phase. All the solvents were of specially pure or chemically pure grade. The development of chromatograms was performed iodine vapor. To determine the concentration of phosphorus present, ammonium molybdate and ascorbic acid (Serva, Electrophoresis GmbH, Germany) and also perchloric acid of chemically pure grade were used to generate a color reaction. A more detailed description of the PL determination method is presented in Ref. [21]. In addition to the quantitative analyzes of the different fractions of PL, the following generalized parameters of lipid composition were also assessed: PL proportion in the total lipid composition (%PL), the ratio of the content of the different fractions in PL and the ratio of sums of the more easily to the more poorly oxidizable PL (ΣEOPL/ΣPOPL), a value of which was calculated by the formula [21]:

$$\Sigma EOPL/\Sigma POPL = (PI + PS + PE + CL + PA)/(LPC + SM + PC),$$

where PI is phosphatidylinositol, PS is phosphatidylserine, PE is phosphatidylethanolamine, CL is cardiolipin, PA is phosphatidic acid, LPC are lysoforms of PL, SM is sphingomyelin, PC is phosphatidylcholine.

All parameters of organs were measured for each animal individually, but under the analysis of PL composition in erythrocytes blood was combined from 2 to 3 species.

The experimental data were processed with a commonly used statistic method by means of statistical package Excel including the method of regression analysis. Differences between parameters were determined by Student's *t*-criterion. The variability of parameters was evaluated as the ratio of the standard deviation to average mean for group expressed as a percentage. The experimental data are presented in figures as the arithmetic means with the indication the standard deviation.

## 18.3 RESULTS AND DISCUSSION

To study the development of the biological consequences under the combined action of the chemical and physical factors on organism it was chosen organs which had the different functions in metabolism and functioning organism. So, spleen is one of the important blood-forming organ, liver is a "chemical laboratory" of organism, and brain is considered to serve for signaling. Recently the blood erythrocyte lipids are supposed to be a convenient model system for estimating of biological consequences under the action of the different factors [22]. The $Pb^{2+}$ concentrations were chosen according to their toxicity for laboratory rodent.

Earlier it was shown that a long-time inhabiting of tundra voles under an increased natural radiation background provokes distortions on a morphophysiological level [10]. Besides, values of the organ indices are sensitive to the action of the ionizing radiation at low doses according to Ref. [23, 24]. Therefore, analysis of the morphophysiological parameters was the fist stage of our investigations. The average magnitudes of the body weight in all groups of mice are done in Table 18.1.

**TABLE 18.1** The Body Weight of Mice in the Dependence on the Experimental Conditions

| Period after action | Body weight (M ± SD), g | | | | | | |
|---|---|---|---|---|---|---|---|
| | A | B | I | II | III | IV | V |
| 1 day | 18.3 ± 1.3 | 19.0 ± 1.3 | - | 18.7 ± 1.0 | 17.1 ± 0.7 | - | 19.1 ±1.2 |
| 30 days | 25.5 ± 1.8 | 24.8 ± 1.8 | 21.4 ± 1.3** | 25.0 ± 0.8 | 20.2 ± 0.9** | 22.7 ± 1.3* | 18.7 ± 1.0** |

*Note*: * Significant differences from the control group ($P < 0.02$). ** Significant differences from the control group ($P < 0.01$).

As seen, while presence of lead in drinking water is practically cause no changes in the body weight within 1 day after action, in remote term it was revealed the substantial diminution of this parameter in groups of mice under the combined action of radiation at low dose and lead independent on their concentration as compared with that both of the control and irradiated groups of mice. Only in group II mice of which receive the lead nitrate at the concentration of 0.01 g/kg the body weight didn't differ from the value in the control and irradiated groups of mice.

The average magnitudes of the spleen and liver indices are presented in Figures 18.1 and 18.2 for groups of mice in early and remote terms after action. As seen, values of the spleen index don't for certain differ in all groups within 1 day after action (Figure 18.1). However, the variability of this parameter is a 1.9 fold increased after irradiation and decreased in groups of mice under combined action of the lead and radiation, especially in group V, compared with control in early period after action. There is a diminution of the value of the spleen index in control group under aging and a growth of its variability simultaneously. Besides, within 30 days it is revealed the substantial increase of the spleen index values in groups

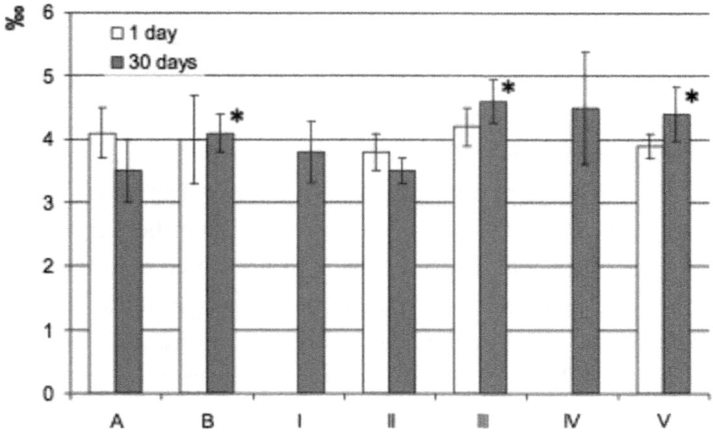


**FIGURE 18.1** The values of the relative weight (index) of spleen in groups of control and experimental mice in early and remote periods after action (Note: * Significant differences from the control group (P < 0.02)).

of nice under the combined action of radiation and lead at concentrations from 0.03 to 0.3 g/kg especially (Figure 18.1). It is need to note, that the variability of values of the spleen index in all experimental groups in addition to group IV are less compared with control, while the variability of this parameter in group IV is a 1.4 fold. As seen from data presented in Figure 18.2, values of the liver index is decreased both in early term under

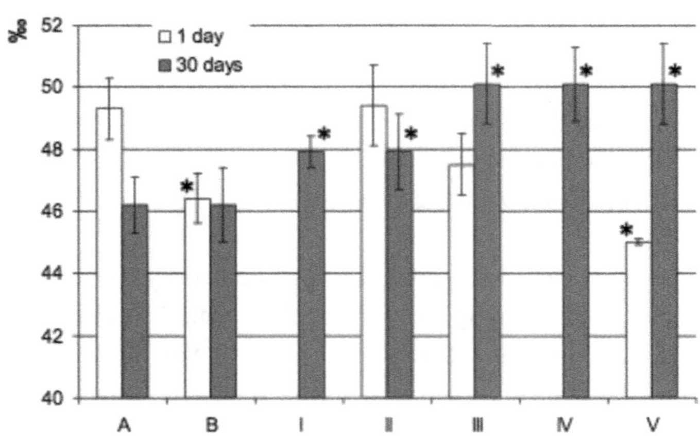


**FIGURE 18.2** The values of the relative weight (index) of liver in groups of control and experimental mice in early and remote periods after action (Note: * Significant differences from the control group (P < 0.02)).

the irradiation of mice at low dose and in group of age control compared with initial control value. Within 1 day it is revealed the diminution of that parameter under the combined action of irradiation and lead depending on their concentration. However, within 30 days after combined action there is for certain enhancement of values of the liver index at all groups which is practically independent on the lead nitrate concentrations (Figure 18.2).

As known, in norm there is a linear interrelation between weights of body and liver for rodent. From analysis of presented data it can see that the organ indices increase under the simultaneous diminution of the body weight. Hence, the combined action of the chronic radiation at low dose and the uptake of $Pb^{2+}$ ion at the different concentrations for a long-time provokes distortions even on a morphophysiological level.

The majority of radiation effects at low doses were found to be induced not by radiation directly, but indirectly through changes in the regulatory and immune system, antioxidant status of organism and genome destabilization [25]. The question of the LPO regulation is of quite high interest as it ranks as one of the earliest regulatory mechanisms in the evolutionary. At present there are much evidences showing that the LPO plays a very important role in the regulation of cellular metabolism in intact animals and under the impact of various damaging factors, including ionizing radiation. Earlier it was shown the existence of physicochemical regulatory system of the LPO both on a cell and organ levels, which is due to the steady state of LPO in the norm [26]. Among parameters of this regulatory system there are the intensity of LPO, the lipid composition, the ratio of sums of the more easily to the more poorly oxidizable PL, which characterizes the ability of lipids to the oxidation, and ratio of the content of the main PL fractions (PC/PE) in the animal organs or SM/PC, which are parameters characterizing the structural state of the membrane system [22, 26].

It is well known that LPO intensity is evaluated by the TBA-reactive substances content in a complex biological system. The influence of the irradiation at low dose and combined action of the radiation and lead nitrate on this parameter in the early period is shown for liver and blood plasma in Figure 18.3. As seen, while the LPO intensity in liver increases, in the blood plasma there is a tendency to the diminution of this parameter under only the radiation action. There is the simultaneous diminution

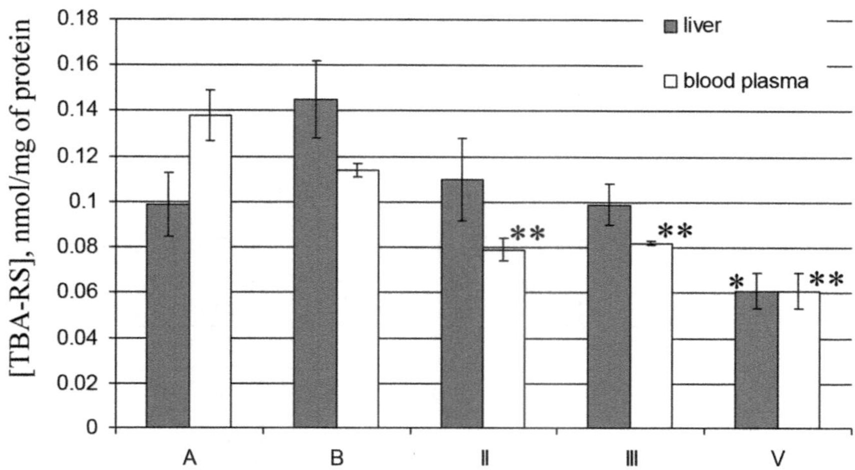


**FIGURE 18.3** The content of the TBA-reactive substances in liver and blood plasma of mice within 1 day after the action in the dependence on experimental conditions (Note: * Significant differences from the control group ($P < 0.02$). ** Significant differences from the control group ($P < 0.01$)).

of their variability in both tissues. However, in the early period after the combined action under the external chronic γ-irradiation at the dose of 1,44–1.6 cGy and the lead nitrate at the concentrations of 0.01 and 0.03 g/kg the LPO intensity in liver isn't differ compared with control (group A) and a 1.7- to 1.75-fold decreases in the blood plasma.

Besides, in the both cases the LPO intensity is independent on the lead concentration. Increasing the lead uptake to 0.3 g/kg results to a significant reduction in the content of TBA-reactive substances by 1.5 and 2.3 fold in the liver and blood plasma, respectively. The absence both of normalization of the LPO intensity and linear relation between the dose of lead nitrate and the LPO intensity in liver and brain of mice is observed within 30 days after action practically at all experimental groups as following from data presented in Figure 18.4. Besides, the catalase activity in liver is 2.6-fold increase in the remote period after action compared with control, but the mean linear relation between the content of TBA-reactive substances and catalase activity which there is in liver of mice in the age control group ($R = 0.64 \pm 0.21$) is absence in liver of mice under only the radiation action. Although the intensity of the LPO in liver of mice under

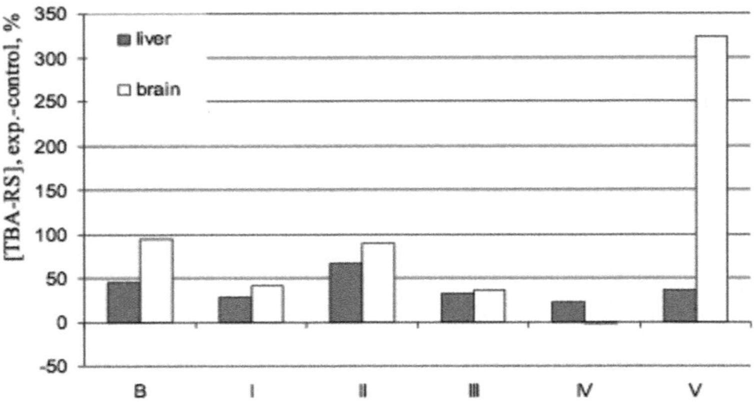


**FIGURE 18.4** The content of the TBA-reactive substances in liver and brain of mice within 30 days after the action in the dependence on experimental conditions (Note: On the ordinate: [TBA-RS], exp.-control, %).

the combined action of radiation and lead nitrate at the range of doses from 0.003 to 0.3 g/kg is higher compared with the control (Figure 18.4), the catalase activity in liver of mice in group I is less compared with that for irradiated mice. Perhaps, a certain reduction of the catalase activity under the combined action of the irradiation and lead nitrate at the dose of 0.003 g/kg is due to a ability of lead to decrease the activity of the antioxidant defense enzymes under its uptake in organism of animals [27]. Thus, both the chronic $\gamma$-radiation at low dose and its combined action with lead nitrate result to disruptions of interrelations between parameters of antioxidant status of tissues which are predominantly more substantial in the remote period after action.

Earlier it was found that lipids of the brain and blood erythrocytes of laboratory mice characterize prooxidat properties, and the liver lipids posses the enough high antioxidant activity in autooxidation reactions [28]. Hence, in early period a response of the LPO regulatory system parameters on the combined action of the irradiation at low dose and lead nitrate is the more pronounced in lipids of tissues with the low antioxidant status. This conclusion is accordance to results of the lipid composition analysis in tissues of mice in control and experimental groups. So, in the early period after action it was only revealed the some reduction the PC share in the liver PL both under the radiation and the combined action of lead nitrate and radiation. Besides, this diminution is increased depending on the lead concentration. The brain PL composition in different groups of mice in early period after the action is presented in Figure 18.5. As seen, the most pronounced changes in quantitative proportion of the PL fractions are revealed in groups on mice under the combined action of investigating factors, especially in group III. Besides, the lowest shares are observed for LPC and PE in group II and PC in group III under the highest shares SM and (CL + PA) sum in group III.

The combined action of the chronic γ-radiation at low dose and lead nitrate at the concentrations from 0.01 to 0.3 g/kg results to the most significant modifying effect on PL composition in the blood erythrocytes of mice when it is revealed by changing shares not only minor but main fractions of phospholipids at all experimental groups of mice in the early period after action (Figures 18.6 and 18.7). Besides, the most changes in proportions of PL fractions are observed for mice of group II under the lead uptake at the dose of 0.01 g/kg. So, these is the substantial growth of the LPC share under the action of only radiation and in groups II and III (it is need to note that this PL fraction is absent in control), the SM share

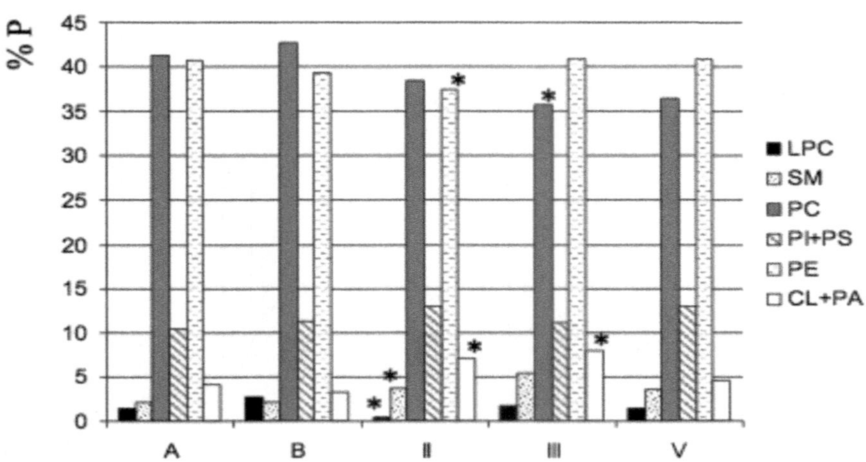


**FIGURE 18.5** The brain phospholipid composition of mice under the radiation at the dose of 1.44–1.6 cGy and combined action of the radiation and lead nitrate at the different concentrations within 1 day after the action (Note: * Significant differences from the control group ($P < 0.02$)).

a 2.5- to 4.8-fold increases in groups B, II and III, and the (CL + PA) sum a 2- to 3.9-fold increases in groups II, II and V (Figure 18.6) under for certain diminution of shares of both main PL fractions practically at all experimental groups simultaneously (Figure 18.7). Substantial changes in the quantitative proportions of PL fractions is due to for certain changes of generalized parameters of the PL composition as following from Figure 18.8. Analysis of these data allow us to make two conclusions both about the absence of linear interrelations of "effect-dose" and the different sensitivity of the LPO regulatory system parameters to the action of investigating factors. So, the highest values are revealed for of SM/PC ratio in group II and PC/PE ration in groups II and III, while the ability of lipids to oxidation (ΣEOPL/ΣPOPL) is a minimum in groups II and III and maximum in group V. Besides, under the chronic γ-radiation at low dose these is for certain increase of the SM/PC ratio and only tendency to increase for PC/PE ratio, but the absence of change for ratio of the more easily to the more poorly oxidizable PL compared with control. It is need to

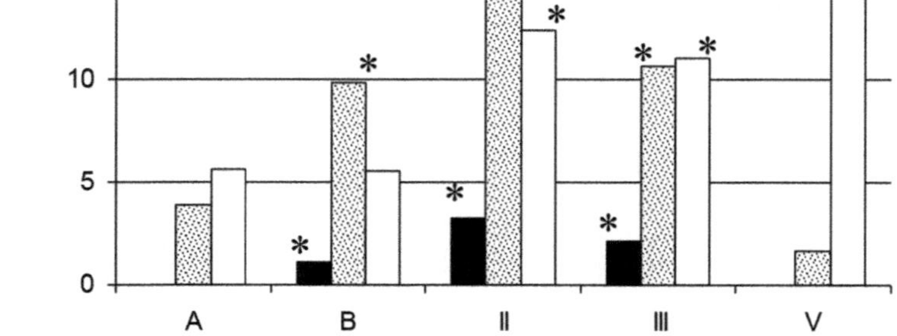


**FIGURE 18.6** The content of minor fractions of phospholipids in the blood erythrocytes of mice under the radiation at the dose of 1.44–1.6 cGy and combined action of the radiation and lead nitrate at the different concentrations within 1 day after the action (Note: * Significant differences from the control group ($P < 0.02$). ** Significant differences from the control group ($P < 0.01$)).

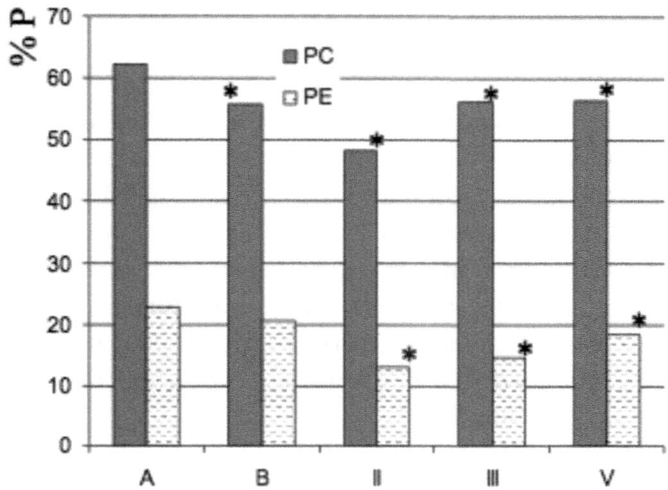


**FIGURE 18.7** The content of main fractions of phospholipids in the blood erythrocytes of mice under the radiation at the dose of 1.44–1.6 cGy and combined action of the radiation and lead nitrate at the different concentrations within 1 day after the action (Note: * Significant differences from the control group ($P < 0.02$)).

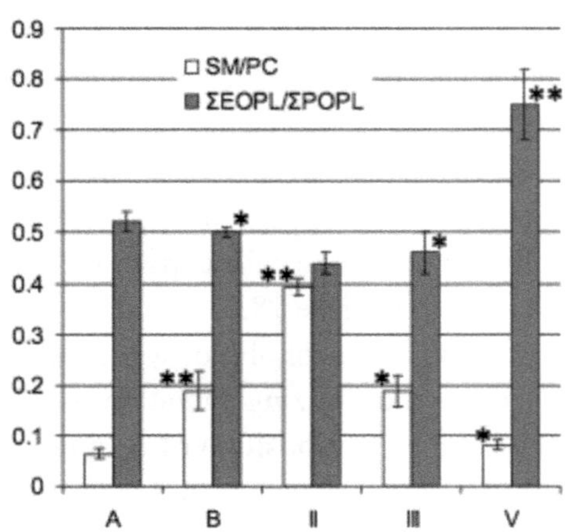


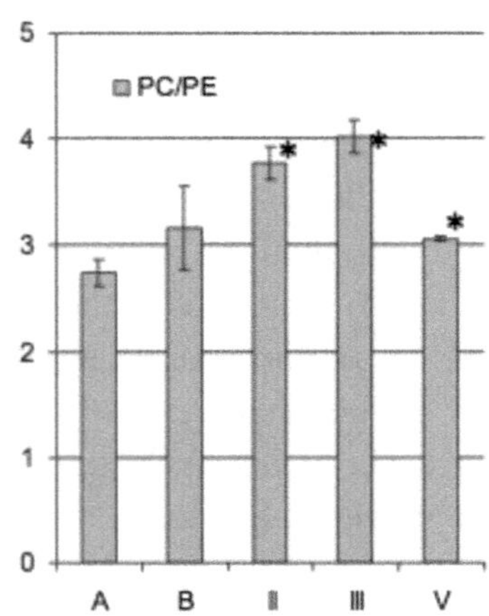


**FIGURE 18.8** Influence of the radiation at the dose of 1.44–1.6 cGy and its combined action with lead nitrate at the different concentrations on the generalized parameters of the phospholipid composition in the blood erythrocytes within 1 day after the action (Note: * Significant differences from the control group ($P < 0.02$). ** Significant differences from the control group ($P < 0.01$)).

emphasize, that the SM/PC and PC/PE ratio characterize a rigidity of the membrane structure. Lysoforms of PL is known to posses the lytic properties [29], and SM prevents a lytic action of lysoforms of PL and hemolysis of erythrocyte [30]. As already marked earlier, the relative content of LPC in PL composition in groups B, II and III is for certain higher compared with control (Figure 18.6). It is also need to emphasize that these fractions of acid PL as PA and especially CL are not typical for the PL composition of erythrocytes in the norm. Thus, in early period the most changes in the lipid composition and the most sufficient rearrangements of the membrane structure are revealed in the blood erythrocytes of mice under the chronic γ-radiation at the low dose and its combined action with chemical toxicant at the different concentrations.

In the remote period the more substantial changes in the PL composition were revealed for the liver mice in experimental groups. Nevertheless, it is need to mark for certain diminution of the PL share in the total lipid composition of mice brain in remote period especially in group IV compared with control and irradiation. While the quantitative proportion of the brain PL fractions had negligible differences from control and values of the generalized parameters of the PL composition were nearly to these, which were found for the irradiated mice, the PL composition in liver is found to have sufficient biochemical differences between the control and at all experimental groups. It is following from data presented in Figure 18.9.

In the remote period under chronic radiation at low dose these is a diminution of the relative content both of the main (PC and PE) and minor (PI + PS) fractions in the liver PL composition under for certain increase of the lysoforms of PL share simultaneously (Figure 18.9).

Earlier an increase of share of lysoforms in phospholipids was revealed in experiments on laboratory mice under the ionizing radiation at sublethal and lethal doses and also in the liver PL composition of wild rodent trapped in areas in 30-kilometer exclusion zone of the Chernobyl accident or inhabiting on the local radioactively contaminated areas in the Komi Republic [9, 10, 31, 32]. This phenomenon is usually considered as a result of the activation of phospholipase $A_2$.

The combined action of the chronic radiation at the low dose and the lead nitrate results to the more substantial changes in the liver PL composition as compared with that for the irradiated mice. It is need to note, the

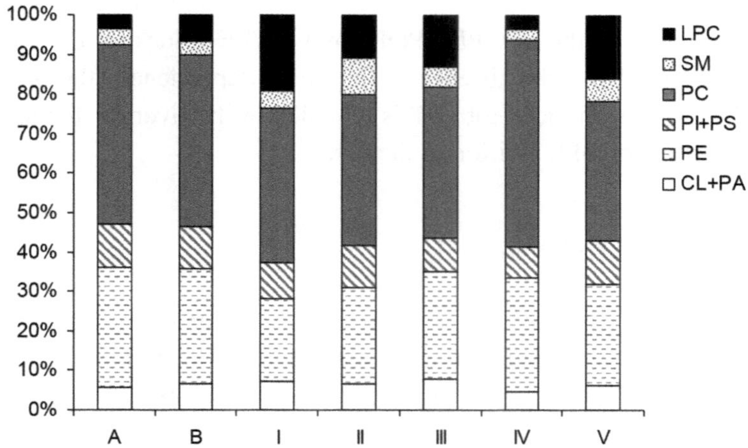


**FIGURE 18.9** The phospholipid composition of the liver mice under the radiation at the dose of 1.44–1.6 cGy and its combined action with lead nitrate at the different concentrations within 30 days after action.

direction of these changes in group IV under the lead nitrate concentration of 0.1 g/kg was an opposite compared with other experimental groups. While in groups I, II, III and V it is found the reduction of the PC and PE share and the rise of the relative content of SM and especially lysoforms of PL, the most high PC share and the least relative content of lysoforms, SM and PI + RS were observed for the liver PL composition in group IV among at all experimental groups. Such profound difference in the liver PL composition between the control and experimental mice causes the sufficient disturbance of the scale and direction of interrelations between the reciprocal parameters of the LPO regulatory system in norm. So, while these is an reverse correlation between the ratio of the sum of the more easily to the more poorly oxidizable PL fraction and the PC/PE ratio in the PL in the liver mice in control which is characteristic of tissues of the intact laboratory animals [33] this correlation is absent in the liver PL both in group of the irradiated mice and under the combined action of the chronic radiation at the low dose and the lead nitrate in groups I, II and V, and it is direct in group III in the remote period after action. As showed, under the combined action of the chronic radiation at the low dose and the lead nitrate at the dose of 0.1 g/kg the interrelation between the ability of lipids to oxidation and the structural state of the membrane system is reduced although the ΣEOPL/ΣPOPL and PC/PE values for certain differ from control. Thus, the absence of the linear dependence "the biological effect – dose of the lead nitrate" is revealed in the liver PL for the long-time after action of the different factors.

## 18.4 CONCLUSIONS

Thus, the data obtained allow us to conclude that the combined action of the chronic γ-irradiation at a low dose and the lead nitrate at the wide range of doses results to the substantial changes of the LPO regulation in the liver, brain and blood corpuscles of CBA mice both in the early and remote periods after action. Fist of all it is need to mark the absence of the linear dependence in the scale and direction of changes of the investigated parameters on the dose of the lead nitrate. Besides, the extent of modifying effect of the lead nitrate is determined by its dose, the duration after action and analyzed tissue and also is observed for any from the studied parameters. In the early period the more pronounced changes are found in tissues with the low antioxidant status, i.e., the blood erythrocytes and brain. In the remote period disturbances of the lipid exchange were predominantly revealed in liver, which characterizes enough a high antioxidant status. In many cases the most profound changes of the LPO regulatory system parameters were obtained under the combined action of the radiation and the lead nitrate at the lower concentrations. In our opinion, it is due to the absence of the reduction process when the dose of the chemical toxicant is too low.

As also found, the combined action of the chronic radiation at low dose and the uptake of $Pb^{2+}$ ion at the different concentrations for a long-time provokes distortions even on a morphophysiological level. It is in accordance with the earlier obtained data about effect of the physicochemical properties of murine liver lipids on interrelations between the lipid composition and the liver index of mice [34].

The data obtained and analysis of literature make it possible to consider the LPO regulatory system parameters in tissues of animals and their interior characteristics and also the changes of interrelations between them as markers for the estimation of the biological consequences under the action of radiation at the low doses and its combined action with the chemical toxicants at the different concentrations.

## ACKNOWLEDGEMENT

This work was supported by the Program of Fundamental Researches of Presidium of Russian Academy of Sciences "Fundamental sciences to medicine" (2012–2014).

## KEYWORDS


- **brain**
- **catalase**
- **erythrocytes**
- **liver**
- **low-intensity $\gamma$-radiation**
- **mice**
- **phospholipids**
- **TBA-reactive substances**


## REFERENCES


1. Petin, V. G., Zhurakovskaya, G. P., Pantyukhina, A. G., Rassokhina, A. V. Low doses and problems of synergistic interaction of environmental factors. Radiation biology. Radioecology. 1999, 39(1), 113–126 (in Russian).
2. Lockitch, G. Perspectives on Lead Toxicity. Clinical Biochemistry. 1993, 26, 371–381.
3. Luo, Z. D., Berman, A. H., The influence of Pb2+ on expression of acetylcholinesterase and acetylcholine receptor. Toxicology and Applied Pharmacology. 1997, 145, 237–245.
4. Minnema, D. J., Michaelson, I. A., Cooper, G. P. Calcium Efflux and Neurotransmitter Release from Rat Hippocampal Synaptosomes Exposed to Lead. Toxicology and Applied Pharmacology. 1988, 92, 351–357.

5. Rosel, E., Costin, K. A., Garcia-Webb, P. Effect of Occupational Lead Expose on Lymphocyte Enzymes Involved in Heme Biosynthesis. Clinical Chemistry. 1990, 36(11), 1980–1983.
6. Albano, E., Bellomo, G., Benedetti, A., Carini, R., Fulceri, R., Gamberucci, A., Parola, M., Comporti, M. Alterations of hepatocyte Ca2+ homeostasis by triethylated lead (Et3Pb+) are they correlated with cytotoxicity? Chemico-Biological Interactions. 1994, 90, 59–72.
7. Ademuyiwa, O., Agarwal, R., Chandra, R., Behari, J. R. Lead-induced phospholipidosis and cholesterogenesis in rat tissues. Chemico-Biological Interactions. 2008, doi: 10.1016/j.cbi.2008.10.057
8. Baranowska-Bosiacka, I., Dziedziejko, V., Safranow, K., Gutowska, I., Marchlewicz, M., Dolegowska, B., Rac, M. E., Wiszniewska, B., Chlubek, D. Inhibituin of erythrocyte phosphoribosyltransferases (APRT and HPRT) by Pb2+: A potential mechanism of leas toxicity. Toxicology. 2009, 259, 77–83.
9. Kudyasheva, A. G., Shishkina, L. N., Zagorskaya, N. G., Taskaev, A. I. 20 Years after the Chernobyl Accident: Past, Present and Future [Eds. Ekena, B. Burlakova, Valeria, I. Naidich]. New York: Nova Science Publishers, 2006, Ch. 17, 303–329.
10. Kudyasheva, A. G., Shishkina, L. N., Shevchenko, O. G., Bashlykova, L. A., Zagorskaya, N. G. Biological consequences of increased radiation natural background for Microtus oeconomus Pall. populations. Journal of Environmental Radioactivity. 2007, 97, 30–41.
11. Evseeva, T. I., Geras'kin, S. A. Combined effect of radiation and non-radiation factors for tradeskantia. Ekaterinburg, 2001, 156 p. (in Russian).
12. Evseeva, T. I., Majstrenko, T. A., Geras'kin, S. A., Belykh, E. S., Kazakova, E. V. Toxic and cytotoxic effects induced in Allium cepa with low concentrations of Cd and 232Th. Cytology and Genetics. 2005, 5, 73–80 (in Russian).
13. Evseeva, T. I., Geras'kin, S. A., Majstrenko, T. A., Belykh, E. S. The Problems of a Quantitative Estimation Combined of Chemical and Radioactive Exposure on Biological Objects. Radiation biology. Radioecology. 2008, 48(2), 203–211 (in Russian).
14. Pozolotina, V. N., Antonova, E. V., Bezel, V. S. Long-Term Effects in Plant Populations from Zones of Radiation and Chemical Pollution. Radiation biology. Radioecology. 2010, 50(4), 414–422 (in Russian).
15. Evseeba, T. I., Geras'kin, S. A., Vakhrusheva, O. M. Evaluation of the Partial Contribution of Naturally Occurring Radionuclides and Nonradioactive Chemical Toxic Elements in Formation of Biological Effects within the Vicia Cracca Population Inhabiting the Area Contaminated with Uranium-Radium Production Wastes in the Komi Republic. Radiation biology. Radioecology. 2014, 54(1), 85–96 (in Russian).
16. Asakawa, T., Matsushita, S. Coloring Conditions of Ttiobarbituric Acid Test for Detecting Lipid Hydroperoxides. Lipids. 1980, 15, 1137–1140.
17. Itzhaki, R., Gill, D. M. A microbiuret method for estimating proteins. Analitical Biochemistry. 1964, 9, 401–410.
18. Korolyuk, M. A., Ivanova, L. I., Maiorova, I. G. Assay of the catalase activity. Laboratory Practice. 1988, 1, 16–19 (in Russian).
19. Kates, M. The Technology of Lipidology (Russian version). Moscow: Mir, 1975, 322 p.

20. Biological Membranes. A practice Approach [Eds. J. B. C. Findlay, W. H. Evans]. Moscow: Mir, 1990, 424 p. (in Russian).
21. Shishkina, L. N., Kushnireva Ye. V., Smotryaeva, M. A. The combined effect of surfactant and acute irradiation at low dose on lipid peroxidation process in tissues and DNA content in blood plasma of mice. Oxidation Communications. 2001, 24, 276–286.
22. Shevchenko, O. G., Shishkina, L. N. Blood erythrocytes lipids as a model for estimating of biological consequences under the action of the physical and chemical factors. Technologies of living systems. 2009, 6(8), 21–32 (in Russian).
23. Klimovich, M. A., Smotryaeva, M. A., Gaintseva, V. D., Shishkina, L. N. The criterion on test for the estimation of the biological consequences of X-ray radiation at low doses under changing dose rate. Radiation biology. Radioecology. 2009, 49, No 4, 473–477 (in Russian).
24. Klimovich, M. A., Kozlov, M. V., Shishkina, L. N. Sensitive Indices of the Murine Liver Lipid to the Action of Low Intensity X-ray Radiation at Low Doses. New Steps in Physical Chemistry, Chemical Physics and Biochemical Physics [Eds. Gennady, E. Zaikov, Elim Pearce, Gerald Kirshenbaum]. New York: Nova Science Publishers, 2013, Ch. 27, 249–263.
25. Mikhailov, V. F., Mazurik, V. K., Burlakova, E. B. Signal function of the reactive oxygen species in regulatory networks of the reaction of cell to damaging effects: contribution to radiosensitivity in genome instability. Radiation biology. Radioecology. 2003, 43(1), 5–18 (in Russian).
26. Shishkina, L. N., Kushnireva, E. V., Smotryaeva, M. A. A new Approach to Assessment of Biological Consequences of Exposure tp Low-Dose Radiation. Radiation biology. Radioecology. 2004, 44(3), 289–295 (in Russian).
27. Saliu, J. K., Bawa-Allah, K. A. Toxicological Effects of Lead and Zinc on the Antioxidant Enzyme Activities of Post Juvenile Clarias gariepinus. Resources and Environment. 2012, 2(1), 21–26.
28. Shishkina, L. N., Khrustova, N. V. The Kinetic Characteristics of Lipids of Animal Tissues in Autooxidation Reactions. Biophysics. 2006, 51(2), 340–346 (in Russian).
29. Gennis, R. Biomembranes: Molecular Structure and Function (Russian version). Moscow: Mir, 1997, 624 p.
30. Nagasaka, Y., Ishii, F. Interaction between erythrocytes from three different animals and emulsions prepared with various lecithins and oils. Colloids and surface B: Biointerfaces. 2001, 22, 141–147.
31. Burlakova, E. B., Archipova, G. V., Shishkina, L. N., Goloshchapov, A. N., Zaslavsky Yu. A. Influence of ionizing radiation on the regulation fumction of Biomembtane. Studia biophisica. 1975, B. 53, S. 67–71.
32. Shishkina, L. N., Polyakova, N. V., Taran Yu. P. Analysis of lipid peroxidation regulatory system parameters in murine organs long time after acute irradiation. Radiation. Radiation biology. Radioecology. 1994, 34(3), 362–367 (in Russian).
33. Shishkina, L. N., Klimovich, M. A., Kozlov, M. V. A New Approach in Analysis of Participation of Oxidative Orocesses in Regulation of Metabolism in Animal Tissues. Biophysics. 2014, 59(2), 310–315.
34. Khrustova, N. V., Shishkina, L. N. Effects of Lipid Physicochemical Characteristics on Interrelations between Lipid Composition and the Mouse Liver Index. Journal of Evolutionary Biochemistry and Physiology. 2011, 47(1), 37–42 (in Russian).

## CHAPTER 19

# RESPONSE OF MARINE FISH LIVER ON ENVIRONMENTAL POLLUTION

IRINA I. RUDNEVA,[1] MARIA P. RUDYK,[2] VICTORIA V. SHEPELEVICH,[2] LARISA M. SKIVKA,[2] NATALIA N. ROSLOVA,[2] EKATERINA N. SKURATOVSKAYA,[1] IRINA I. CHESNOKOVA,[1] VALENTIN G. SHAIDA,[1] and TATIANA B. KOVYRSHINA[1]

[1]*Kovalevski Institute of Marine Biological Research RAS, Nahimov av., 2, Sevastopol, 299011, Russia, Tel. +7 (8692) 54-41-10, E-mail: svg-41@mail.ru*

[2]*Biological Institute of Taras Shevchenko Kiev National University, Akademic Glushkov av., 2, Kiev, 03022, Ukraine*

## CONTENTS



### ABSTRACT

The response reactions of fish *Scorpaena porcus*, caught in Sevastopol bays, with different degrees of chemical pollution, to the unfavorable ecological factors were studied. The significant increase of the number of melanomacrophage centers (MMCs) in the liver of fish from the most polluted site was shown. Increase of oxidized products and chemiluminescence values in the liver extracts of fish from the contaminated bay was the result of oxidative stress in the animals. Induction of antioxidant enzyme activities in the liver demonstrated the response of Scorpion fish to pollution. Defense mechanisms of resistance activation of the organism to the consequences of oxidative stress and adapt it to the unfavorable environmental conditions were shown. Therefore, the anthropogenic impact on the fish provoked pathological changes of tissues in the liver as well as oxidative stress, what induced the increase of the activity of antioxidant and immune systems. These parameters could be used as informative indicators for the development of monitoring programmes.

## 19.1 INTRODUCTION

At the last decades researchers take their attention to the consequences of the environmental pollution and negative effects on the living organisms in ecosystems and biosphere. Long-term and large-scale monitoring studies indicate the changes of anthropogenic impact on the water ecosystems, which can be chronically stressed by multiple environmental factors The key idea of the investigations is the development of the methodology of identification and evaluation of the damages in ecosystems, including water bodies, caused polluted compounds and the search criteria of the rate of anthropogenic impact on them. Indicators of negative effects allow the direct determination of pollutant impact on living organisms in aquatic systems. Marine environment and especially the coastal waters play an important role in human activity because more than, 2,000,000,000 people live in the coastal part of the sea and ocean [1]. Fish are very important for commercial purposes, including fishery and aquaculture. Additionally, they play the key role in ecosystem because they belong to the vertebrates, which are at the top of the food chain.

Usually the investigators demonstrate the elevated levels of the man-made pollution and its negative effects on marine organisms in all levels of their biological organization Environmental pollutants transfer to fish organism and accumulate in it, caused early biological effects especially in cell and molecular levels. Accumulation of these alterations provokes damage of organ structure and function, tissues and systems disfunction, and during some time the pathologies and morphological anomalies are documented.

For the analysis of negative factors on the organism of many different criteria are used. Indicators of different biological levels (biochemical, physiological, cell, morphological, and population) could be grouped in three series [2–4]: (i) direct indicators (biomarkers) of exposure of unfavorable factors (stressors) in the environment; (ii) direct indicators (bioindicators) of the effects of these stressors on the organisms, and (iii) indirect indicators exposure and effects. The most informative biomarkers are the parameters of oxidative stress and antioxidant activities in the tissues of aquatic organisms [2, 5]. Toxicants, distributed in the environment, transfer into the aquatic organisms, accumulate in tissues and organs and cause oxidative stress, characterizing the increase of reactive oxygen species (ROS), accumulation of the oxidized biomolecules in tissues and the induction of the components of antioxidant defense system [6–8].

Histopathological analysis of various organs and tissues is an important tool of environmental monitoring of water pollution, which allows assessing of structure changes, and lesions that caused by environmental toxicants and various negative factors. The most common pathological changes in the liver of fish as the primary organ for xenobiotic accumulation and detoxication include alterations of the mural architecture (degenerative hepatocyte lesions: hepatocyte hypertrophy, pyknotic nuclei, nuclear pleomorphism and peripheral nucleoli, focal necrosis of hepatocytes that can be noticed by the presence of darkly stained eosinophilic debris as a result of cellular component disintegration). Extensive dilation of sinusoids with blood congestion, hypertrophy and hyperplasia of bill duct cells, fibrosis of blood vessels, increased numbers of Kupffer cells are also often occur in fish inhabiting polluted environments [9]. Among the others pathological alterations it needs to be noted a loss of lipid vacuoles in hepatocytes, pancreatic atrophy characterized by decreased number of acinous and appears as remnants around hepatic blood vessels [10]. Subcapsular edema and fat accumulation in liver is a common problem in aquaculture that can be induced by parasites, inappropriate nutrition, pesticides and toxins [11] as well as steatosis – fatty degeneration [12]. Histopatological lesions also include lymphocytic infiltration, hypertrophied hepatocyte nuclei with coarse chromatin, hepatocellular and nuclear pleomorphism [12, 13], presence of syncitia characterized by multiple nuclei and granular cytoplasm [14].

Biological markers allow the direct determination of pollutant impact on living organisms in aquatic systems. Firstly the biomarkers come in ecotoxicology from the clinic diagnostic and they were adopted for the analysis of biochemical processes, which take place in natural populations, impacted by negative factors. They are used as tests usually as the indicators of the "early warning system" in the water bodies impacted the increasing anthropogenic activity. Their application helps to improve the management of the environment and to develop the monitoring programs, to protect the living organisms health and the total ecosystem.

Anthropogenic activity, including industry, agriculture, fishery, mariculture, tourism, petroleum and gas production, coastal domestic infrastructure, maritime transport impact on the habitat of the Black Sea. The specificity of Sevastopol bays including their geographical position and different level of pollution allow to understand the main trends of anthropogenic impact on fish health and to define the biomarkers on stress in the existence conditions of different species.

The aim of the present study was to determine the hystopathological and biochemical indicators of the liver of Scorpion fish caught in two bays near Sevastopol with different levels of anthropogenic pollution – Streletskaya Bay and Alexandrovskaya Bay.

## 19.2 MATERIALS AND METHODOLOGY

Scorpion fish *Scorpaena porcus* is highly distributed benthic species in Black Sea (Figure 19.1). Fish were caught in spring period 2012–2014 in two Sevastopol bays Streletskaya Bay ($n$=17), Alexandrovskaya Bay ($n$=18) characterizing different levels of anthropogenic impact and pollution. The animals were immediately placed in the aerated tank and subjected to anetezia.

The tested bays are differed each from other by hydrological characteristics, hydrochemical parameters and the level of anthropogenic impact

**FIGURE 19.1** Scorpion fish Scorpaena porcus.

and pollution. Streletskaya Bay is the most polluted site caused active maritime transport and recreation. There are petroleum moorage and ship-repairing yard, two domestic sewage treatment enterprises on the coast of this bay and the effluents enter into the aquatoria and contaminate water and sediments [15, 16]. Many buildings and hotels, increasing the number of moorages for small ships continue to pollute the water. Concentration of petroleum hydrocarbons in 2003–2009 was increased in 1.2–1.5-folds as compared with the values of 1990 [17]. There are no any direct pollution sources in Alexandrovskaya Bay; however, the rain sewage of the city domestic treatment enterprise enters into its aquatoria. In addition, close position of this bay promotes accumulation of xenobiotics in the bottom sediments. However, the content of petroleum hydrocarbons in the sediments of this site was approximately 7–4 lower as compared with the corresponding values in the sediments of Streletskaya Bay and their level continues to decrease [18].

Therefore, it could be proposed that Streletskaya Bay is the most polluted site and the living conditions in it are not preferable for aquatic organisms, while ecological status of Alexandrovskaya Bay is better. Pollution level is differed in the bays, especially in the bottom water in which the Scorpion fish inhabits.

The liver was removed from fish captured in both bays for histochemical studies. Liver samples were fixed in 10% neutral buffered formalin folowing by dehydration in a graded series of ethanol solutions of increasing concentrations. Fixed specimens were processed to paraffin embedded blocks by accepted, routine methods and sections were cut at 6 μm on a sledge microtome. Tissue sections were stained with Boehmer's haematoxylin and 1% eosin by standard method and mounted into Canadian balm. Samples were examined using light microscope and MICROmed digital camera. To count melanomacrophage centers (MMCs), 8–10 fields were randomly selected on each slide, captured using the camera, and readings were performed at 200X magnification. After each field of liver tissue samples had been photographed, the area of MMCs was measured ($\mu m^2$) (MICROmed digital camera software). Quantitative changes in the numbers and size of the liver MMC were determined by counting the mean number of MMCs per field and by measuring of the average absolute ($\mu m^2$) and relative (% of the field area) size of single MMC (the percentage of organ area occupied by MMC) and total MMCs per field [19–21].

For biochemical determinations liver samples were washed several times by cold 0.85% solution of NaCl, homogenized and centrifuged at 8000 g during 15 min at cool conditions. In obtained extracts thiobarbituric acid (TBA)-reactive products concentration was determined used Spectrophotometer Specol–211 (Carl Zeiss, Iena, Germany) [22]. Parameters of spontaneous chemiluminescence (SChL) and inducible by $H_2O_2$ and $FeSO_4$ chemiluminescence (IChL) were measured used Luminometer 2010 (LKB, Sweden) [6]. The value of chemiluminescence was estimated in arbitrary units per protein concentration which was determined used the kit of Filicit (Ukraine).

The analysis of the concentration of oxidized proteins was processed according the method of Dubinina et al. [22]. Optical density of carbonil groups reacted with 2,4-dinitrophenil hydrazones was assayed at the following wavelength 346, 370, 430, 530 nm used the spectrophotometer Specol–211 (Carl Zeiss, Iena, Germany).

Antioxidant activities in fish liver extracts were determined according the methods, which we described previously with small modifications. Catalase (CAT) was measured by the method involving the reaction of hydroperoxide reduction [24]. Superoxide dismutase (SOD) was assayed on the basis of inhibition of the reduction of nitroblue tetrasolium (NBT) with NADH mediated by phenazine methosulfate (PMS) under basic conditions [25]. Peroxidase (PER) activity was detected by spectrophotometric method using benzidine reagent [26]. Glutathione reductase (GR) was determined according the method of NADH degradation and Glutathione-S-transferase (GST) activity was assayed by the method of by following the increase in absorbance at 340 nm due to the formation of the conjugate 1-chloro-2,4-dinitrobenzene (CDNB) using as substrate at the presence of reduced glutathione (GSH) spectrophotometrically [27].

Aminotransferases (ALT and AST) activity was determined spectrophotomtrically with 2,4-dinitrophenylhydrazine at 500–530 nm used the standard kit (Felicit – Diagnosis, Ukraine) [28].

The results were processed to statistical evaluation with Student's tests. All numerical data are given as means ± SEM [29]. The significance level was 0.05.

## 19.3 RESULTS AND DISCUSSION

Fish liver tissue that does not exhibit histopathological changes displays only few MMCs in the parenchyma or their absence [9]. Polygonal shaped or classical hexahedral hepatocytes with large spherical nucleolus and variable amount heterochromatin were detected in samples of fish liver (Figures 19.2 and 19.3). Hepatocytes formed hepatic cell cords locating among blood capillaries (hepatic sinusoids). Erythrocytes were mainly observed in the lumen of sinusoids. Kupffer cells (tissue-resident macrophages with large processes and bean-shaped nucleus) were registered in the space between hepatocytes and on the luminal surface of the sinusoid endothelium.

At the same time, significant distinctions were detected in liver histological parameters between the fish captured in the different examined bays (Figure 19.4). The mean number and area of MMCs were statistically greater in the liver of the fish from Streletskaya Bay. Liver MMCs were not detected in 60% (3/5) of fish from Alexandrovskaya Bay and occurred very rarely in the rest 40% of fish, in which they were relatively small and associated with the blood vessels. The MMC size in

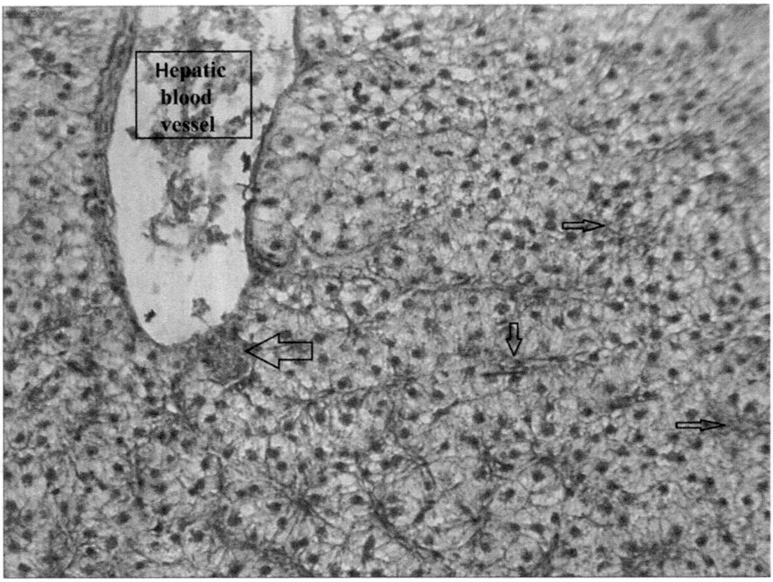


**FIGURE 19.2** Section through liver of Black Sea Scorpion fish (Scorpaena porcus) captured in Alexandrovskaya Bay showing melanomacrophage center (big arrow), hepatic sinusoids (small arrows) and hepatic blood vessel. H&E, 200X, Bar = 25 μm.

fish from Streletskaya Bay was 3.8 times greater than that in fish from Alexandrovskaya Bay. In fish from Streletskaya Bay, numerous MMCs were adjacent to the blood vessels or bile duct cells and they were pale colored. Large amount of MMCs were infiltrated by lymphocytes, which surrounded MMCs in some cases and induced granuloma-like structure formation.

It is necessary to point out a significant individual variability of MMC area in the liver of the fish captured in different bays. Thus, the results of hystochemical studies showed the significant differences between the liver samples of the Scorpion fish caught in the bays with different level of pollution. The goal of our further investigations was to study the response of the animals inhabited two tested bays used the biomarkers of oxidative stress and antioxidant activity. The results observed the increase of TBA-reactive products concentration in the liver of the fish from Streletskaya Bay more than 2-fold as compared with the corresponding values of the animals from Alexandrovskaya Bay (Figure 19.5).

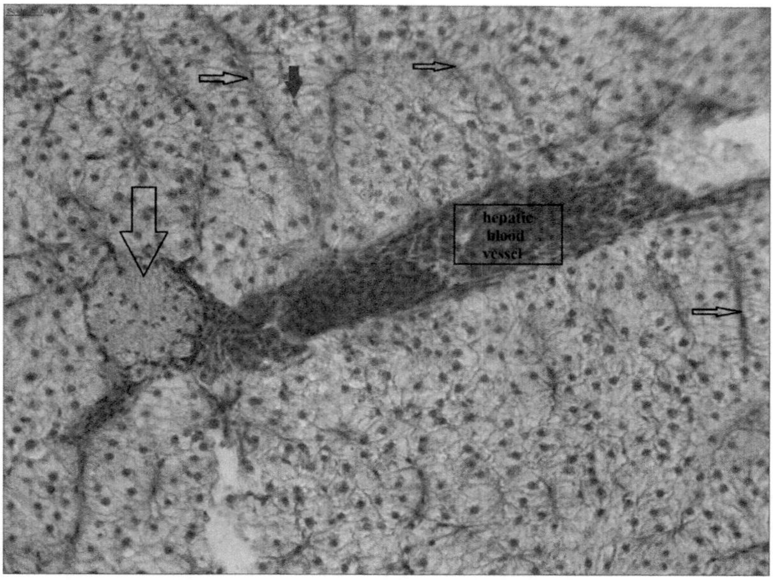


**FIGURE 19.3** Section through liver of Black Sea Scorpion fish (Scorpaena porcus) captured in Streletskaya Bay showing the large melanomacrophage center (big arrow) adjacent to the blood vessel, dilated hepatic sinusoids (small arrows), congested blood vessel (hyperemia) with perivascular lymphocytic infiltration and sign of fibrosis around endothelial cells, increased Kupffer cells (red arrow). H&E, 200X, Bar = 25 μm.

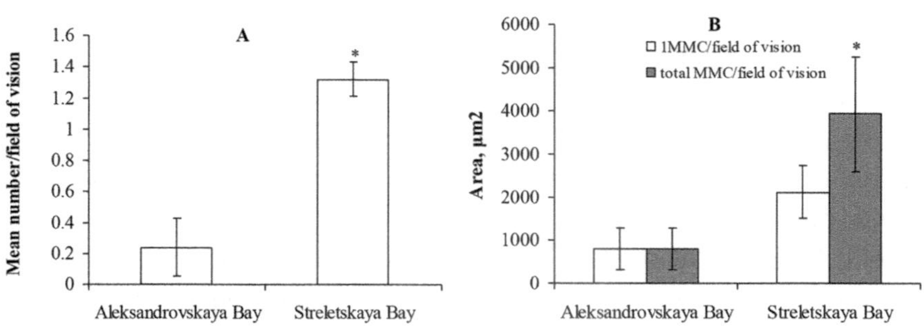


**FIGURE 19.4** Qualitative evaluation of liver melanomacrophage centers (MMCs) in Black Sea Scorpion fish (Scorpaena porcus) captured in two bays of Sevastopol. A – the mean number of MMCs; B – size of MMCs area (μm2); * – P<0.05 compared with the corresponding values of fish captured in Alexandrovskaya Bay.

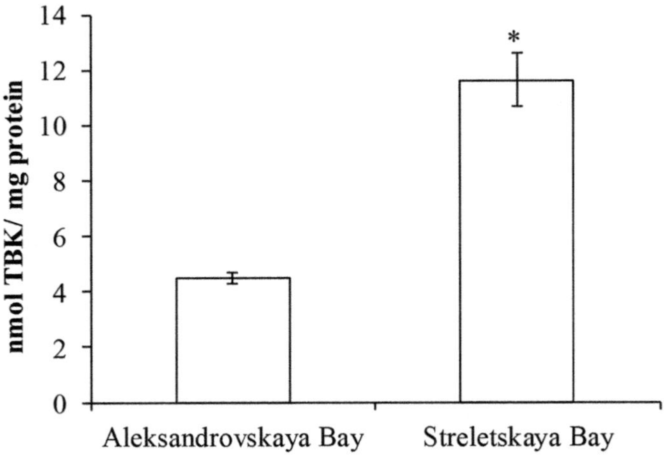


**FIGURE 19.5** Concentration of TBA-reactive products in the liver of Scorpion fish (Scorpaena porcus), captured in two bays of Sevastopol; * – significant difference, $P < 0.05$.

The values of spontaneous chemiluminescence (SChL) in the liver extracts of fish from both tested bays did not show significant differences, while in this case we can mark the increase of these parameters in fish from Streletskaya Bay (Figure 19.6). However, the inducible chemiluminescence (IChL) was higher approximately 10-fold in the liver of Scorpion fish from the polluted site.

Concentration of oxidized proteins in the liver extracts of the fish from the polluted bay was significant higher than in the non-polluted site (Figure 19.7).

Significant differences of antioxidant enzyme activities were shown also in the liver of fish from two tested bays (Figure 19.8). SOD activity tended to increase in the liver of fish from Streletskaya Bay, while the activity of CAT, PER, GR and GST was significantly greater ($P<0.05$) more than 2–3 fold as compared with the values of fish from Alexandrovskaya Bay.

Aminotransferase activity in the liver of fish from both bays is present in Figure 19.9. ALT activity tended to decrease in the liver of the animals from Streletskaya Bay, while the AST activity significantly decreased.

Therefore, the parameters of oxidative stress and antioxidant activities were differed in the liver of Scorpion fish captured in both Sevastopol bays and they were significantly higher in the animals from the polluted site.

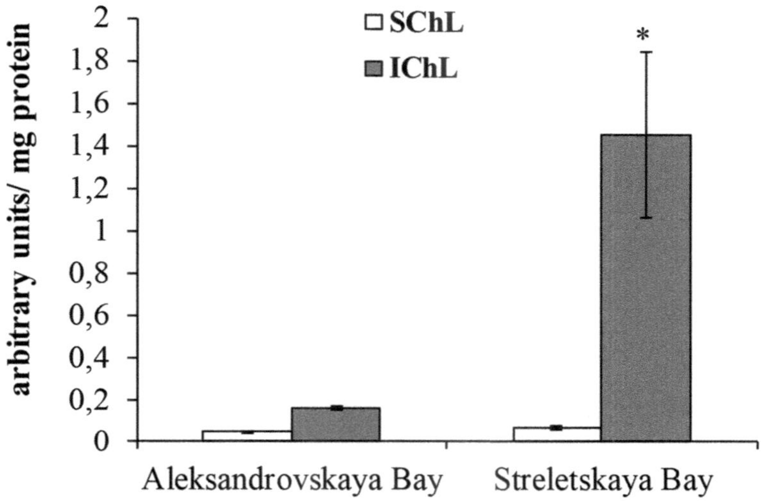


**FIGURE 19.6** Chemiluminescence level in the liver of Scorpion fish (Scorpaena porcus), captured in two bays of Sevastopol; * – significant difference, $P<0.05$, SChL – spontaneous chemiluminescence, IChL – inducible chemiluminescence.

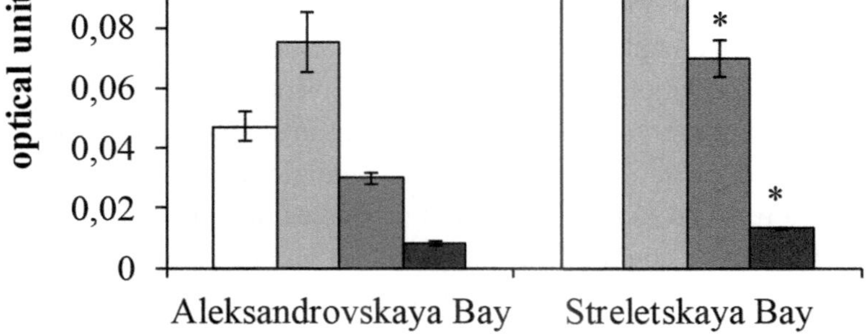


**FIGURE 19.7** Concentration of oxidized proteins in the liver of Scorpion fish (Scorpaena porcus), captured in two bays of Sevastopol; * – significant difference, $P<0.05$.

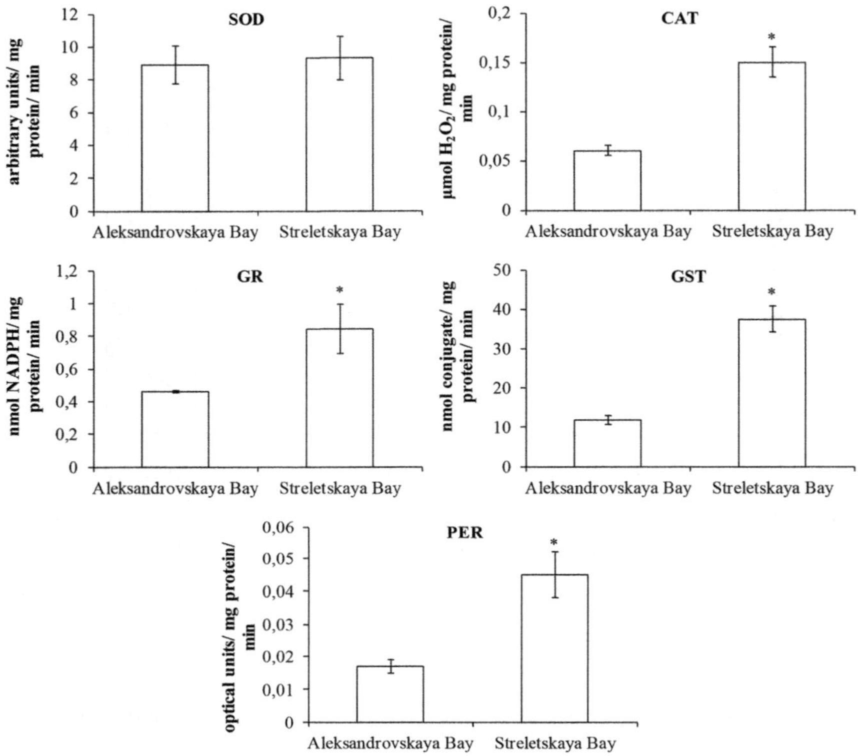


**FIGURE 19.8** Antioxidant enzyme activities in the liver of Scorpion fish (Scorpaena porcus), captured in both bays of Sevastopol; * – significant difference, $P<0.05$.

The results of hystopathological and biochemical studies of the liver of Scorpion fish, captured in two Sevastopol bays, characterizing different level of pollution, demonstrated significant impact of the negative ecological conditions on fish health. Taking into account that the liver is the important metabolic organ, in which xenobiotics are accumulated and detoxified resulted biotransformation processes, parameters of the evaluation of its status could reflect the pollution level of environment and its harmful for living organisms. The data obtained are important for the environmental monitoring programs, ecological risk assessment and validation of the criteria of ecological rate of anthropogenic impact.

As we described previously [5], Scorpion fish is highly distributed benthic settled life species inhabiting coastal waters of Black Sea. Its

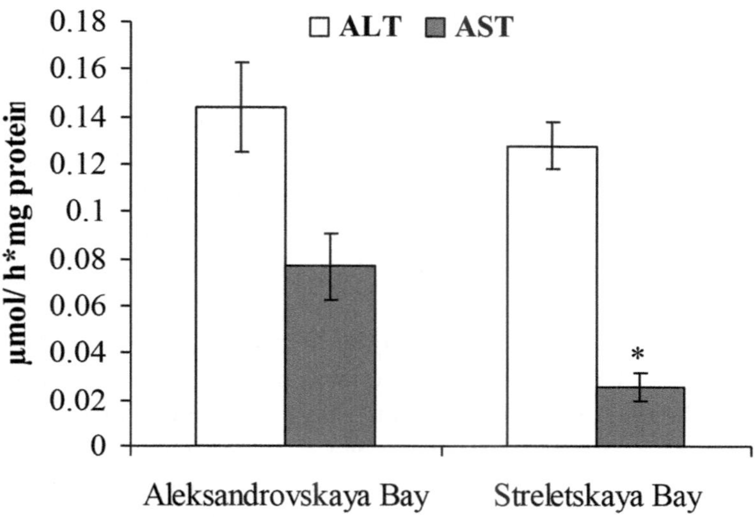


**FIGURE 19.9** Aminotransferase activity in the liver of Scorpion fish (Scorpaena porcus), captured in two bays of Sevastopol; * – significant difference, $P < 0.05$.

biology and ecology is well known, and thus it can be used as biomonitor species in the Black Sea coastal waters for monitoring purposes. Our findings of hystochemical and biochemical investigations demonstrated the responses of Scorpion fish liver to long-term chemical pollution of environment caused man-made activity on the coastal area of the tested bays.

It is well known that MMCs in fish tissues can contain four types of brown pigments: melanin (catabolic product of fatty acids degradation at low temperatures), lipofuscin (product of the destruction of cellular membranes and the end product of lipid peroxidation which is utilized in the lysosoms), ceroid and hemosiderin (product of red blood cells degradation) [19, 20]. Morphologic features of MMCs vary markedly depending on fish species, organ of their location and physiological status (age, starvation, tissue type, iron and hemoglobin metabolism, anatomopathological conditions). The composition and functions of liver MMCs are differ from spleen or kidney MMCs because liver is the key organ for the metabolic state that plays important role in the processes of cell renovation and storage as well as in detoxication and biotransformation processes [20].

The researchers had been shown that the number of MMCs was significantly higher in the liver of Sea bass (*Dicentrarchus labrax*) exposed to the various environmental pollutants. It can be used as a bioindicator of environmental stress. Strong correlation between the appearance of MMCs and high level of resistant intracellular bacteria and parasites had been demonstrated. MMCs play important role as the main locations for long-term pathogen retention [21, 30]. It corroborates usefulness of the bioindicator for potential monitoring of fish health and their environment.

Histopathological alterations in the liver of Black Sea Scorpion fish were mainly characterized by the dilated hepatic sinusoids and marked focal infiltration by lymphocytes, increase in the number of Kupffer cells, hyperemia appeared as congested hepatic blood vessel and sinusoids. Moreover, vacuolization in the cytoplasm of hepatocytes, which was presumably caused by fat accumulation, was noted. Similar patterns of histopathological changes were observed in many fish species exposed to the toxicants or captured from polluted waters [31, 32].

The level of water pollution significantly influences on the patterns of fish liver pathology. Lesions in the liver of Black Sea Scorpion fish captured in Alexandrovskaya Bay mainly included extensive dilation of hepatic sinusoids, hyperemia, and degenerative changes in hepatocytes. Marked focal infiltration by lymphocytes in liver tissue possibly indicating infection process was observed and it was associated with increase in the number of Kupffer cells and in the number and size of MMCs. At the same time, fish without significant signs of hepatic pathology and low MMCs count were detected in that bay. However, large MMCs were found in liver of all animals captured from Streletskaya Bay as well as focal infiltration by lymphocytes and hyperemia were detected indicating marked pathological changes in liver and plausible focal inflammatory lesions.

In conclusion, detected histopathological changes in the liver of Black Sea Scorpion fish captured in Alexandrovskaya and Streletskaya Bay pointed out unfavorable environmental conditions due to water and sediment pollution in the tested sites, which were more prominent for Streletskaya Bay. Our further studies of the parameters of oxidative stress and antioxidant system validated the conclusions of the obtained results of hystopathological investigations.

High concentration of xenobiotics in the bottom sediments and water in which Scorpion fish inhabits, causes oxidative stress in fish resulted their transfer via water and food in the organism. Liver is the main organ of pollutants accumulation and biotransformation caused ROS-generation and accumulation of their metabolites [8]. Our findings demonstrated significant increasing (more than 2-fold) the level of TBA-reactive products in the liver of Scorpion fish caught in the most polluted Streletskaya Bay as compared with the corresponding values of the animals from the less polluted Alexandrovskaya Bay. Identical trends were shown for the parameters of chemiluminescence, which were associated with the increasing of lipid peroxidation in the liver of fish inhabiting polluted environment. Concentration of oxidized proteins in tissues is informative indicator of the chronic pollution impact on fish organism resulted oxidative stress also. In our studies we showed significant increase of oxidized proteins concentration in the liver of fish from the most contaminated Streletskaya Bay. Growth of lipid and protein peroxidation production in tissues of the animals in contaminated habitats are postulated as the non-specific response of the organism to the stress effects and it was documented at the case of environment pollution by various chemical agents and their complex [33–36]. In polluted areas the exposure of aquatic organisms to xenobiotics results to interaction between these compounds and biological systems, which may give elevation to biochemical and physiological damage or/and adaptive mechanisms via the induction of defense immune and antioxidant systems [36]. Thus, biochemical and physiological parameters are used as biomarkers for contaminants and could be applied for evaluation of environmental stress and its after-effects. However, biomarkers exposure to environmental stressors varied widely depending on the type of anthropogenic activity involved [2].

Antioxidant enzyme activity in the liver of fish from high-contaminated Streletskaya Bay was also significantly higher which demonstrated the fish response to pollution and its resistance against the consequences of oxidative stress and adapt to the unfavorable environmental conditions.

The similar responses of enzyme antioxidant system were documented in the tissues of fish affected different levels of pollutants in the laboratory and field conditions: heavy metals [37], petroleum and oil compounds resulting oil spills [34], organic chemicals (benz[α]pyren [35]), etc. However, we could note the possibility of the specificity response of antioxidant enzymes in fish liver to different toxicants expose in the water and sediments as well as synergic and antagonistic effects at the case of interactions between xenobiotics and with the water. These interactions could modulate organism response and result not only adaptive effect, but toxic one also characterizing the inhibition of defense systems, including immune and antioxidant components. Thus, in some cases in the tissues of fish inhabiting high polluted water bodies the researchers documented the decrease of the activity of antioxidant and other key enzymes which characterized clearly toxic effect caused high level of heavy metals a and chlororganic compounds accumulation, which we described previously [5]. At present study we demonstrated significantly decrease of aminotrasferase activity in the liver of fish from Streletskaya Bay, wich was connected with the great intoxication resulted accumulation in the organ high concentration of toxicants.

## 19.4 CONCLUSIONS

Hystopathological and biochemical biomarkers apparently are useful indicators of the health status of fish in monitoring studies. The obtained results can be applied for development monitoring management and they are the perspective for conservation ecology and biodiversity in impacted on aquatic ecosystems. The analysis of biomarkers in fish liver is important tool for the evaluation of fish abilities to protect against chemical pollution and keeping life in the polluted environments. In contaminated areas the exposure of aquatic organisms to xenobiotics are by results to interaction these compounds and biological systems which may give elevation to biochemical and physiological damage or/and adaptive mechanisms via the induction of defenses immune and antioxidant systems.

## KEYWORDS


- **anthropogenic impact**
- **antioxidant enzymes**
- **Black Sea**
- **fish Scorpaena porcus**
- **melanomacrophage centers**
- **oxidative stress**
- **Sevastopol bays**


## REFERENCES


1. The World Health Report. World Health Organization, Geneva, Switzerland. 2002, 162 p.
2. Adams, S. M. (2005). Assessing cause and effect of multiple stressors on marine system, Marine Pollution Bulletin, , 51(8–12), 649–657.
3. Galloway, T. (2006) Biomarkers in environmental and human health risk assessment, Marine Pollution Bulletin, 53, 606–613.
4. Ababouch, L. (2006). Assuring fish safety and quality in international fish trade. Marine Pollution Bulletin, 53, 561–568.
5. Rudneva, I.I., Skuratovskaya, E.N., Dorohova, I.I., & Kovyrshina, T.B. (2012). Application fish blood biomarkers in evaluation of marine environment health, Asian Journal of Experimental Biological Science, , 2(7), 19–25.
6. Vladimirov, J. A. (2001). Activated chemiluminescence and bioluminescence as an instrument in medicine-biological studies. Soros Educational Journal, 7, 16–20 (in Russian).
7. Van der Oost, R., Beyer, J., & Vermeulen, N. P. E. (2003). Fish bioaccumulation and biomarkers in environmental risk assessment: a review, Environmental Toxicology and Pharmacology, 13, 57–149.
8. Sole, M., Rodriguez, S., Papiol, V., Maynou, F., & Cartes, J. E., (2009). Xenobiotic metabolism markers in marine fish with different trophic strategies and their relationship to ecological variables, Comparative Biochemistry and Physiology, 149 C (1), 83–89.
9. Abdel-Warith, A. A., Younis, E. M., Al-Asgah, N. A., & Wahbi, O. M. (2011). Effect of zinc toxicity on liver histology of Nile tilapia, Oreochromis niloticus, Scientific Research and Essays, , 17, 3760–3769.
10. Kaewamatawong, T., Rattanapinyopituk, K., Ponpornpisit, A., Pirarat, N., Ruangwises, S., & Rungsipipat, A. (2013). Short-term exposure of Nile Tilapia (Oreochromis niloticus) to mercury histopathological changes, mercury bioaccumulation,

and protective role of metallothioneins in different exposure routes, Toxicologic Pathology, 41, 470–479.
11. Pietsch, C., Schulz, C., Rovira, P., Kloas, W., & Burkhardt-Holm, P. (2014). Organ damage and hepatic lipid accumulation in carp (Cyprinus carpio L.) after feed-borne exposure to the mycotoxin, deoxynivalenol (DON), Toxins, 6, 756–778.
12. Feist, S. W., Stentiford, G. D., Kent, M. L., Ribeiro Santos, A., & Lorance, P. (2015). Histopathological assessment of liver and gonad pathology in continental slope fish from the northeast Atlantic Ocean, Marine Environmental Research, 106, 42–50.
13. Al-Mamoori, A. M. J., AlZubaidy, F. M., Abd Al-Rezzaq, A. J., Adil Hadi, M., Hass, & M.J. (2014). Biomarkers of Chlorfos toxicity in Common Carp Cyprinus carpio, IOSR-JESTFT, , 1, 109–112.
14. Way, K., Bark, S. J., Longshaw, C. B. et al. (2003). Isolation of a rhabdovirus during outbreaks of disease in cyprinid fish species at fishery sites in England, Dis. Aquat. Org., 57, 43–50.
15. Ovsyanyi, E. I., Romanov, A. S., Minkovskaya, R. J. et al. (2001). Main sources of pollution of marine environment of Sevastopol region. Proceedings of MHI NAS of Ukraine: Ecological safety of the coastal and shelf zone and the complex use of shelf resources. Sevastopol: ECOSY-Hydrophysica, 138–152 (in Russian).
16. Sergeeva, N. G., Kolesnikova, E. A., & Mazlumyan, S. A. (2010). Taxonomic biodiversity of meiobenthos in Sevastopol bays (Crimean coastal line of Black Sea). Biodiversity and sustainable development: Proc. Int. Scientific-practical Conference, May 19–22, Simpheropol, 113–115 (in Russian).
17. Osadchaya, T. S. (2013). Petroleum hydrocarbons in the bottom sediments of the coastal aquatoria of Sevastopol (Black Sea), Scientific research and their practical application. Modern State and Ways of Development. URL: http://www.sworld.com.ua/index.php/ru/conference/the-content-of conferences/archives-of-individual-conferences/oct-2013 (in Russian).
18. Kopytov, J. P., Minkina, N. I., & Samyshev, E. Z. (2010). Pollution level of the water and bottom sediments in Sevastopol bay (Black Sea), Systems of the control of environment, NAS Ukraine. MHI: Proceedings. Sevastopol, 14, 199–208 (in Russian).
19. Kranz, H., (1989). Changes in splenic melano-macrophage centers of dab Limanda limanda during and after infection with ulcer disease, Diseases of aquatic organisms, 6, 167–173.
20. Manrique, W. G., da Silva Claudiano, G., Petrillo, T. R., et al. (2014). Response of splenic melanomacrophage centers of Oreochromis niloticus (Linnaeus, 1758) to inflammatory stimuli by BCG and foreign bodies, Journal of Applied Ichthyology, 5, 1–6.
21. Kurtovic, L. B., Teskeredzic, E., & Teskeredzic, Z. (2008). Histological comparison of spleen and kidney tissue from farmed and wild European Sea bass (Dicentrarchus labrax, L.), Acta Adriatica, 49(2), 147–154.
22. Dubinina, E.E., & Shugalei, I.V. (1993). Oxidizide modification of the proteins, Achievements of the Modern Biology, , 113(1), 71–81 (in Russian).
23. Asatiani, V. S. (1969). Methods of enzymatic analysis. Moscow, Nauka, 611 p. (in Russian).
24. Nishikimi, M., Rao, N. A., & Jagik, K. (1972). The occurrence of superoxide anion in the reaction of reduced phenazine, Biochim. Biophys. Res. Com., , 46(2), 849–854.

25. Litvin, F.F. (1981). Manual book of physical-chemical methods in biology. Moscow, MSU Ed, 86–87 (in Russian).
26. Pereslegina, I. A. Antioxidant enzyme activity of the salva of healthy children, Laboratornoe Delo, 1989, 11, 20–23 (in Russian).
27. Katalog of instructions Kits for clinic, biochemical and microbiological analysis. Dnepropetrovsk: OOO NPP "Filicit-diagnostica," 2005, 199 p. (in Russian).
28. Lakin, G. F. Statistical analysis. Moscow, Vysshaya shkola, 1990, 352 p. (in Russian).
29. Zaki, M. M., Eissa, A. E., Saeid, S. Assessment of the immune status in Nile Tilapia (Oreochromis niloticus) experimentally challenged with toxogenic/septicemic bacteria during treatment trial with florfenicol and enrofloxacin, World Journal of Fish and Marine Sciences, 2011, 3(1), 21–36.
30. Biyasheva, Z. M., Kenzhebayev, N. A., Ibragimova, N. A., Omar, A. E. Environment and life quality by the example of surface water in the area of HPS-2 influence (Almaty), IJAIR, 2014, 4, 622–627.
31. Ferreira, C. M. H. Can fish liver melanomacrophages be modulated by xenoestrogenic and xenoandrogenic pollutants? Experimental studies on the influences of temperature, sex, and ethynylestradiol, using the platy fish as the model organism, 2011, 53 p.
32. Hansson, T., Schiedek, D., Lehtonen, K. K., Vuorinen, P. J., Liewenborg, N., Noaksson, E., Tjarnlund, U., Hansson, M., Balk, L. Biochemical biomarkers in adult female perch (Perca fluviatilis) in a chronically polluted gradient in the Stockholm recipient (Sweden), Marine Pollution Bulletin, 2006, 53, 451–468.
33. Napierska, D., Barsiene, J., Mulkiwicz, E., Podolska, M., Rybakovas, A. Biomarker responses in flounder Platichthys flesus from the Polish coastal area of the Baltic Sea and applications in biomonitoring, Ecotoxicology, 2009, 18, 846–859.
34. Martinez-Gomez, C., Fernandez, B., Valdes, J., Campillo, J. A., Benendicto, J., Sanchez, F., Vethaak, A. D. Evaluation of three-year monitoring with biomarkers the Prestige oil spill (Spain), Chemosphere, 2009, 74 (5), 613–620.
35. Martinez-Porchas, M., Hernandez-Rodriguez, M., Davila-Ortiz, J., Vila-Cruz, V., Ramos-Enriquez, J. R., A preliminary study about the effect of benzo[α]pyrene (BaP) injection on the thermal behavior and plasmatic parameters of the Nile tilapia (Oreochromis niloticus, L.) acclimated to different temperatures, Pan-American Journal of Aquatic Sciences, 2011, 6(1), 76–85.
36. Liu, H., Wang, W., Zhang, J., Wang, X. Effects of copper and its ethylenediaminetetraacetate complex on the antioxidant defenses of the goldfish, Carassius auratus, Ecotoxicological Environmental Safety, 2006, 65(3), 350–354.
37. Goksoer, A., Beyer, J., Egaas, E., Grosvik, B. E., Hylland, K., Sandvik, M., Skaare, J. U. Biomarker responses in flounder (Platichthys flesus) and their use in pollution monitoring, Marine Pollution Bulletin, 1996, 33, 36–45.

# GLOSSARY

**Activation energy ($E_a$)**

Activation energy is the minimum amount of energy (heat, electromagnetic radiation, or electrical energy) required to activate atoms or molecules to a condition in which it is equally likely that they will undergo chemical reaction or transport as it is that they will.

**Alanit**

Alanit is the natural zeolite-containing clay from Northern Ossetia, harvested in the Terek, contains the following elements: silicon; aluminum; iron; calcium; potassium, phosphorus, manganese, sulfur, magnesium.

**Alluvium**

Alluvium is a silty stream deposits made during flood stage overflow of normal channels.

**Apoptosis**

Apoptosis is the process by which a cell dies in a programmed way, or in other words, kills itself; it is the most common form of physiological (as opposed to pathological) cell death; it is an active process requiring metabolic activity by the dying cell; often characterized by shrinkage of the cell and cleavage of the DNA into fragments.

**ATP-ases**

ATP-ases are a class of enzymes that catalyze the decomposition of ATP into ADP and a free phosphate ion. This dephosphorylation reaction releases energy, which the enzyme (in most cases) harnesses to drive other chemical reactions that would not otherwise occur. These include: adenylpyrophosphatase, ATP monophosphatase, triphosphatase, SV40 T-antigen, adenosine 5'-triphosphatase, ATP hydrolase, complex V (mitochondrial electron transport), ($Ca^{2+}$ +$Mg^{2+}$)-ATPase, $HCO_3^-$-ATPase, adenosine triphosphatase.

**Baykal EM-1**

Baykal EM-1 is effective biological preparation contains a stable community of physiologically compatible and complementary useful microorganisms.

**Bioremediation**

Bioremediation is the complex methods of water purification, soil and the atmosphere using metabolic potential of biological objects – plants, fungi, insects, worms and other organisms. According to the United States Environmental Protection Agency (EPA or sometimes USEPA) is an agency of the U.S. federal government which was created for the purpose of protecting human health and the environment by writing and enforcing regulations based on laws. Bioremediation is a "treatment that uses naturally occurring organisms to break down hazardous substances into less toxic or non toxic substances."

**Bioutilization**

Bioutilization of sewage sludge can be successfully solved by the creation on their basis of valuable fertilizing means, such as composts, organic-mineral fertilizers, including pelleted ones. During the processing pathogenic microflora and other pathogenic organisms in them are destroyed. By reducing humidity and granulation one can transport them over long distances.

**Boreal ecosystem**

Boreal ecosystem is an ecosystem with a subarctic climate in the Northern Hemisphere, roughly between latitude 45° to 65° N.

**Boreal forest>>>Taiga**

**Boreal forests** are also known as the **taiga**, particularly in Europe and Asia. A different use of the term taiga is often encountered in the English language, with "boreal forest" used in the United States and Canada to refer to only the more southerly part of the biome, while taiga is used to describe the more barren areas of the northernmost part of the biome approaching the tree line and the tundra biome.

***Bromus inermis***

It is a species of the grass family (Poaceae). This bunchgrass is native to Europe. The plant is characterized by an erect, leafy, long-lived perennial, 46 to 91 cm (1 1⁄2 to 3 ft.) tall, rhizomatous, and commonly with dense rhizome.

**Carbon dioxide ($CO_2$)**

It is a colorless, odorless gas vital to life on Earth. This naturally occurring chemical compound is composed of a carbon atom covalently double bonded to two oxygen atoms.

**Carbon monoxide (CO)**

It is a colorless, odorless, and tasteless gas that is slightly less dense than air. It is toxic to hemoglobic animals (including humans) when encountered in concentrations odorless, and tasteless gas that is slightly less dense than above about 35 ppm.

**Carotenoids**

Carotenoids are organic pigments that are found in the chloroplasts and chromoplasts of plants and some other photosynthetic organisms, including some bacteria and some fungi. They are essential for plant growth and photosynthesis.

**Catalase (CAT)**

Catalase is a common enzyme that is active in cells and tissues throughout the body, where it breaks down hydrogen peroxide ($H_2O_2$) molecules into oxygen ($O_2$) and water ($H_2O$). Hydrogen peroxide is produced through chemical reactions within cells. At low levels, it is involved in several chemical signaling pathways, but at high levels it is toxic to cells. If hydrogen peroxide is not broken down by catalase, addition reactions convert it into compounds called reactive oxygen species that can damage DNA, proteins, and cell membranes.

**Cesium (Cs)**

Cesium is a chemical element with atomic number 55.

**Chemicalization**

Chemicalization is accumulation of chemical compounds in soils due application of agricultural chemicals.

**Chemiluminescence**

Chemiluminescence is the emission of light (luminescence), as the result of a chemical reaction. There may also be limited emission of heat. It was used as the method of the evaluation of ROS production generation in living organisms.

**Chlorophyll**

Chlorophyll is any member of the most important class of pigments involved in photosynthesis, the process by which light energy is converted to chemical energy through the synthesis of organic compounds. Chlorophyll is found in virtually all-photosynthetic organisms, including green plants, prokaryotic blue-green algae (cyanobacteria), and eukaryotic algae. It absorbs energy from light; this energy is used to convert carbon dioxide to carbohydrates.

**Chromium, chrome (Cr)**

It is a chemical element with atomic number 24.

**Closed canopy forest>>>Taiga**

**C-mitosis**

It is a metaphase that is visible under the microscope after treatment of object by colchicine

**Cobalt (Co)**

It is ninth element (the old classification – secondary subgroup eighth of group) of the fourth period of the periodic table of chemical elements of Mendeleev, atomic number – denoted by 27. Co (lat. Cobaltum).

**Compost**

is organic matter that has been decomposed and recycled as a fertilizer and soil amendment.

**Concentric Double Membrane System**

It is formation in cells of the double concentric membranes, that is the universal answer of cells on the influence of substances of different nature, toxic already action. Species specificity of process appears in speed of formation of internal membrane, thickness and period of dissipative destructions. An external membrane is conservative according to its thickness. The formation of the double concentric membrane system is accompanied with the row of successive changes in its structure, composition and functions. Content of the basic classes of adaptive lipids in membranes increases. The general regularity for the actions of all investigated factors is the increase of content of triacylglycerols and phospholipids.

**Copper (Cu)**

It is chemical element with the atomic number 29, atomic weight 63.546.

**Crop Rotation**

It is the practice of growing a series of dissimilar/different types of crops in the same area in sequenced seasons. It also helps in reducing soil erosion and increases soil fertility and crop yield.

**Cytochrome oxidase (CO)**

It is a large transmembrane protein complex found in bacteria and the mitochondria of eukaryotes. It is the last enzyme in the respiratory electron transport chain of mitochondria or bacteria located in their membrane. It receives an electron from each of four cytochrome *c* molecules, and transfers them to one oxygen molecule, converting molecular oxygen to two molecules of water. In the process, it binds four protons from the inner aqueous phase to make water, and in addition translocates four protons across the membrane, helping to establish a transmembrane difference of proton electrochemical potential that the ATP synthase then uses to synthesize ATP.

**Deoxyribonucleic acid (DNA)**

It is a molecule of genetic information, a nucleic acid that contains the genetic instructions used in the development and functioning of all known living organisms.

**Diacylglycerol (DAG)**

It is glycerol substituted on the 1 and 2 hydroxyl groups with long chain fatty acyl residues. DAG is a normal intermediate in the biosynthesis of phosphatidyl phospholipids and is released from them by phospholipase C activity is a glyceride consisting of two fatty acid chains covalently bonded to a glycerol molecule through ester linkages. In biochemical signaling, diacylglycerol functions as a second messenger signaling lipid, and is a product of the hydrolysis of the phospholipid phosphatidylinositol 4,5-bisphosphate by the enzyme phospholipase C. The production of DAG in the membrane facilitates translocation of protein kinase C (PKC) from the cytosol to the plasma membrane.

**Dialbekulit**

It is zeolite-containing natural clay. Accumulations is in the foothills of the North Caucasus.

**Ecotoxicology**

It is the study of the effects of toxic chemicals on biological organisms, especially at the population, community, ecosystem level. Ecotoxicology is a multidisciplinary field, which integrates toxicology and ecology.

**Eluvial Accumulative Coefficient (EAC)**

It is ratio of algebraic sum of bringing or bringing out oxide to its original content. EAC is used for estimation of degree of bringing out or accumulating of any elements in some horizons of soil profile with respect to mother rock.

**Eluvium**

Eluvium or eluvial deposits are those geological deposits and soils that are derived by in situ weathering or weathering plus gravitational movement or accumulation. A soil horizon formed due to eluviation is an eluvial zone or eluvial horizon.

**Fertilizers**

It is any material of natural or synthetic origin, that is applied to soils to supply one or more plant nutrients essential to the growth of plants. This also depends on its soil fertility.

**Gley**
Gley is reduced gray soil due to poor drainage and high water table

**Glutamate dehydrogenase (NADH/NADPH-GDH)**
It is an extremely important enzyme which links carbohydrate and nitrogen metabolism and the physiological roles of GDHs as anabolic and/or catabolic enzymes are generally defined by the nature of their cofactor specificity. NADP(H)-specific GDH enzymes usually catalyze the assimilation of ammonia by reductive amination of a-ketoglutarate to form *L*-glutamate (anabolic), while NAD(H)-dependent GDH enzymes catalyze the reverse reaction (catabolic).

**Glutathione peroxidase (GPx)**
It is the general name of an enzyme family with peroxidase activity whose main biological role is to protect the organism from oxidative damage. The biochemical function of glutathione peroxidase is to reduce lipid hydroperoxides to their corresponding alcohols and to reduce free hydrogen peroxide to water.

**Gmelin salt**
It is $K_3Fe(CN)_6$ >>> potassium ferricyanide

**High boreal>>>Taiga**
**Heavy metals (HM)**
It the term refers to any metallic chemical element that has a relatively high density and is toxic or poisonous at low concentrations

**Horizon A**
*Horizon A* is surface soil: layer of mineral soil with the most organic matter accumulation. This layer is depleted of iron, clay, aluminum, organic compounds, and other soluble constituents. The A horizon is the topmost mineral horizon, often referred to as the 'topsoil.' This layer generally contains enough partially decomposed (humified) organic matter. The A horizons are often coarser in texture, having lost some of the finer materials by translocation to lower horizons and by erosion. This layer is known as the zone in which the most biological activity occurs. AB-horizon – transition

horizon between A and B master horizon, wherein the dominant horizon is placed before the non-dominant horizon. Ap – surface soil horizon: layer of ineral soil with the most organic matter accumulation; index p indicates that there is plowing or other disturbance

**Horizon B**

*Horizon B* is subsoil: subsurface layer reflecting chemical or physical alteration of parent material. This layer accumulates iron, clay, aluminum and organic compounds, a process referred to as illuviation. The B horizon is commonly referred to as the "subsoil." In humid regions, B horizons are the layers of maximum accumulation of materials such as silicate clays, iron and aluminum oxides, and organic material. These materials typically accumulate through a process termed illuviation, wherein the materials gradually wash in from the overlying horizons. Accordingly, this layer is also referred to as the "illuviated" horizon or the "zone of accumulation."

**Horizon O**

*Horizon O* is organic matter: surficial organic deposit with litter layer of plant residues in relatively non- decomposed form. It is a surface layer dominated by the presence of large amounts of organic material derived from dead plant and/or animal residues which is in varying stages of decomposition. O horizons may also be divided into three subordinate O horizons denoted as: $O_i$, $O_e$, and $O_a$.

**Hydrogen chloride (HCl)**

It is a colorless, thermally stable gas (under normal conditions) with a pungent odor, fuming in moist air, easily soluble in water.

**Hydrogen ion** ($H^+$)

It is a general term for all ions of hydrogen and its isotopes.

**Illuvium**

**Illuvium** is material displaced across a soil profile, from one layer to another one, by the action of rainwater. The removal of material from a soil layer is called **eluviation**. The transport of the material may be either mechanical or chemical. The process of deposition of illuvium is termed illuviation. It is a water-assisted transport in a basically vertical direction, as compared to alluviation, the horizontal running water transfer.

**Irlits**

Irlits are zeolite-containing natural clays. Accumulations are in the foothills of the North Caucasus.

**Iron (Fe)**

It is a chemical element with symbol Fe from Latin: ferrum, has atomic number 26. It is a metal in the first transition series.

**Jaccard index**

Jaccard index also known as the Jaccard similarity coefficient (originally coined coefficient de communauté by Paul Jaccard), is a statistic used for comparing the similarity and diversity of sample sets. The Jaccard coefficient measures similarity between finite sample sets, and is defined as the size of the intersection divided by the size of the union of the sample sets.

**Lead (Pb)**

It is a chemical element in the carbon group with symbol Pb (from Latin: plumbum) and atomic number 82.

**Labile part of total carbon ($C_{lab}$)**

Labile part of total carbon ($C_{lab}$) is a quickly reactive labile organic matter, which provides energy and nutrients for soil microorganisms and releases part of the nutrients for plant usage. Its half-life is between days and few years. It provides short-term organic matter turnover during the year.

**Lichen woodland>>>Taiga**

**Lysophospholipids (LPL)**

They are the class of signaling lipids, which include sphingosine 1-phosphate (S1P) and lysophosphatidic acid (LPA). These lipids have many important actions mediated by G protein-coupled receptors. Lysophospholipids have been accepted as intermediate products of the phospholipid deacylation–reacylation pathway.

**Malondialdehyde (MDA) >> TBA-reactive products**
**Manganese (Mn)**
It is a chemical element with atomic number 25.

**Maslova's method**
It is a method for determination of potassium in the soil based on the extraction of soil exchangeable potassium by 1-normal solution of ammonium acetate at 1:10 ratio soil: solution and one hour shaking on a rotator.

**Maximum speed of penetration ($V_{max}$)**
It is the maximum initial velocity of an enzyme-catalyzed reaction, ie. at saturating substrate levels.

**Melanomacrophage centers**
It also known as macrophage aggregates, are distinctive groupings of pigment-containing cells within the tissues of heterothermic vertebrates. In fish they are normally located in the stroma of the haemopoietic tissue of the spleen and the kidney, and they are also found in the liver. They may develop in association with chronic inflammatory lesions elsewhere in the body. In higher teleosts, they often exist as complex discrete centers, containing lymphocytes and macrophages, and may be primitive analogs of the germinal centers of lymph nodes. Melanomacrophage centers usually contain a variety of pigments, including melanins. Melanomacrophage centers act as focal depositories for resistant intracellular bacteria, from which chronic infections may develop.

**Membrane fluidity**
It refers in biology to the viscosity of the lipid bilayer of a cell membrane or a synthetic lipid membrane. Lipid packing can influence the fluidity of the membrane. Viscosity of the membrane can affect the rotation and diffusion of proteins and other bio-molecules within the membrane, thereby affecting the functions of these molecules.

**Michaelis (Michaelis-Menten) constant ($K_m$)**
It is equation derived from a simple kinetic model of enzyme action that successfully accounts for the hyperbolic (adsorption isotherm) relationship between substrate concentration S and reaction rate V • $V = V_{max} \cdot S/(S + K_m)$.

**Middle boreal >>> Taiga**
**Mitotic index**
It is the fraction of cells undergoing mitosis in a given sample defined as the ratio between the number of cells in a population undergoing mitosis to the number of cells. The purpose of the mitotic index is to measure cellular proliferation.

**Mitotic pathologies >> Pathologic mitosis.**
**Molybdenum (Mo)**
It is a chemical element with atomic number 42.

**Mother rock**
Mother rock is altered geologic material from which the soil is presumed to have formed

**Natural clays**
They are zeolite-containing – dialbekulit, irlits, alanit, are mined in the foothills of the North Caucasus and along the banks of the river Terek river, contains the following minerals: silicon, iron, calcium, cobalt, zinc, nickel, and phosphorus, and other in different ratios. The clays are used to enrich by minerals of arable the soils at the process bioremediation.

**Nemoral species**
Nemoral species are pertaining to or living in a forest or wood.

**Nickel (Ni)**
It is a chemical element with atomic number 28.

**Nitrogen (N)**
It is a chemical element with symbol N and atomic number 7.

**Nonetherified fatty acids (NEFA)**
It is a class of compounds containing a long hydrocarbon chain and a terminal carboxylate group (-COOH). Fatty acids belong to a category of biological molecules called lipids, which are generally water-insoluble but highly soluble in organic solvents. Fatty acids function as fuel molecules and serve as components of many other classes of lipids, including

triglycerides (commonly known as "fats") and phospholipids, which are important building blocks of biological membranes.

**Nucleoli Persistent**

They are modified material of nucleoli, is observed between chromosomes at metaphase, anaphase and telophase of mitosis are round or drop-shaped nucleolus-like structures connected with the chromosomes at metaphase, anaphase and early telophase of mitosis, when the nuclear membrane is absent and chromosomes are significantly reduced. Phenomenon of the appearance of the persistent nucleoli at such mitosis phases as metaphase, anaphase and telophase is not typical for the normal course of division, because, as a rule, the nucleolus disappears.

**Nucleolus**

It is tight formation, was detected in interphase nuclei of eukaryotic cells, which is formed in specific loci of chromosomes. An RNA-rich, intranuclear domain found in eukaryotic cells that is associated with the nucleolus organizer and is the site of pre-ribosomal RNA synthesis and processing and of ribosomal particle assembly. The nucleolus is composed of the primary products of the ribosomal RNA genes and a variety of proteins, including RNA-polymerases, ribonucleases, molecular chaperones, helicases, ribosomal proteins, and proteins of unknown function. rRNA genes and their nascent transcripts were first seen as Miller trees in nucleoli from salamander oocytes.

**Nucleus**

It is the spheroidal, membrane-bounded structure present in all eukaryotic cells. Nucleus contains DNA, usually in the form of chromatin.

**Oil**

It is natural oily flammable liquid and with strong odor consisting essentially of a complex mixture of hydrocarbons of various molecular weights and other chemical compounds.

**Para-aminobenzoic acid (PABA)**

It is 4-amino-2-oxibezoic acid ($C_7H_7NO_2$) with molecular mass of 137.1 is soluble in the hot water (80–90°*C*), benzol, ethanol, acetic acid. PABA represents fine crystalline powder of white or slightly cream color. It belongs to the group of vitamins B (vitamin H), is a part of folic acid, is the antagonist of novocain, sulfonamide.

**Pathologic mitosis (mitotic pathologies)**
They are any cell division that is atypical, asymmetric, or multipolar and results in an unequal number of chromosomes in the nuclei of the daughter cells. Level of mitotic pathologies is the ratio of cells with impaired dividing the total number of cells in percentages. Spectrum of mitosis pathologies – the proportion of each type of cell division disturbances of their total number in percentages.

**Perennial rye-grass**
has common name *Lolium perenne*, or English ryegrass or winter ryegrass, is a grass from the family Poaceae.

**Pheophytin (Pheo)**
It is a chemical compound that serves as the first electron carrier intermediate in the electron transfer pathway of photosystem II (PS II) in plants, and the photosynthetic reaction center (RC P870) found in purple bacteria. Pheophytin is a form of chlorophyll *a* in which the magnesium ion is replaced by two hydrogen ions. It participates in the crucial step of converting light energy to chemical energy.

**Phospholipid (PL)**
It is class of lipids that are a major component of all cell membranes as they can form lipid bilayers. Most phospholipids contain glyceride, a phosphate group, and a simple organic molecule such as choline; one exception to this rule is sphingomyelin, which is derived from sphingosine instead of glycerol. The structure of the phospholipid molecule generally consists of hydrophobic tails and a hydrophilic head. Biological membranes in eukaryotes also contain another class of lipid, sterol, interspersed among the phospholipids and together they provide membrane fluidity and mechanical strength.

**Phosphorus (P)**
It is a chemical element with automatic number 15.

**Podzolic soil**
This soil usually forming in a broadleaf forest and characterized by moderate leaching, which produces an accumulation of clay and, to some degree, iron that have been transported (eluviated) from another area by water. The humus formed produces a textural horizon (layer) that is less than 50 cm from the surface. Podzolic soils may have laterite (a soil layer cemented together by iron) in place of the humic horizon or along with it.

**Potassic fertilizer**
It is a variety of fertilizer that has potassium as a primary ingredient.

**Potassium (K)**
It is a chemical element with symbol K (derived from Neo-Latin, kalium) and atomic number 19. It was first isolated from potash, the ashes of plants, from which its name is derived.

**Potassium ferricyanide ($K_3Fe(CN)_6$)**
>>>potassium geksatsianoferriat, Gmelin salt, red blood salt

**Potassium geksatsianoferriat**
>>> Gmelin salt, red blood salt

**Prologed trial**
It is studies conducted in the field on a dedicated site for two or more turns of the rotation.

**Reactive oxygen species (ROS)**
They are chemically reactive molecules containing oxygen. They include peroxides, superoxide, hydroxyl radical, and singlet oxygen. In living organisms ROS are formed as a natural byproduct of the normal metabolism of oxygen and have important roles in cell signaling and homeostasis. However, ROS levels can increase dramatically in the organisms, impacts

different stressors. This may result in significant damage to cell structures, known as oxidative stress.

**Red blood salt**

>>> potassium ferricyanide, potassium geksatsianoferriat, Gmelin salt

**Red clover**

(*Trifolium pretense*) is a herbaceous species of flowering plant in the bean family Fabaceae, native to Europe, Western Asia and northwest Africa, but planted and naturalized in many other regions.

**Red fescue**

(*Festuca rubra)* is a species of grass known by the common name red fescue or creeping red fescue. It is widespread across much of the Northern Hemisphere and can tolerate many habitats and climates.

**Revolution per minute (rpm)**

It is a measure of the frequency of rotation, specifically the number of rotations around a fixed axis in one minute.

**Ruderal species**

Ruderal species is a plant that grows, as a rule, along fences and roads, at garbage dumps, and in other waste places. Ruderals have various adaptations (poisonous substances, thorns, stinging hairs) for protection from destruction by man and animals. Ruderals and segetal plants make up the weed group.

**Segetal species**

Segetal species are plant species growing in fields of grain.

**Selenium (Se)**

It is a chemical element with atomic number 34.

**Sewage sludge**

It is the residual semi-solid material that is produced as a by-product during sewage treatment of industrial or municipal wastewater. The term septage is also referring to sludge from simple wastewater treatment but is connected to simple on-site sanitation systems such as septic tanks.

**SH-group**
It located at the active centers of enzymes, activated by the formation of hydrogen bonds with neighboring functional groups

**Silicon**
It is a chemical element with symbol Si and atomic number 14. It is a tetravalent metalloid, more reactive than germanium, the metalloid directly below it in the table.

**Snow forest>>>Taiga**
**Sod gleyic drained clay loam**
It is soil which has blue-bluish or greenish color caused by the presence of divalent iron that is formed in a stagnant waterlogged and characterized by low acidity.

**Sod-podzolic heavy loami soil**
It is zonal soil formed under forest which consists mainly of fine silica and has a heavy mechanical composition; it occupies an intermediate position between clay and sandy soils.

**Sod-podzolic loami sand soil**
It is the soil in which an admixture of sand more than in loamy soil.

**Software program (STRAZ)**
It is abbreviation of Russian words which characterize the program and mean: old, traditional, functioning, very simple, disused

**Soil amendments**
It include a wide range of fertilizers and non-organic materials fiber flax – a special additive to fertilizer before planting potatoes.

**Soil horizon**
Soil horizon is a layer of soil, approximately parallel to the surface, whose physical characteristics differ from the layers above and beneath. Each soil

type usually has three or four horizons. Horizons are defined in most cases by obvious physical features, chiefly color and texture. Each main horizon is denoted by a capital letter, which may then be followed by several alphanumerical modifiers highlighting particular outstanding features of the horizon. Horizon nomenclature consists of two major parts, the master horizon and the subordinate designation:

**Soil layer**
Soil layer is a layer in the soil deposited by a geologic force (wind, water, glaciers, oceans, etc.) and not relating to soil forming process.

**Soil profile**
Soil profile is a vertical section of the soil extending through all its horizons and into the parent material.

**sparse taiga>>>Taiga**
***Stevia rebaudiana* Bertoni**
It is a natural sweetener content stevoside and other diterpene glycosides, nicotinic acid, flavonoids, aminoacids, pectins, volatile oils, beta-carotene, mineral elements and vitamins.

**Strontium (Sr)**
It is a chemical element with symbol Sr and atomic number 38.

**Subordinate horizons**
Subordinate horizons may occur within a master horizon and these are designated by lowercase letters following the capital master horizon letter (e.g., $A_p$, $O_i$). Since the nature of a master horizon is only generally described by the capital master horizon letter, the lowercase letter symbols following the master horizon designation is often used to indicate more specific horizon characteristics, for example: e – organic material of intermediate decomposition (humic material); used with O-horizon; h – illuvial accumulation of organic matter; I – slightly decomposed organic material (fabric material); used with O-horizon; p – plowing or other disturbance; used only with surface horizon.

**Succession**

Succession is a process by which, in a specific area, species replace each other in a directed way, for instance the re-establishment of a forest after clear-cutting.

**Succinate dehydrogenase (SDH)**

It is an enzyme complex, bound to the inner mitochondrial membrane of mammalian mitochondria and many bacterial cells. It is the only enzyme that participates in both the citric acid cycle and the electron transport chain. In step 6 of the citric acid cycle, SQR catalyzes the oxidation of succinate to fumarate with the reduction of ubiquinone to ubiquinol. This occurs in the inner mitochondrial membrane by coupling the two reactions together.

**Superoxide dismutase (SOD)**

It is an enzyme that alternately catalyze the dismutation (or partitioning) of the superoxide ($O_2^-$) radical into either ordinary molecular oxygen ($O_2$) or hydrogen peroxide ($H_2O_2$). Superoxide is produced as a by-product of oxygen metabolism and, if not regulated, causes many types of cell damage. Hydrogen peroxide is also damaging, but less so, and is degraded by other enzymes such as catalase. Thus, SOD is an important antioxidant defense in nearly all-living cells exposed to oxygen.

**Taiga Also Known As Boreal Forest or Snow Forest**

It is a world's largest terrestrial biome characterized by coniferous forests. There are two major types of taiga. The southern part is the *closed canopy forest*, consisting of many closely spaced trees with mossy ground cover. The other type is the *lichen woodland* or *sparse taiga*, with trees that are farther-spaced and lichen ground cover; the latter is common in the northernmost taiga. In Canada, Scandinavia and Finland, the boreal forest is usually divided into three subzones: The *high boreal (north taiga)* or taiga zone; the *middle boreal (middle taiga)*; and the *southern boreal (southern taiga)*, a closed canopy boreal forest with some scattered temperate deciduous trees among the conifers, such as maple, elm and oak. This southern boreal forest experiences the longest and warmest growing season of the biome, and in some regions (including Scandinavia, Finland and western Russia) this subzone is commonly used for agricultural purposes.

**TBA-Reactive Products or Malondialdehyde (MDA)**
They are the organic compounds with the formula $CH_2(CHO)_2$. The structure of this species is more complex than this formula suggests. These compounds react with 2-thiobarbituric acid and they are the markers for oxidative stress.

**Transitional layers**
Transitional layers are between the master horizons. They may be dominated by properties of one horizon but also have characteristics of another. The two master horizon letters are used to designate these transition horizons (e.g., AB, BC), wherein the dominant horizon is placed before the non-dominant horizon.

**Triacylglycerol (TAG)**
It is an ester derived from glycerol and three fatty acids. There are many triglycerides: depending on the oil source, some are highly unsaturated, some less so.

**Triglyceride (TG)>>> Triacylglycerol**
**Unesterified Fatty Acids (UFA)**
It is class of compounds containing a long hydrocarbon chain and a terminal carboxylate group (–COOH). Fatty acids belong to a category of biological molecules called lipids, which are generally water-insoluble but highly soluble in organic solvents. Fatty acids function as fuel molecules and serve as components of many other classes of lipids, including triglycerides (commonly known as "fats") and phospholipids, which are important building blocks of biological membranes.

**Unit of Illuminance**
Unit of illuminance in photometry, illuminance is the total luminous flux incident on a surface, per unit area, the SI unit of illuminance and luminous emittance, measuring luminous flux per unit area.

**Vegetation zone**

Vegetation zone is a region on a defined latitude and altitude with a typical vegetation type > major life zone > biome.

**Zinc (Zn)**

Zinc in commerce also spelter is a chemical element with atomic number 30.

# INDEX

**B**

**F**



**G**

## N

## O



## P

## Q



## R

## T

For Product Safety Concerns and Information please contact our EU representative GPSR@taylorandfrancis.com Taylor & Francis Verlag GmbH, Kaufingerstraße 24, 80331 München, Germany

**Batch number: 10725800**

Printed by Printforce, the Netherlands